LEGAL AND ETHICAL DILEMMAS IN OCCUPATIONAL HEALTH

RMCOEH

LEGAL AND ETHICAL DILEMMAS IN OCCUPATIONAL HEALTH

Edited by

JEFFREY S. LEE, MPH, PhD
WILLIAM N. ROM, MD, MPH

ANN ARBOR SCIENCE
THE BUTTERWORTH GROUP

300406

PREFACE

The field of occupational and environmental health is becoming increasingly complex due to a multitude of inherent legal and ethical issues. It is often impossible to separate basic science in the tumultuous arena also inhabited by politics, ethics and philosophy.

This text brings together discussions from leading experts, with a wide variety of expertise, from government, labor and academia. Several of the papers are highly controversial, as evidenced by the printed discussions that follow them. The papers will challenge and stimulate the thinking of not only practicing occupational and environmental health professionals, including industrial hygienists, occupational medicine physicians, occupational health nurses and safety professionals, but also that of lawyers, ethicists and industrial managers.

Several forces within our society bear ultimately on protection of workers' health. For example, the authors discuss balancing costs vs benefits; individual freedoms and rights to privacy vs scientific needs and employer liability; economic and technical feasibility vs prudence in setting occupational health standards; and the advantages vs disadvantages of the workers' compensation system. The text does not provide definitive solutions to issues, but rather presents various points of view, accompanied by the latest precedent cases and relevant science and, as such, documents a point in our evolution toward a safe and healthy workplace.

Jeffrey S. Lee
William N. Rom

ACKNOWLEDGMENTS

The Third Annual Park City Environmental Health Conference was sponsored by the Rocky Mountain Center for Occupational and Environmental Health at the University of Utah; The Arizona Center for Occupational Safety and Health at the University of Arizona; and the University of Texas Occupational Health Program. The Conference Chair was Jeffrey S. Lee, PhD; John Repko, PhD, from Arizona, and Glendel J. Provost, JD, MPH, from Texas, served as Assistant Chairs.

We wish to thank the Planning Committee, especially Ms. Mary Alexander, for their efforts. Special credit goes to Ms. Katharine Blosch, Director of Continuing Education for the Rocky Mountain Center.

We are particularly indebted to Ms. Sheran Smith for her editorial assistance in the preparation of these conference proceedings; to Ms. Jackye Brooks and Ms. Robyn Drew for typing; and to Ms. Sara Shopkow for her bibliographical assistance.

Lee   **Rom**

Jeffrey S. Lee is Director of Industrial Hygiene for the Rocky Mountain Center for Occupational and Environmental Health and is Assistant Professor in the Department of Family and Community Medicine, School of Medicine, University of Utah. Dr. Lee spent over nine years as an industrial hygienist with NIOSH and its predecessor organizations, and three years with the OSHA Health Response Team. He began the Industrial Hygiene Program at the University of Utah in 1979 and has served as its director since its inception. He is Board-Certified in the Comprehensive Practice of Industrial Hygiene and is a registered Professional Engineer in Safety. Dr. Lee received his MPH and PhD degrees in industrial hygiene from the University of California at Berkeley.

William N. Rom joined the faculty of the University of Utah School of Medicine in 1977. He is Director of the Rocky Mountain Center for Occupational and Environmental Health, which is one of 12 NIOSH-funded Educational Resource Centers. He has a joint appointment in Pulmonary Disease of the Department of Internal Medicine and Occupational Medicine in the Department of Family and Community Medicine. His research interests are in occupational lung disease, energy-related developments, mining and reproductive toxicology.

He was a research fellow in environmental and pulmonary medicine at Mt. Sinai School of Medicine (New York), and received a MPH degree in environmental health from Harvard School of Public Health. He completed an internal medicine residency at the University of California, Davis, and graduated from the University of Minnesota Medical School. He also studied electives at the Mayo Clinic, Dartmouth and the University of Edinburgh.

He was a wilderness canoe guide in the Boundary Waters of Minnesota–Ontario for eight summers, leading three canoe expeditions into the Canadian Barren Lands of the Arctic and Hudson Bay. He is an avid skier and mountain climber, having climbed Mt. McKinley in Alaska. He is also an honors graduate of the University of Colorado in political science with a commitment to protecting the environment in the West. Dr. Rom is co-editor of *Health Implications of New Energy Technologies,* published by Ann Arbor Science.

AUTHORS

Mary Alexander (Zerwas), MPH
President
Alexander Associates
270 Aldrin
Ambler, Pennsylvania 19002

Frederick R. Anderson, Jr., JD
Professor, College of Law
University of Utah
Room 207, College of Law
Salt Lake City, Utah 84112

Nicholas A. Ashford, PhD, JD
Associate Professor of Tech-
   nology and Policy
Assistant Director, Center for
   Policy Alternatives
Massachusetts Institute of
   Technology
E40-250
Cambridge, Massachusetts 02139

Frederick M. Baron, JD
Baron and Cowley
8333 Douglas, Suite 1400
Dallas, Texas 75225

Leslie I. Boden, PhD
Assistant Professor of
   Economics
Occupational Health Program

Harvard School of Public
   Health
665 Huntington Ave.
Boston, Massachusetts 02115

Richard F. Boggs, PhD
Director, Occupational Safety
   and Health
Organization Resources Coun-
   selors, Inc.
Suite 802
1625 I Street, N.W.
Washington, DC 20006

Carin Clauss , JD
College of Law
Room 604
University of Wisconsin
Madison, Wisconsin 53706

Timothy F. Cleary, JD
Commissioner
U.S. Occupational Safety and
   Health Review Commission
1825 K Street, N.W.
Washington, DC 20006

Kenneth S. Cohen, PhD, PE,
   CIH
Industrial Hygiene/Toxicology
Consulting Health Services

P.O. Box 1625
El Cajon, California 92022

Merl Coon, PhD, MPH
Western Institute for Occupa-
tional and Environmental
Sciences, Inc.
2520 Milvia
Berkeley, California 94704

W. Clark Cooper, MD
2150 Shattuck Ave.
Suite 401
Berkeley, California 94704

Thomas H. Corbett, MD
Anesthesia Service Associates of
Toledo
Sylvania, Ohio 43560

Molly Joel Coye, MD, MPH,
MA
Medical Officer
National Institute for Occupa-
tional Safety and Health
50 United National Plaza
San Francisco, California 94102
and Chief, Occupational Health
Clinic
San Francisco General Hospital
1001 Potero Ave.
San Francisco, California 94110

Morris E. Davis
Presiding Official
United States Merit Systems
Protection Board
San Francisco Field Office
525 Market St., 24th Floor
San Francisco, California 94105

Larry C. Drapkin, JD, MPH
Legal Coordinator

Labor Occupational Health
Program
University of California at
Berkeley
Berkeley, California 94720

Stanley W. Eller, MS
Consultant
Workers' Institute for Safety
and Health
Washington, DC 20006
and Former Director, Health
and Safety Department
International Chemical Workers'
Union
206 W. Hilcrest
Dayton, Ohio 45405

Willie Hammer
12361 E. Droxford Place
Cerritos, California 90701

Charles Henri Hine, MD, PhD
Clinical Professor and Chief of
Division
Occupational Medicine
School of Medicine
University of California at San
Francisco
400 Parnassus
San Francisco, California 94123

Joseph T. Hughes, Jr.
Research Associate
Institute for Southern Studies
P.O. Box 531
Durham, North Carolina 27702

Robert L. Jennings, Jr., JD
Baskin and Sears
10th Floor, Frick Building
Pittsburgh, Pennsylvania 15219

and Former Special Assistant to
Assistant Secretary of Labor
Eula Bingham
Occupational Safety and Health
Administration
Washington, DC

Paul Kotin, MD
Senior Vice President
Health, Safety and Environment
Department
Johns-Manville Corporation
Denver, Colorado 80217

Richard A. Lemen, MS
Director, Division of Criteria
Documentation and Standards
Development
National Institute for Occupa-
tional Safety and Health
Rockville, Maryland 20857

Marvin Lieberman, JD, PhD
Executive Secretary for Com-
mittee on Medicine in Society
New York Academy of Medicine
2 E. 103 D Street
New York, New York 10029

Miles W. Lord, LLB
Chief Judge
U.S. District Court
District of Minnesota
1100 S. Fourth St., Room 684
Minneapolis, Minnesota 55401

William F. Martin, MSEHE, PE
Deputy Director, DTS
National Institute for Occu-
pational Safety and Health
Room 230 Taft Center
4676 Columbia Parkway
Cincinnati, Ohio 45226

Robert D. Moran, JD
Attorney at Law
Vorys, Sater, Seymour and Pease
1828 L Street, N.W.
Washington, DC 20036

Douglas G. Mortensen, JD
Clyde, Pratt, Gibbs & Cahoon
200 American Savings Plaza
77 W. 200 South
Salt Lake City, Utah 84101
and Energy Law Fellow
University of Utah
College of Law
Salt Lake City, UT 84132

K. W. Nelson, MS
Vice President, Environmental
Affairs
ASARCO Incorporated
3422 South 700 West
Salt Lake City, Utah 84119

Phillip L. Polakoff, MD, MPH,
MS
Director
Western Institute for Occupa-
tional/Environmental Sciences
2520 Milvia
Berkeley, California 94704

Otto P. Preuss, MD
Medical Director
Brush Wellman, Inc.
17876 St. Clair Ave.
Cleveland, Ohio 44110

Richard W. Prior, MD
Corporate Medical Director
General Motors
3044 West Grand Boulevard
Detroit, Michigan 48202

Mary E. Rahjes, MPH
Director
Community Health Services
    Division
Colorado Department of Health
4210 E. 11th Ave.
Denver, Colorado 80220

Attilio D. Renzetti, Jr., MD
Professor of Medicine
University of Utah
50 North Medical Drive
Salt Lake City, Utah 84132

Leonard A. Sagan, MD
Program Manager
Electric Power Research
    Institute
P.O. Box 10412
Palo Alto, California 94303

Marvin A. Schneiderman, PhD
Senior Science Advisor
Clement Associates
Washington, DC 20007

John J. Sheehan, MA
Assistant to the President and
    Legislative Director
United Steelworkers of America
Suite 706
815 Sixteenth St., N.W.
Washington, DC 20006

Theodor D. Sterling, PhD
University Research Professor
Computing Science Department

Simon Fraser University
Burnaby, British Columbia
    V5A 1S6
Canada

Howard Walderman, LLB
Senior Attorney
Office of General Counsel
U.S. Department of Health and
    Human Services
Parklawn Building
5600 Fishers Lane
Rockville, Maryland 20857

Donald Whorton, MD
Clinical Associate Professor
School of Public Health
University of California
Berkeley, California 94720
and Senior Medical Associate
Environmental Health Asso-
    ciates, Inc.
Berkeley, California 94704

Steven H. Wodka
International Representative
Health and Safety Department
Oil, Chemical and Atomic
    Workers International Union
1120 19th St. N.W., Suite 620
Washington, DC 20036

John D. Yoder, ScD
Director, Environmental Health
I.T.T. Corporation
320 Park Ave.
New York, New York 10022

DISCUSSION PARTICIPANTS

Herbert K. Abrams, MD, MPH
Director, Arizona Center for
  Occupational Safety and
  Health
University of Arizona Health
  Sciences Center
Tucson, Arizona 85724

Victor E. Archer, MD
Rocky Mountain Center for
  Occupational and Environ-
  mental Health
University of Utah
Building 512
Salt Lake City, Utah 84112

Robert S. Bernstein, MD, PhD
Center for Disease Control
National Institute for Occu-
  pational Safety and Health
944 Chestnut Ridge Road
Morgantown, West Virginia 26505

MacDonald Caza, MD
Noranda Mines Limited
240 Hymus Blvd.
Pointe-Claire, Quebec H9R 1G5
Canada

David R. Denton, PhD
Occupational Health and Safety

Magazine
P.O. Box 7573
3700 W. Waco Drive
Waco, Texas 76710

Owen B. Douglass, Jr.
NL Industries, Inc.
1230 Avenue of the Americas
New York, New York 10020

Marie Geraci
Rocky Mountain Center for
  Occupational and Environ-
  mental Health
University of Utah
Building 512
Salt Lake City, Utah 84112

Tee Lamont Guidotti, MD
San Diego State University
Graduate School of Public
  Health
San Diego, California 92182

Michael Holthouser, MD
Rocky Mountain Center for
  Occupational and Environ-
  mental Health
University of Utah
Building 512
Salt Lake City, Utah 84112

Ana Kimball
Rocky Mountain Center for
  Occupational and Environ-
  mental Health
University of Utah
Building 512
Salt Lake City, Utah 84112

John J. Mattick
Reynolds Electrical and
  Engineering
P.O. Box 14400, M/S 706
Las Vegas, Nevada 89114

Russell McIntyre
Rutgers Medical School
University Heights
Piscataway, New Jersey 08854

Janet Miller
Pepperidge Farm, Inc.
Box 238
Richmond, Utah 84333

Glendel J. Provost, JD, MPH
Conference Assistant Chair
Assistant Professor, Occupa-
  tional Health
Occupational Health Program
University of Texas School of
  Public Health
P.O. Box 20186
Houston, Texas 77025

William N. Rom, MD, MPH
Director, Rocky Mountain Cen-
  ter for Occupational and
  Environmental Health
Associate Professor, Department
  of Family and Community
  Medicine
Chief, Division of Occupational

and Environmental Health
Associate Professor, Department
  of Internal Medicine, Division
  of Respiratory Critical Care
  and Occupational (Pulmonary)
  Medicine
University of Utah College of
  Medicine
Salt Lake City, Utah 84132

Mark A. Rothstein
West Virginia University
College of Law
Morgantown, West Virginia
  26506

Joyce A. Spencer
University of California at
  Los Angeles
405 Hilgard Avenue
Los Angeles, California 90024

James H. Spraul, MD
Monsanto Company
800 North Lindbergh
St. Louis, Missouri 63166

Jon R. Swanson
Swanson Environmental, Inc.
29623 Northwestern Hwy.
Southfield, Michigan 48034

Craig Swick
University of Utah
College of Law
1972 South 8th East
Salt Lake City, Utah 84105

Stephen P. Teret, JD, MPH
Johns Hopkins University
615 North Wolfe Street
Baltimore, Maryland 21205

Mark Veckman
International Paper Company
77 West 45th Street
New York, New York 10036

William L. Wagner, MS
Rocky Mountain Center for
   Occupational and Environ-
   mental Health
University of Utah
Building 512
Salt Lake City, Utah 84112

J. D. Watts
Texaco, Inc.
P.O. Box 52332
Houston, Texas 77052

Michael Zacks
Office of the Ombudsman/
   Ontario Canada
125 Queens Park
Toronto, Ontario M5S 2C7
Canada

CONTENTS

Section 3: Workers' Compensation

Section 4: Job Discrimination

Section 5: Ethics

CHAPTER 1

INTRODUCTION

Jeffrey S. Lee, PhD and William N. Rom, MD
Rocky Mountain Center for Occupational
and Environmental Health
University of Utah
Salt Lake City, Utah

Legal and ethical issues have been inherent in the field of occupational safety and health for centuries. As early as the 1700s Ramazzini recognized the forces that exist between preservation of worker health and economic development. Ramazzini addressed "risk acceptance" and spoke of workers in certain occupations "that derive from them grave injuries, so that where they hoped for a subsistence that would prolong their lives and feed their families they [were] too often repaid with the most dangerous diseases and finally, uttering curses on the profession to which they had devoted themselves, they desert their post among the living" [1].

Society's position on these issues is constantly in a state of flux. Political views are conditioned by one's background, source of income and personal mores. The conference on Legal and Ethical Dilemmas in Occupational Health was held following a radical redirection of American politics toward a more conservative, free-enterprise approach. Occupational Safety and Health Administration (OSHA) regulatory and enforcement policies were radically being redirected. The current economic stagnation and inflation are, in part, being blamed on the overenthusiastic regulatory apparatus that has evolved over the past decade. Yet, public support for a safe and healthful workplace has not diminished. It was at this crucial turning point in public policy that this conference was convened.

Perspectives, comments and criticisms expressed at the conference represented the full political spectrum of our society. The responsibilities and ethics of the corporate enterprise were discussed. The role of regulators, the ethics involved in conducting occupational health research and the newly defined legal responsibilities of occupational safety and health professionals were reviewed. Questions such as: "Who is responsible for occupational disease compensation? Are remedies adequate? What benefits must be shown before standards are promulgated? What investigative authority should government have?" were discussed thoroughly at the conference.

Individuals representing the worker's perspective stated that any occupational disease and injury that occurs is, by definition, not ethical and represents a failure to provide a safe and healthy workplace. Industry supporters responded that zero risk is not feasible.

The Workers' Compensation system was recognized as outdated and unresponsive. Other compensation programs, such as Social Security, are inappropriately bearing a significant portion of the cost of occupational disease. Benefits are woefully inadequate. The Workers' Compensation system does not provide to industry the necessary economic incentives to prevent occupational disease. The burden of proof is placed on the worker, who is most often least able to bear it.

The concept of occupational safety and health standards was endorsed by all parties, with opinions differing only on what was "feasible." Uniform standards accepted as "safe" and applied consistently from state to state were supported by industrial representatives. Maintaining exposures below such standards partially protects companies from unjustified liability and Workers' Compensation claims. If standards are uniformly applied, the cost of compliance is theoretically borne by all companies, thus removing the potential unfair advantage of a firm that does not invest the necessary funds to control exposures. Labor union representatives were also supportive of standards that protect their workers. Yet, they cannot assume responsibility for health and safety in a plant without assuming liability as well. A union's role should include ensuring that management conforms with standards.

Certainly workers are interested in stringent standards that protect their health, but also will not cost them their jobs. All workers—the pregnant woman, the hypersensitive, the smoker—desire a safe workplace. Job discrimination based on the fact that certain worker groups are at increased risk was discussed. The level of "risk acceptance" is not insulated from economic and political pressures. A person is more inclined to risk his health by working with a toxic chemical if alternative employment is unavailable. Similarly, the lessons learned from the saccharin case and from unregulated tobacco usage demonstrate that individuals consciously trade off risks for "benefits."

Cost-benefit approaches are increasingly being used in public decision-making. However, conference participants argued that one person's benefits cannot unilaterally be traded for another's costs.

The prevention of occupational disease and injury has become increasingly important for industry. Related social and ethical issues are high-priority concerns that affect corporate productivity and liability as well as human resources. The benefits accrued by a healthy workforce, although difficult, if not impossible, to quantify, are overcoming traditional apathy. The "hidden costs" of occupational disease and injury are well recognized as enormous, and are ultimately borne by all of society.

It is easy to say that workers should be informed of the hazards and risks of their occupations, and that they and government representatives should have access to their medical records. However, the issue becomes obscure when scientists disagree about the level of risk, or the reliability of predictive tests, or the implications of exposure to the pregnant female (or reproductive male). Informing workers of a risk with a low probability of occurrence or an abnormality that scientists themselves cannot interpret may unduly alarm them and may be a disservice.

Occupational safety and health law is in its infancy, with precedent-setting cases with far-reaching implications; but, in general, the courts have upheld standards that will protect the occupational safety and health of workers. The appropriate balance between worker rights and employer rights is being argued on many fronts and from many different perspectives. Many of these cases are discussed in this book.

Dilemmas and issues in the field of occupational safety and health need to be discussed in an open, constructive manner. Areas of uncertainty need to be identified and addressed with cooperative approaches toward research. The villains are excess disease and ignorance. Our field is replete with tragedies; however, to dwell on them or to try to establish blame for them is more often unproductive, indeed, destructive.

Advocacy points of view should be respected, and all advocates should be allowed an opportunity to constructively debate their views. The Third Annual Park City Conference attempted to provide such a forum which we consider to be a basic responsibility of academia.

REFERENCE

1. Ramazzini, B: De morbis artificum (Diseases of workers), G Rosen, trans. Hefner Publishing Co., New York (1964), p. 7.

CHAPTER 2

KEYNOTE ADDRESS: CORPORATE ETHICS AND ENVIRONMENTAL POLLUTION*

Miles W. Lord, LLB
U.S. District Court
District of Minnesota

I have to give a couple of caveats as I start, or else I am liable to be attacked right here on the lectern.

First of all, when you go to church on Sunday, you are all called sinners, aren't you? You don't really mind because the preacher says he is a sinner himself and we are all working for our salvation. I feel much that way as I talk about the environment and various actors who are working to desecrate it, if not to destroy it. I am going to confess to you.

When I was the Attorney General of Minnesota, I conceived the very good idea that the metropolitan area of Minneapolis and St. Paul would be better off without mosquitos, so we organized a Metropolitan Mosquito Control District, and we spread DDT on everybody and everything. We put thousands of pounds of DDT everywhere. We thought we were doing pretty well. We didn't do too much to the mosquitos, and I don't know what we did to our progeny. Ultimately, we learned better and we stopped.

During the same period, as attorney general I went to Washington, DC, with a delegation and literally wrestled about $25 or 30 million out of the Department of Agriculture so that we could start an atomic energy plant just about 15 miles north of the Twin Cities' water intake on the

---

*Editor's note: This address was not intended to be an article or a treatise. Its syntax reflects that fact.

5

Mississippi River to produce cheap electricity. The plant leaked and generated almost no electricity. They are still cleaning it up. I sinned on that one.

My father-in-law was an iron mine owner, and I sometimes worked with him. We filled lakes; we pumped refuse into the streams; we did as we pleased without interference — all in the name of jobs and progress. We learned better, but it took time and education.

So now when I talk to you about what mistakes we make, I am not pointing my finger at you; I am speaking from my own observations and experiences. We are all guilty by virtue of being alive. Individual life, in any form, creates some pollution. Why, then, should we concentrate on corporate conduct? The answer is that corporations are the principal actors. Corporations create 80% of our gross national product. They, of all the entities working, have the most potential for good or evil in our society. However, it is not the individual, the corporate person, about which I am talking. It is the role that the corporation itself plays. When an individual works for a corporation, he gets into the corporate mold. He must behave and move along in the corporate manner.

The inner workings of a corporation, with all of its people, its money and privacy, is almost an impregnable citadel; wherein the planning and execution of moves and countermoves that affect our lives and the future of our world take place in relative secrecy.

Corporations are always far ahead of the government. They know what they are doing, what they plan to do and they know the effects of what they are doing. Whether or not they always tell the government about that is another thing. However, there are some good legal reasons why the corporation poses special problems and is very difficult to deal with.

By and large, we can control the conduct of individuals or families or small groups by the coercive force of law. We can just threaten to put them in jail or fine them. The larger the group that is involved, of course, the less responsibility each individual member has for the actions of his group. It then becomes more difficult to enforce the law by these coercive means. So, generally, it is thought that a larger group creates a greater threat to society. Let us look at a common example from our system of justice.

In the eyes of the criminal law, the individual is strictly accountable for his misdeeds. In fact, the law would prefer that all misdeeds be done by a single actor so that the individual who, for instance, builds an illegal bonfire of leaves would be caught and punished. However, this individual, without criminal liability, could plan to build a fire, prepare for it and contemplate doing it, but to be criminally liable for it, he would actually have to burn the leaves. However, people acting in groups pose an added threat under our criminal system. Therefore, if you get two or more people together and they plan a crime and pile up those same leaves, they never need to strike the match to be criminally liable. Merely getting together and planning an illegal act and any member then taking one act in furtherance thereof constitutes the crime of conspiracy. This concert of action among the larger group members represents a greater threat to society. So, those individuals have committed a crime simply by agreeing to get together to do the crime. If they make one overt act in furtherance thereof, they need not complete the contemplated crime. Well, that seems to be appropriate.

We have perceived over the centuries that this greater danger is posed by people working together. If two or more individuals or families or friends or buddies or rascals get together to carry out a crime and make the slightest move toward it, the law jumps on them. So, you cannot prepare to commit crimes in small groups. That is a conspiracy.

You cannot pollute in small groups either, or they will catch you. If you must pollute or commit a crime, you should form a corporation. The reason is that a corporation, in the eyes of the law, is looked at as an individual entity.

Take an outdoor biffy located on running streams — detection, conviction and punishment are fairly sure and rather swift.

We will now go to the other extreme. Take the metropolitan sewage waste disposal system. If you lived where I live, the system can overrun its capacity, distributing raw effluent to the people down in New Orleans, with relatively no fear of punishment. They can continue the process long enough to wait for federal grants, or if they are a big enough city in a small enough state, they won't have to do that. They can get a permit to continue it.

So, I would say the lesson we learn from this is that it is a crime to pollute in miniscule amounts. You should gather great lumps of it together and put it in a river as an unwelcome addition to the diets of those millions of people downstream.

Those two examples about the privy on the creek and the sewer disposal on the river have parallels throughout our society. In many states you are not allowed to burn leaves in your yards even though you love the smell in the fall. If you are the guy burning leaves and you live next door to a giant smoke stack that spews forth its pollution and potential for acid rain 600 or 700 feet high, you will be the one nailed, not the corporation.

Sometimes I think the reason we prohibit people from doing small, little, silly things like burning leaves is just to make people hate the environmentalists. You get a bunch of people in an environmental committee, including fellows representing industry, and the first thing they will do is prohibit leaf burning or dusting or washing on Thursday or something. Why does this imbalance in enforcement exist?

As I said, if you are going to pollute, form a corporation. You can have 60 or 100,000 employees. Very seldom do the government or the law enforcement people accuse the individual within a corporation of the act of conspiracy. They say no. It is the act of the corporation. The corporation is a "person" and can only act through its agents. Therefore, the actors are all one person and are not charged as conspirators. It makes for a disparity under the law, because we find the government proceeding against all small groups all over for the crime of conspiracy, but they do not go after the corporations. Even when several corporations work together, there is often immunity from the conspiracy claims.

I suppose most of you remember in your economics courses in high school or college about John Locke and Adam Smith and how they indicated that each of us should be able to utilize our property to its fullest extent, maximize the profits, and an unseen hand would guide us to our natural destiny; everybody would be better off as a result of each person exploiting his property to the fullest. Well, those fellows are still very much alive, and they have even become more alive since the last election. The current thought is that you should maximize profit; you should buy cheap and sell dear. The country, and the well-being and the very existence of the government can depend on how we do on Wall Street. Buy as cheap as you can and sell as dear as you can, and have something left for yourself. That is the American way. I have done all right by it. But it is a kind of fancy way of saying that our country is run by selfishness and greed. If you just look at it in its barest elements, that is what it is.

Now, is that a very good system to turn loose to control the long-term destiny of mankind? Having selfishness and greed control our environment, control our work places, decide that which we are going to breathe, what we are going to eat, what we are going to drink?

You read horror stories daily about the unfit and unsafe foods, drugs, air and water to which we are exposed. Who is doing these things and why are they being done? By and large, it is corporate activity.

The corporation that is doing most of the acting is not ordinarily trying to attract the philosopher or the scholar. They see the corporate goal as making money, while the average corporate leader looks disdainfully at the cry of those who do not have a job—the Chicano, the Black or the poor minorities of any kind. You know, these minorities have a little slogan: they say, "We would like to be excused for crimes of survival." These poor people say that if they have to commit a crime to get groceries or clothing or to heat their houses, they should be forgiven. However the cry from that strata of society is by and large ignored by the courts or the government or the businessman. The leaders give little heed to those unfortunates who choose to exist by committing a crime, even though it be a crime of survival. Yet, as we come around to the end of the story, we see that time and again the corporations are making that same cry—the cry that does not work for the poor works well for the corporations.

The corporations say, "We cannot stay in business unless we do this thing. If we did a wrong thing in terms of turning out bad products for people to eat, or a drug that poisoned instead of curing them, we cannot be responsible. We have done so much good for this country; we provide for so many people. We cannot survive unless we get away with this." It is a crime of survival. So we come around full circle.

The people who can best afford to pay for the crimes of survival are getting away with it. The people who need the crimes of survival are not getting away with it. I am not sure either of them should get away with it.

Unfortunately, in some cases we find people who are turning out foods or fibers or drugs and selling a product that is harmful, and we cannot find out why they are doing it. Let us attempt to ascertain who is responsible. If you go to the research department, you might find that somebody tested it on rats and wrote it up in a report. Where the report went, you will never find out. Certainly, the boys in the production department did not know about it when they began the production. The fellows in the sales department were never told. And it would be unthinkable to believe that one of the higher corporate officials got hold of information to the effect that what they are selling as a cure was actually poisoning people. The board of directors is the last to know. There is something wrong with the system. Their officers are not expected to compute the incalculable damage to the world as a part of the cost of production.

But, they say, the corporation does what it is assigned to do. It maximizes a return on the investment. It uses its political muscle and its economic power to achieve a maximum return on the investment—not next year, not the year after, not ten years from now, but today. The name is instant profits.

For example, you could set up a little plant right here in Park City. You could manufacture something with 10 or 15 men on the payroll and make $50,000 a year, spew poisonous waste into the earth or into the air. That would cause ultimate damages in the billions, but the profit system says you have a right to do that without government control. We should not be able to produce anything that we cannot handle, but we do.

Let us get back to the balance sheet. Within the narrow confines of the balance sheet, the corporation is doing well. That is all that counts. If the product is unsafe or actually poisonous, it may be known by others, but not by the corporation officials. Suppose one of the men in the industry learns what is happening. He does not like the fact that it is happening, so he goes out and tells somebody. He might tell a government official or a newspaperman about what he found. Now, what do you think *his* ultimate fate will be? He will be hounded from that company. "He's disloyal." "He's a trouble-maker." "He's a squealer." In industry, you are expected to follow the code of the underworld: *omerta* — silence. The code of the underworld requires that you never squeal, no matter what happens. You go to jail, you do your time, you take your punishment, but you do not squeal. To me, it is rather sickening to see that code enforced in business circles. If a man is run out of an industry because he squealed, ordinarily he cannot find a job in another similar industry. In addition, he is often foreclosed from positions on college faculties. He violated the code. The clout that big industry has on colleges, in addition to the few well-placed men on the board of regents, is enough to keep those troublemakers off the staff of the universities and colleges. The universities themselves are unwilling captives of this segment of our economy. They depend on the grants from the government and big industry to finance much of the work of their professors in research and study.

I recently heard a case involving the Chemistry Department at the University of Minnesota. The case itself involved alleged discrimination against women, but we covered the grant procedures in the department. Of the 40 chemists in that department, there was not one ecologist, and all grants were received to invent new polymers or to enhance production; not one grant was dedicated, nor was one penny spent, to determine the effect of the products on mankind. Not one penny. Turn it out; if it is useful, use it. Let somebody else worry about how mankind should protect itself from it. There is a great deal of orientation in academia towards the profit motive.

In the old days — you know golden olden days — if you killed somebody, if you produced something that would hurt somebody, you were stopped. If you poisoned their cattle, you were stopped. If you burned all of the surface off their lands, you were stopped. Not today. Today, we have the cost-benefit analysis, where you weigh how much a human life is worth. I always thought life was a sacred and priceless thing.

The cost-benefit analysis is a wonderful tool. You can get an economist to put a price on a human life and then you weigh that against the benefit. When you put a price on the priceless, all is lost. Economists can do anything they wish with figures. If the pollution is pervasive and many people are killed or injured, the corporation asks the government to step in and take over. They can have a Black Lung Bill passed as was done for coal or they can have a White Lung Bill as is about to be done for asbestos. They are now talking about Brown Lung for cotton. So, you may go ahead and poison, and, if it gets bad enough, the federal government will take action. If it does not get bad enough, you can deny all responsibility. So we do not have much sense about how we do this cost-benefit analysis. There is no rhyme or reason to it. We sort of play Russian roulette. This sentence of death and debilitation is not imposed on us by some foreign power. We are doing it to ourselves, by the cost-benefit analysis approach. We are ashamed to acknowledge the stark reality of it. What we are doing is to trudge along with no clear plan as to where we are going, or which individual will be sacrificed.

You know, even if a group of people were alone in a lifeboat and had to sacrifice one person to save water and get a little raw meat, they would probably pick on some old doodler, who had lived beyond his time anyway. Even Hitler, when he was butchering people, articulated a reason to his madness. We do not even do that. We go along and make those sacrifices in easy stages. We poison them in the womb, and in the milk of their mother's breast and the air that they breathe, the water that they drink, the food they eat. It is random selection imposed by the adults on the young and unborn generations.

Will the calls of those who give us warning go unheeded? There is no profit in following their advice. The corporate officials who are responsible as the actors in the main are decent people. It seems to be the system that they get caught up in that causes them the trouble. They are kind and honorable in their personal relationships, their personal mores. They attend church; they do charitable deeds. Almost any corporate president or official would walk miles to help a little child who is hungry or injured, or who is hurting. But, he could then walk back to the office and approve a plan that would dump tons of poison into the drinking water of that same child.

There is an additional reason why corporate conduct will become even more difficult to control. The laws once prohibited election contributions by corporations. There was the fear that our political system could not handle that kind of thing. So, they did not let the corporations contribute. This prohibition was founded in the same fears of group action that caused conspiracy to be made a crime. Large numbers of people working together, as in a corporation, pose a threat to the freedom of our election process. Recent cases involving the alleged civil rights of corporations have changed this concept substantially. They now say that since a corporation is a "person" in the eyes of the law, that it — the corporate "person" (a piece of paper) — has the constitutional right to freedom of speech and thus can use its corporate war chest to influence public opinions and elections. They can advertise and use corporation profits for any purpose that furthers their economic interests. It puts them into the middle of politics. They are, by court decision, allowed to have political action groups in which they collect a percentage of the salaries of officers and employees and use it to finance the campaigns of candidates favorable to their cause. I can tell you that some of the recent Supreme Court decisions and some of the laws that have opened up the way for the corporations have changed our way of doing politics in America. Now, that may be good from the point of view of some, and it may be bad according to others. I personally feel that it bodes evil for our individual freedom, our environment and our way of life. Yes, even for democracy as we know it.

Corporations are now free of search and seizure. They have a right not to have their speech repressed; they have a right to do many of the things that individual people can do. The evolution of a corporation resulting in the granting of personal rights began with the then-amazing concept that a corporation could own land the same as an individual owns land. Many people objected to that, and some philosophers could not understand the concept. Thereafter, corporations were allowed to do other things the same as individuals were, and now we have finally come to the point where the corporation is an individual. Corporations have rights never originally conceived of as belonging to them. They have utilized their corporate clout to make corporations into something they were never intended to be—a person, in the eyes of the law, in all respects that are helpful to them and in none of the respects that are harmful to them. When trouble comes along for corporations, they have a way of banding together like a herd of musk oxen being attacked by a wolf. It is rumps together and horns out. It makes the corporations rather difficult to handle, particularly in times like these.

Recently, the government itself has taken an about-face. In times past, according to the law, government employees and agents would be imprisoned for disclosing the contents of files to special interest groups with which the government was dealing. What have we done recently? We have put the special interest groups in charge of the agency itself. Not only do we give them the keys to their own files, but also to those of everyone else who is in a similar position. We actually have government officials who claim they are polluting in the name of God. It occurs to me that the church might develop an ethic which would say that pollution, contamination and desecration of our environment are a sin, and that we should not look to the Second Coming immediately, but should plan for future generations. Of course, if we build enough neutron bombs, we need not worry about whether or not Christ comes.

The corporation can outlast you, or the politician. Their perpetual existence makes it possible for them to outwait anyone who stands in the way. It also makes it possible for them to await the opportune time. It appears that they have now arrived at that time. They have waited since the time of Teddy Roosevelt, two or three times the public life span of any official, and they see this as the time to move.

When I watch the wonderful progress President Reagan is able to make, I think it is remarkable how much easier it is to destroy than it is to build and preserve.

When I see the people who are now in charge of governmental agencies, I despair.

We have been moving along over the past 10 or 15 years, starting with Earth Day, initiated by Senator Gaylord Nelson. We have begun to develop an environmental ethic and to start to try to protect our people from drowning in their own sewage and effluent.

After moving along for several years, all at once we have elected a leader of our country who turns the opposite way, who says, echoed by his advisors, "Well, we must give some breaks to the rich. You haven't seen any breaks for the rich lately," they say. "The rich haven't been getting any breaks out of social welfare." I agree with them that we must mine the coal. We must cut some of the forest. We must develop the oil. We have to live and we have to grow. All these things are necessary, to a degree, but strict controls should be imposed by the government to minimize the damage to the environment. Under the present administration, it is concluded that business knows what is best for our people and that the competition for resources can dictate their preservation.

Now, I have talked a lot about corporations, and a lot of labor people here may be feeling pretty good about this. I do not think you should feel too good about it. In my own experience I found some of the nation's largest labor unions willing to sell some segments of their labor force to death and debilitation and ill health to keep an industry going.

Historically, the political system evolved so that there was a kind of division — labor was on one side, management on the other — to form a kind of balance. Industry alone is a formidable adversary for those who would preserve the environment. It generally throws its balance against the environment.

The corporations' pervasive influence penetrates the executive, legislative and judicial departments of government. When labor and management stand opposed, the scales are often balanced politically, but the recent phenomenon that I have witnessed is indeed cause for alarm. When labor lends its clout to a polluting industry, all is lost. It is an invincible combination, often controlled by the national management of the labor union involved. Labor unions that have traditionally expressed great concern for a healthy environment in which to work have sold out on some segments of their membership and have allowed the workers and the public to be unnecessarily exposed to death and debilitation. Labor unions are not a democracy. If the local unions had their way, the pollution would be stopped. So, they will sell off segments of their work force, of their own members, because they need to keep those other people working. They do their own cost-benefit analysis without any benefit or supervision from the court or the government.

Jobs and the people who hold them are the most potent hostages available to industry. Environmental impact statements were designed to prevent the commencement of industries that have as their ultimate result the desecration of the environment. The impact statement looks at the long-term damage to the environment. The alternative, which works much better for the industry and which can frustrate governmental control, is to start slowly, hire ten workers, and thereafter use these few jobs and the prospect of more as leverage on the public. As soon as ten men are employed, the stage is set. It is literally impossible to stop or to control them. "It will cost jobs and depress the economy," they proclaim.

They call an economist. The payroll is hypothetically recycled through each of the local business enterprises in the community, and projections of prosperity and affluence created by these ten jobs will go a long way toward paying off the state debt.

For example, if you pay $100 to the grocery store operator for groceries, he'll pay it to his suppliers, who buy gas and tires and so on. This $100 keeps changing hands until it is $50,000 worth of benefits. With ten jobs gone, you have lost hundreds of thousands of dollars right there. All they are doing after all is pumping stuff in this little creek or killing a few babies, and we have too many babies anyway.

Even Workers' Compensation and the immunity it furnishes the employer for his own negligence protect polluters. When the employer does not personally pay for the workman's injury, it removes his incentive to make a safe workplace. Although I hate to see the premiums escalate, I would suggest that an even greater escalation of the premium than we now have, where you have an imbalance of too many injuries, might be a good cure for that. The employer should be penalized for having an unsafe workplace.

President Reagan said the other day that even coal has its civil rights. I am trying to think how that works.

Well, coal has veins; maybe it is something like a person. But now, he said the other day, do you realize that if we enforce these environmental standards and do not let them tear up the sod and the topsoil, that some coal might not be mined? Can't you see that little piece of coal down there? He says, "Gosh, President Reagan is going to let me be mined."

We are all for the environment. Every one of us. We talk about it and we think about it, we act on it, but one of the things you affect when you start to move on the polluters is jobs. Jobs will be lost.

I have been around a long time. It is not easy to tell anyone they cannot have a job because they are producing poison. It is not easy to tell a bank robber that he has to change his way of making a living.

You are probably saying to yourself right now, "What kind of bird is this speaker? He's really anticorporation, isn't he?" If you are saying this, let us examine, not my thinking, but your thinking.

You undoubtedly applauded when the President and the Chief Justice of the Supreme Court announced that we have to crack down on criminals who are making our country a less safe place in which to live. People are being mugged, slugged and drugged by criminals. These bad actors must all be put in jail. We must stop giving them the right not to be searched. We must agree to allow as evidence formerly illegally seized evidence against them. We must streamline the jailing of individual criminals.

While you applauded these efforts by the President or the Chief Justice, you felt proud and you did not think that they were un-American. You did not think of them as being anticriminal, but rather pro-people—"stand up for the good guys." Merely denouncing these actors who kill, maim, cheat, rob and steal does not earn us the title of being anti- anything. However, if you call attention to the fact that a great deal of this kind of conduct is carried on in the corporate form, you may find yourself being called anticorporation, antibusiness or worse. The mere fact that you are an individual who may work for a corporation does not mean that you are being denounced any more than being a member of the public makes you a criminal. What is being denounced here are those people and institutions who are doing wrong. If the shoe fits, I might suggest that it is perfectly appropriate for you to wear it so long as you recognize your reason for being so alarmed.

As I told you, some of our finest people are selling poison, but they should not be allowed to do it. The corporation should not be allowed to do it. One of the ways to stop them is to publicize it.

In my experience, the public is shielded from accusations and disclosures of corporate wrongdoing. Such stories are all too often handled by the financial writers for the newspaper. On a local basis, the writers are much more concerned with protecting the industry, its stock prices and the local tax base from exposure than they are interested in disclosing to the public the true nature of the evil practices that are perpetrated to make those profits. Most newspapers, insofar as local stories are concerned, believe that corporate crime or white collar crime cannot exist unless the actors are Italians and associated with the Mafia.

I am firmly convinced that if the newspapers followed corporate civil trials as closely as they follow the trial of an individual charged with a crime, that the disclosures made in the course of this coverage would change the manner in which business is done in America. These practices can only be carried on by management, shrouded in secrecy. The stock-owners themselves would often rebel if they knew of some of the dirty tricks being done in their names for the sake of dividends.

The man in the research department is just one of the many actors. He did his thing. The man in the sales department noticed that a few workers were dying, keeling over, but did not think much of it. Maybe it was something they ate. And the president and the board of directors are the last ones to know. So, you move in. What are you going to do? You are a prosecutor or a governmental official or an adversary in a civil case. Do you arrest the man in the lab who found it but did not fully report it? Do you arrest the production manager? Do you arrest the salesperson? No, you go to the board of directors, and they are completely innocent. You cannot put them in jail; you cannot fine them. If you fine them, they are already covered by a policy of insurance that says they will not suffer any harm whatsoever. Nobody gets fired. So, what you do then is to take the corporation as a whole, not any one individual, and fine it. Just lay a big lump on it. See, and they say, well, this corporation is not owned by the people who owned it when these things were done. This corporation is now owned by a whole new set of shareholders that just bought in on Wall Street. If you penalize those new shareholders, punish them for ancient crimes, that is not fair. And the board of directors, of course, is just working for the shareholders to do what is good for them so the board shouldn't be punished. Some shareholders are new, so they should not be punished. Just turn us loose and let us go home.

I have had these arguments made by some of the finest lawyers in America. We talk the way we are talking here. They say, who are our shareholders? These are the people that are going to be hurt. Do you know who it is? It is the school teachers and the teamsters and the steel workers, the health and welfare funds. It is the state employees' pension fund. It is all of these funds really owned by the labor fund. This is not a reliable statistic, but approximately 30% of shares of stock in America, in the bigger companies, are owned by these funds. I am afraid to give a definite statistic, but I know it is a tremendous amount handled by eight banks in New York, which represent billions and billions of dollars of clout in the investment world.

The average investor is a 28 year old graduate of Radcliffe—a young woman who sits in an office and looks at certain parameters and does the investing. In any event, they tell you now that levying a fine will not punish the corporation. They say, indeed, you are going to punish the school teachers of this world, you are going to punish the teamsters or the steel workers, whomever it may be. And when you say yes, they come to their rationalization.

You cannot punish management because it did not know. Management acted for the shareholders; there is insurance to cover. The middle management did not have the whole picture; they can easily be replaced. In any event, shareholders, teachers and others similar should not be punished. So you get a system where there can be a theoretical punishment but nothing to stop the predation on society. Some of the ideas that you might give some thought to would be to bring the attorneys in on some kind of case similar to an antitrust action.

When we see the courts being caught up in this dilemma and giving heed to these arguments, we must recognize it for what it is. It is a plea to be forgiven for crimes of survival. They had to break the law to meet competition, to stay in business, to keep their people working, to meet the economy. Thus, the crimes that are necessary for the survival of business often go unpunished while the crimes of survival that the poor people commit are disallowed.

The law contemplates that the government is simply not capable of effectively enforcing the antitrust laws without the assistance of the private lawyers who represent the individual litigant. That way the lawyer gets his attorney's fees and the litigant gets triple damages, and by that means they hope to keep our competitive system free and open. We are having some success at that. In like fashion, maybe we should give substantial attorney's fees to those who would come in and force an industry to clean up. In this manner, we would put real teeth into the antipollution laws.

And there again you get your old selfishness and greed. The lawyer will do it, perhaps, maybe not quite so much for his altruistic view of the matter, but to make a buck. But we have to have those forces at work. We do have greater tax benefits for the people who clean up their workplace and prevent the poisoning of their employees, but the award of attorney's fees would put a carrot on one end of a stick and a whip on the other end.

As I have said, I worship the First Amendment. I love newspapers even when I am maligned by them. If newspapers would do their bit to expose what they see going on, it could change our way of doing business. The methods that we have used so far are ineffective. We have got to have better systems of rewards for those within corporations who correct the situation. Possibly we need to have more government inside, more government inspectors in there. We have to have people designated by the corporation to watch things, people who have a responsibility to the public and who are required, under oath, to report what they find.

There has got to be a way to get the public more involved in the corporate decisions, and I do not mean the public as individuals because many of them would not understand or could not care less, but elected public officials and their representatives.

SECTION 1

OCCUPATIONAL SAFETY AND HEALTH ACT ISSUES

CHAPTER 3

THE RIGHT OF WORKERS TO REFUSE TO WORK UNDER HAZARDOUS CONDITIONS: IMPACT OF THE *WHIRLPOOL* SUPREME COURT CASE

Mary Alexander, MPH

Alexander Associates
Ambler, Pennsylvania

The workers at the household appliance company in Marion, Ohio, had worried for some time about the accidents that were occurring in the plant. The plant had 13 miles of overhead conveyers to transport appliances and parts. Objects often fell off the conveyor belt. The employer had installed a wire screen about 20 feet above the floor to protect employees from these falling objects. The maintenance workers routinely placed paper on the wire mesh to catch oil drips and removed fallen parts from the screen. However, several people fell through the mesh and were injured, and a newer screen was installed in some areas. One person fell completely through to the floor but survived. Then, on June 28, 1974, a man working on the screen slipped through and fell to his death. Following the incident, the employer, Whirlpool, made some repairs and issued an order that maintenance employees were not to step onto the screen.

One night, just 12 days after the fatal accident of their fellow employee, Virgil Beemer and Thomas Cornwell came on duty for the night shift. The two employees had asked the night before for the telephone number of the Occupational Safety and Health Administration (OSHA) office from the plant's safety director, who made veiled threats about such activities. The foreman ordered them onto the old part of the screen to perform their usual maintenance duties. Because one of their fellow workers had lost his life only a few days before doing similar work, Virgil and Thomas refused to go out on the screen. They were sent home without pay for the balance of the shift, and official written reprimands were subsequently placed in their personnel files.

Virgil and Thomas contacted OSHA, and the Secretary of Labor brought suit on their behalf in the District Court [1]. The District Court denied relief, stating that the work refusal regulation exceeded OSHA's authority. The Sixth Circuit reversed [2] and the Supreme Court granted *certiorari* [3].

The Supreme Court held unanimously that workers have the right to refuse to work in the face of serious injury or death. It upheld the validity of an Occupational Safety and Health Act (OSHAct) [4] regulation [5] regarding a worker's right to refuse to work in hazardous conditions.

To fully understand this controversial issue and the Whirlpool case, we must review those statutes that give workers the right to refuse to work in the face of imminent danger. These include Section 7 of the National Labor Relations Act (NLRA) [29 U.S.C. 157 (1976)], Section 502 of the Labor Management Relations Act (LMRA) [29 U.S.C. 143 (1976)], and Section 11(c) of the OSHAct [29 U.S.C. 660(c)(1) (1976)] and the regulation promulgated under it [29 U.S.C. 1977.12 (1979)].

SECTION 7 OF THE NATIONAL LABOR RELATIONS ACT

In *NLRB v. Washington Aluminum* [370 U.S. 9 (1962)], the Supreme Court upheld the right of a group of workers to refuse to work in severe cold in their machine shop. On a day when the furnace was not working, seven of eight workers walked out and were subsequently terminated by their employer. Their right to walk out was upheld by the court under Section 7 of the NLRA. Section 7 says that workers have the "right to...engage in...concerted activities for...mutual aid or protection...." [6] The court's test for the adequacy of the workers' decision to stop work was not based on objective standards. The court stated that "the reasonableness of the workers' decisions to engage in concerted activity is irrelevant...." [7] Thus, the crucial point is not whether there are actual unsafe conditions, but whether the workers *perceive* the conditions to be unsafe.

An additional requirement of Section 7 is that the activity be "concerted." This means that an individual cannot exercise his or her rights under the section. However, the test has been expanded to include individual actions that tend to advance group interests. The National Labor Relations Board has found "implied" concert in certain situations. For example, an employee may be implied to be upholding a health and safety clause of a union contract [8]. Or, as in *Alleluia Cushion Co.* [221 N.L.R.B. 999 (1975)], concert may be implied if, in the absence of a labor contract, an employee seeks to enforce safety and health statutory provisions designed to protect all workers. In *Alleluia Cushion Co.,* it was held that, in those instances where one person is acting for the group, the activity is implied to be concerted in the absence of any evidence that the fellow employees disavowed the representation.

However, in *NLRB v. Bighorn Beverage Co.* (103 L.R.R.M. 3009), the 9th Circuit refused to extend the concept of implied concerted activity to situations that did not involve collective bargaining and reversed the NLRB. It was alleged that cement trucks operating inside a warehouse caused an accumulation of carbon monoxide, and many employees developed headaches. Several workers left work, and one was fired after calling the state health and safety authority. The court said it would not extend *Alleluia Cushion Co.* to this situation where there was no union and held that there was no concerted activity. Since safety statutes had not been achieved through collective bargaining but, rather, were imposed on the workplace, their enforcement by an individual could not constitute concerted activity. However, the NLRB decision that there had been illegal retaliation due to the employee's union organizing activity was upheld.

SECTION 502 OF THE LABOR MANAGEMENT RELATIONS ACT

Section 502 of the Labor Management Relations Act (Taft-Hartley Act of 1947) [29 U.S.C. § 143 (1976)] provides "nor shall the quitting of labor by an employee or employees in good faith because of abnormally dangerous conditions for work at the place of employment...be deemed a strike under this Act." Thus, employees have a right to walk out if they believe, in good faith, that their work involves abnormally dangerous conditions. They will not be considered to be engaged in a strike and, therefore, may be protected under that act.

While Section 7 applies to nonunionized workers, Section 502 is applicable to union workers. However, Section 502 does not protect workers who strike en masse. If an employee is covered by a no-strike clause that does not exempt safety and health matters, the employee may not engage in any concerted work refusal that might be considered to be a strike. Section 502 exempts certain work refusals from consideration as a strike.

In *Gateway Coal Co. v. UMWA* [415 U.S. 368, 38 L. Ed. 2d (1974)], coal miners refused to work in a mine in which a ventilation system had collapsed. The Supreme Court held that the arbitration clause of their collective bargaining agreement gave rise to an implied no-strike obligation. This obligation was found to support a court-ordered injunction against the work stoppage. The workers were ordered back to work and were required to go to arbitration. The court also required that "ascertainable, objective" evidence be introduced to prove the workers' case.

Thus, an employer can obtain an injunction to stop strikes and can submit the dispute to arbitration unless the walkout falls under Section 502. If there are no contract provisions stating otherwise, a no-strike clause will preclude a lawful walkout over unsafe conditions, unless the conditions are abnormally dangerous. If the degree of hazard is not known or if it cannot be determined whether the hazard is abnormally dangerous, Section 502 is of little protection for the worker.

Grievance and arbitration mechanisms in union contracts provide an alternative to regulation for worker self-help. Safety and health issues are subject to arbitration unless specifically exempted in a union contract. Since *Gateway Coal Co.*, an employer may obtain injunctive relief to enforce an implied no-strike obligation in instances where the workers refuse to work in hazardous conditions that have not been abated or are not based on objective reasons. If the refusal meets Section 502's "abnormally dangerous" test, it will not be enjoined. For a union contract to be useful in this area, it must specifically reserve the right to strike over safety disputes and the right to refuse to work in hazardous conditions [9].

OSHAct, SECTION 11(c) AND
THE REGULATION SECTION 1977.12

Congress enacted the Occupational Safety and Health Act of 1970 "to assure so far as possible every working man and woman...safe and healthful working conditions..." [10]. One of the rights afforded workers by the Act is freedom from discharge or discrimination resulting from "the exercise...of any right afforded by [OSHA]" [29 U.S.C. § 660(c)(1)]. However, the right to refuse work in the face of death or serious bodily injury was not a right expressly granted by the act, as is acknowledged in the regulation [11]. The act does give an employer the "general duty" to furnish the employees a workplace free from hazards likely to cause death or serious physical harm [12]. The Secretary of Labor is given the authority by the act to promulgate regulations consistent with the act. The Secretary promulgated in 1973 a regulation (§ 1977.12) that states that among the rights the act protects is the right of an employee to choose not to perform an assigned task because of a reasonable apprehension of death or serious injury, coupled with a reasonable belief that no less drastic alternative is available [13]. The Secretary acknowledged that the act did not specifically create a right to refuse to work in hazardous situations; but implied from Section 11(c) of the act that a worker could refuse to work under certain circumstances without retaliation [14].

Section 1977.12 has given rise to a great deal of controversy. In *Marshall v. Daniel Construction Co.* [563 F. 2d 707 (CA 5 1977)], the Fifth Circuit disagreed with the Secretary's interpretation of 11(c). An ironworker was fired following his refusal and that of others to return to work when so ordered. The job required fitting steel beams 150 feet above the ground with the aid of a crane. The day was so windy that the worker determined it threatened his life, and he came down from the worksite, as did the rest of his crew. A foreman ordered him to return to work. He refused and was fired.

The court held that the regulation according the worker the right to refuse to work (Section 1917.22) was invalid because it was inconsistent with congressional intent in passing the act. The court said Congress did not intend to grant workers the right to walk off the job when faced with dangerous conditions. It referred to the legislative history of the act [15]. Representative Daniels of New Jersey had sponsored one of several House bills. The Daniels' Bill contained a section entitled the "strike with pay" provision [16]. The provision provided a long process in which employees could request Department of Health, Education and Welfare (HEW) examination of the toxicity of any substances in the workplace. If any substance was found to have "potentially toxic or harmful effects in such concentration as used or found," the employer was to be given 60 days to correct the potentially dangerous situation [3]. After that period, the employers could not require that an employee be exposed to toxic concentrations of the substance unless the employee: (1) was told the health hazards and symptoms of the substance; (2) was told what the proper handling procedures were; and (3) was furnished personal protective equipment. If these conditions were not met, an employee could "absent himself from such risk of harm for the period necessary to avoid such dangers without loss of regular compensation for such period" [13]. Representative Steiger of Wisconsin introduced a bill that did not contain a "strike with pay" provision. Ultimately, the House passed the

Steiger Bill. The Senate passed a bill introduced by Senator Williams that did not contain a "strike with pay" clause, but did give employees the right to request OSHA inspections. Both the Williams Bill and the Steiger Bill were submitted to conference committee, with the House accepting the inspection request provision included in the Senate bill [16].

Provision for administrative shutdown authority in the case of imminent danger was included in the Daniels Bill but not in the successful Steiger Bill. The Steiger Bill gave the courts exclusive authority to restrain imminent dangers. It was the decision of the majority of the court in *Daniels* that, based on the eventual exclusions (both the imminent danger and the "strike with pay" provisions) in the final Williams-Steiger Act, Congress had not intended that workers should have the right to refuse hazardous work as provided by the regulation. The court held that the Act does not provide workers with an express or implied right to strike when faced with a hazardous work environment.

Similarly, in *Marshall v. Certified Welding Corp.* (CA-10, 1978, 1979 OSHD ¶ 23,257), the Tenth Circuit said the regulation providing the right to refuse hazardous work was not valid in a situation where a worker refused to work around a pipe leaking nitric acid. However, the Sixth Circuit in *Marshall v. Whirlpool* [593 F. 2d 715 (1979)] upheld the validity of Section 1977.12 and was, therefore, directly in conflict with the *Daniel* and *Certified Welding* decisions. Three other district courts upheld the validity of the section on similar grounds: *Marshall v. Seaward Construction Col. Inc.* (DC NH, 1979), 1979 OSHD ¶ 24,433; *Marshall v. Halliburton Services, Inc.* (DC W.Va., 1979), 1979 OSHD ¶ 23,409; and *Usery v. Babcock & Wilcox,* 424 F. Supp. 753 (DC Mich., 1976), 1976-1977 OSHD ¶ 21,330, enf denied (DC Mich., 1979), 1979 OSHD ¶ 14,110.

WHIRLPOOL: THE SUPREME COURT UPHOLDS THE REGULATION

Because the decisions in the Fifth Circuit *(Daniels)* and Tenth Circuit *(Certified Welding)* were in conflict with the Sixth Circuit's decision in *Marshall v. Whirlpool,* the Supreme Court granted *certiorari* [3]. The issue before the Court was whether the regulation, Section 1977.12, authorizing workers' "self-help" was consistent with the act, given the act's language, structure and legislative history. The Court held in a unanimous opinion written by Justice Stewart that the regulation was valid on two grounds: first, that the regulation conforms to the act's purpose "to assure as far as possible every working man and woman in the nation safe and healthful working conditions..." [25.415 U.S. 368, 38 L. Ed. 2d 583 (1974)]. The section was found to be consistent with the act's "general duty" clause (Section 5(a) requiring an employer to furnish a workplace free from recognized hazards that are causing or are likely to cause death or serious injury to the employees [17]. Secondly, the Court felt that the fact that Congress considered and rejected the "strike with pay" provision of the act is not inconsistent with administrative interpretation by the Secretary.

The Sixth Circuit had outlined the requirements that a worker must meet to invoke the protection of the regulation and to avoid disciplinary action. The worker must show:

1. that he or she acted in good faith;
2. that he or she had no reasonable alternative;
3. that a reasonable person would have the same apprehension of death or serious physical injury; and
4. that there was insufficient time to use the regular statutory enforcement procedures [18].

The Supreme Court did not address the issue of what requirements a worker must meet to invoke the protection of the regulation. The Supreme Court stated only that "any employee who acts in reliance on the regulation runs the risk of discharge or reprimand in the event a court subsequently finds that he acted unreasonably or in bad faith [19].

The legislative history arguments made by Whirlpool, the petitioner, were rejected by the Court. The opinion stated that, when Congress rejected the Daniels Bill's "strike with pay" provision, it was "not rejecting a legislative provision dealing with the highly perilous and fast-moving situations covered by the regulation..." (100 S. Ct., 893). The provision provided a long-term process for evaluating hazards in the workplace, but did not cover immediate threats of death or severe injury. Furthermore, when Congress rejected the "strike with pay" concept, the intent had been to reject a law unconditionally requiring employers to pay workers who walked out over safety problems. The regulation "does not require employers to pay workers who refuse to (work) in the face of imminent danger" (100 S. Ct., 894). The regulation merely prohibits an employer from discriminating against a worker. Discrimination was defined to be "when he treats that employee less favorably than he treats others similarly situated" (100 S. Ct., 894).

The Court also disagreed with the argument that the administrative plant shutdown provision rejected by Congress is similar to the regulation. The regulation does not authorize government officials or private individuals to close plants. Nor can the employees order the employers to correct the hazardous condition (100 S. Ct., 895).

The Court said that "the regulation on its face appears to further the overriding purpose of the Act...," and it should be upheld, since safety legislation in general "is to be *liberally* construed to effectuate the Congressional purpose" (100 S. Ct., 890, emphasis added).

With respect to the issue of consistency with the general duty clause of the act, the U.S. Court of Appeals for the District of Columbia Circuit ruled on March 12, 1981, that the Secretary of Labor must offer evidence to show that a specific means of abatement is feasible in order to sustain a finding of violation under Section 5(a)(1), the general duty clause, in *Whirlpool Corporation v. Occupational Safety and Health Review Commision and Ray Marshall* (Nos. 79-1692 and 80-2426). This decision vacated a decision of the Review Commission that had affirmed the citation for violation of the general duty clause. The Court said the evidence presented by the Secretary had not affirmatively indicated that abatement by means of installing heavy-duty wire screens was appropriate at the Whirlpool plant. The employee had claimed that replacement of all but one-third of the screen was architecturally unfeasible. The Court stated that the Secretary's witnesses had not addressed "the feasibility of complete screen panel replacement" [20].

The *Whirlpool* court left to the lower court the decision, on remand, as to whether the employees should receive pay for the six hours they did not work [3]. Thus, the Supreme Court gave no indication as to whether back pay should be ordered; but the District Court, on remand, held that the employees were entitled to back pay [21].

In *Marshall v. N. L. Industries, Inc.* [618 F. 2d 1225 (1980)], a post-*Whirlpool* case, the Seventh Circuit felt that the Supreme Court's refusal to decide on the back-pay issue did not proscribe judicially ordered back-pay and held that back pay was "an appropriate remedy for a discriminatory and therefore unlawful discharge" [22]. *N. L. Industries, Inc.* involved an employee who refused to continue operating a payloader that had no windshield or enclosed cab. His refusal followed an incident in which, while dumping lead scrap into a melting kettle, the dross carrying the molten lead in the kettle separated. A similar accident a week earlier resulted in the molten lead exploding. At that time, the worker would have been injured had he not had a payloader with a windshield and enclosed cab. When the worker stopped work, he was suspended and subsequently fired [22]. The Court held that N. L. Industries could be held liable if it could be shown that the employee had a reasonable and good faith belief that conditions leading to his refusal to perform certain job-related duties were dangerous and that back pay is an appropriate remedy under the OSHAct [22]. Thus, the Court set precedent for judicially ordered back pay in those cases where there is unlawful discharge or suspension of an employee for engaging in a protected activity.

EXPOSURE TO HAZARDOUS SUBSTANCES AND WHIRLPOOL

Whirlpool deals with refusal to work under imminent threat of death or serious bodily injury. The question arises as to whether a right to refuse to work exists if the hazard is exposure to a toxic substance or carcinogen that may result in occupational disease.

The *Whirlpool* decision could be expanded to allow workers to refuse to work with toxic substances if: (1) the hazard is very clear, that is, the substance is of high toxicity, its health effects are well known and the workplace exposure levels are high; and (2) the symptoms of exposure are obvious and undeniable. (As examples: In exposure to carbon tetrachloride, the acute effects are well recognized; and in carbon monoxide poisoning, symptoms are seen immediately.) If a worker is exposed to a carcinogen and the cumulative effect may result in disease, can the worker refuse to work with the substance if he or she believes in good faith that it is extremely dangerous? It may be argued that, to the worker, the hazard is as clearly obvious as was the hazard of falling through the wire mesh; and, therefore, he or she has reasonable fear of serious injury and may refuse to work without protection under *Whirlpool*.

But what about the employee who faces a one-time exposure to a carcinogen for which no "safe" level has been determined. Can such an exposure be compared to walking on a wire mesh, even though the injury does not occur until years later? It is scientifically difficult to prove a cause-and-effect relationship of a one-time exposure. Furthermore, the worker does not generally have the access or the ability to evaluate scientific data. Yet, the workers may be faced with making decisions that scientists have been unable to make.

Short-term exposure to toxic substances has been demonstrated to lead to serious occupational disease and death. Asbestos workers, for example, exposed for under three months have demonstrated deaths from lung cancer at 3.5 times their expected rate [23]. This indicates that even shorter exposures may be harmful. This becomes a complex issue when the substance is one for which the health hazard is clear, but the level at which it has its effect is controversial. Suppose, for example, the worker is exposed to 7 ppm of benzene. The standard was upheld at 10 ppm by the Supreme Court following OSHA's attempts to lower it to 1 ppm [24]. If the worker feels, in good faith, that 7 ppm is a serious threat to his or her health, may he or she refuse to work?

A worker may *reasonably* fear that the exposure to a toxic substance is a serious threat. When this occurs, the facts must be considered in a court of law and a determination made as to the reasonableness of the fear according to the objective standard.

The Supreme Court held in *Whirlpool* that "safety legislation is to be liberally construed to effectuate the Congressional purpose" [25], and in *N. L. Industries, Inc.* the Court said that the Secretary of Labor's interpretations of the OSHAct are "entitled to great weight and are controlling if reasonable" [22]. These inclinations toward liberal interpretation and deference to the Secretary's approach could support more liberal decisions in the future and could lead to decisions upholding the right of workers to refuse to work with toxic substances under a reasonable threat of injury, including occupational disease, even if the "injury" is far in the future.

This reasoning breaks down somewhat when considering that *Whirlpool* is based on an imminent danger requiring immediate action. However, for the worker faced with chronic exposure to toxic substances, there is time to petition for administrative action, including calling in an OSHA inspector to solve the problem.

Perhaps a July, 1980, California decision by the San Jose Labor Commission gives indication of the future [26]. A group of Pacific Gas and Electric employees were told to clean up a polychlorinated biphenyl (PCB) spill (the PCB had leaked from a transformer on a pole onto the ground). The workers refused to respond, alleging they had not been given proper training in cleanup procedures, nor had they been given adequate personal protective equipment. The employer was cited by Cal-OSHA for violations (issued July 24, 1980, for 8CAL 3382, 3384 and 5144), including failure to provide face protection, impermeable foot protection or respirators [26]. The California Labor Code, §37, §6310, and §6311, gives workers the right to refuse to do unsafe work and is similar in scope to the federal regulation. The Commission heard testimony regarding the health effects of PCB, including the recommendation from the NIOSH criteria document, which based recommendations on animal studies that showed PCB to be carcinogenic and teratogenic [26]. The Commissioner concluded that the "complainants' testimony as to their fears of contamination in working with PCB spills (was) reasonable" and said "the instant spill and clean-up at hand is concluded to be reasonably found as hazardous by complainants" [26]. The company was ordered to compensate the six complainants for back pay and benefits for the hours suspended and to remove the letters of reprimand or suspension from their records. Thus, in this instance, *Whirlpool* seems to extend to refusal to work with toxic substances, at least in those instances in which the court finds that the substance is particularly hazardous and back pay is an appropriate remedy.

CONCLUSION

Whirlpool has upheld the validity of the regulation that gives workers the right to refuse to work under conditions without discrimination that they reasonably believe pose a threat of death or serious injury. The decision is a narrow one and not a "broad mandate for self-help reasons" nor "open season" under the regulation, as some have suggested [27]. As the Court points out, a worker who relies on the regulation runs the "risk of discharge or reprimand in the event a court subsequently finds that he acted unreasonably or in bad faith [19]. It does provide another avenue to "self-help" in addition to Section 7 and Section 502. In upholding the validity of the Section 1977.12 regulation, the Court has provided workers with additional ways to exercise their right to refuse work under hazardous conditions, but workers are still entitled to exercise Section 7 and Section 502 rights simultaneously with their rights under the regulation [28].

The regulation is perhaps more favorable for the workers than Section 502, since the regulation only questions the "reasonableness" of the workers' fear of death or serious injury, while Section 502 is concerned with the objective "dangerousness" of the hazard. The regulation is also broader than the Section 7 standard because it does not require "concerted activity." *Individuals* can refuse to work in hazardous conditions without showing concerted activity of their fellow employees. Other differences include the requirement for 30 days to file charges for 11(c) complaints and the regulation, while the NLRB requires 180 days. Furthermore, the NLRB is experienced in hearing unfair labor practice charges. Federal court judges hear complaints under the regulation and such cases are rare.

The Supreme Court has given the worker the right to refuse to work in hazardous conditions, but has not decided two issues. First, the Court has left open the question of back pay and discrimination as it applies to employer withholding pay from workers who refuse to work. The Supreme Court's failure to give guidance in this area is seen by some labor leaders as a setback for the OSHA administration, but the decision does signal that withholding pay may, in *some* circumstances [21,22], be illegal discrimination, and the *Whirlpool* District Court and the *N. L. Industries, Inc.* Court granted back pay [29]. These circumstances are yet to be defined, but may exist if workers are not provided alternative work assignments. Failure to provide alternative work assignements that are as safe as those of other employees, coupled with withholding of pay, may constitute discrimination.

Second, the Court has not defined the requirements to be met by the worker to invoke the regulation to ensure that he or she does not abuse the power it gives [1]. The requirements listed by the Sixth Circuit will likely be used until further litigation, and perhaps the Supreme Court, clarifies the issue.

In summary, *Whirlpool* validated the 1977.12 regulation giving workers the right to refuse to work under conditions they reasonably believe pose a threat of death or serious injury. It provides another, per-haps broader, right to "self-help" in addition to the NLRA Section 7 and Section 502. However, there are still issues to be resolved, including the right to pay while refusing to work, what is required of a worker who exercises the right, and how far the courts will extend the right to refuse in situations of exposure to toxic substances. Perhaps the impact of *Whirlpool* is best summarized by former Assistant Secretary of Labor, Eula Bingham, who said, "[B]y sending an unequivocal signal to employ-ers and workers alike that the law protects employees who refuse to work when forced to choose between their jobs or their lives, the Court has given renewed meaning to the fundamental purpose of the OSH Act" [27].

REFERENCES

1. *Usery v. Whirlpool Corp.,* 416 F. Supp. 30 (1976).
2. *Marshall v. Whirlpool Corp.,* 359 F. 2d 715 (1979).
3. *Whirlpool Corp. v. Marshall,* 100 S. Ct. 883 (1980).
4. 29 U.S.C., §651–678 (1970).
5. PL 91–596, §511, Dec. 29, 1970, 84 stat.
6. 29 U.S.C., §157 (1976).
7. *NRLB v. Washington Aluminum,* 370 U.S. 9,16 (1962).
8. *Interboro Contractors, Inc.,* 388 F. 2d 495, 500.
9. Drapkin L: The right to refuse hazardous work after Whirlpool. Ind. Relations Law J. 14:54 (1980).
10. 29 U.S.C., §651(b) (1976).
11. 29 C.F.R., §1977.21(b)(1) (1979).
12. 29 U.S.C., §654(a) (1976).
13. 29 C.F.R., §1977.12, 38 Federal Register 2681, 2683 (1973), as corrected 38 Federal Register 4577 (1973).
14. 29 U.S.C., §660(c)(1) (1970).
15. Rothstein, M: Occupational Safety and Health Law (1978).

16.   H.R. Rep. No. 1765, 91st Cong., 2d Sess., 37–38 (1970), Leg. Hist. 1190–1191. See 29 U.S.C., §657(f).
17.   29 U.S.C., §654(1)(b).
18.   *Marshall v. Whirlpool,* 593 F. 2d 735; U.S.C., §1977.12(b).
19.   100 S.Ct. 890; Employer-employee relations—OSHA. Am J Trial Advocacy 3:568 (1980).
20.   Occupational Safety and Health Reporter, March 19, 1981:1346.
21.   *Marshall v. Whirlpool Corp.,* 9 O.S.H. Cases 1038 (N.D. Ohio, 1980).
22.   *Marshall v. N.L. Industries, Inc.,* 618 F. 2d 1224 (1980).
23.   Boland, BM, Ed.: Cancer and the Worker New York: New York Academy of Sciences, 1977:9.
24.   *Ind. Union, AFL-CIO v. Amer. Petro.,* 100 S. Ct. 2844 (1980).
25.   100 S.Ct., 890. *United States v. Bacto-Unidisk,* 394 U.S. 784,798, 89 S.Ct. 1410, 1418, 22 L.Ed. 2d 762; *Lilly v. Grand Trunk R. Co.,* 317 U.S. 481, 486, 63 S.Ct. 347, 351, 87 L.Ed. 411.
26.   *Neal et al v. Pacific Gas and Electric,* Case No. 12–12015–(1–6) EMM (1980).
27.   Occupational Health Reporter, March 6, 1980:927.
28.   Kirschner, K: Workers in a whirlpool: employee's statutory rights to refuse hazardous work. Lab L. J. 31:293, 1980.
29.   Meehan, M: Your job or your life! The Sixth Circuit's endorsement of the OSHA imminent danger regulation in *Marshall v. Whirlpool Corp.* receives Supreme Court approval. U. Toledo L. Rev. 11:632, 1980.
30.   593 F. 2d 715 (1978); Employer-employee relations—OSHA. Am J Trial Advocacy 3:564–569, 1980.

DISCUSSION: CHAPTER 3

Glendel J. Provost, Mary J. Alexander,
Nicholas A. Ashford and Michael Holthouser

MR. PROVOST: Based on information we have, it would appear that the new administration will eliminate a number of existing regulations. We have already heard about the walk-around pay provision. Now we hear talk about the possible withdrawal of the regulations dealing with access to medical records, and I assume there is a possibility, at least, that the regulation dealing with worker's right to refuse to work under hazardous conditions may also be withdrawn. If that occurs, based on your review of the case law, what sort of remedies do you think will be available to individual workers?

MS. ALEXANDER: It appears that the test put forth by *Whirlpool* is the reasonableness of the worker's fear. I think it may allow the employee to rely on self-help by invoking this reasonableness standard—a quasi-objective standard. Perhaps it may be an avenue beyond that of regulation in which the employee can turn to the courts for assistance.

DR. ASHFORD: It is fairly easy to meet the concerted action requirement in Section 7 if an employee refuses hazardous work with the welfare of his fellow employees in mind. As I read the case law under the Labor Law, it is not necessary for two employees simultaneously to refuse work or have an intent to do so. As welcome as the Supreme Court decision was in *Whirlpool,* the right to refuse hazardous work under Section 502 is limited to those hazards that are likely to cause death or serious bodily harm. For a lot of toxic exposures we are not clear if the toxic material in question is of that nature or if an existing standard provides reasonable protection.

I think there is a promising use of Section 7 rights rather than Section 502 under the Labor Law. If the labor unions decide to exclude from either collective bargaining, mandatory arbitration, or the no-strike provision, the right to refuse hazardous work, the individual union workers can use the Section 7 rights. *Washington Aluminum* is quite broad. It did not depend on the likelihood of death or serious bodily harm. It provides a much more protective right, and I would suggest that Section 7 under the Labor Act is probably a much more promising way to go. Would you agree or disagree? What are your views on that?

MS. ALEXANDER: Of course, all three sections, 7, 502 and the regulation, are available to the employee. I think the significance of the *Whirlpool* decision is greatest in that it was a *unanimous* decision, that it did not take this avenue of self-help away from the employee. Perhaps in certain situations your point is well taken, and Section 7 may be the best alternative. What I would like to suggest is that the reasonableness standard that *Whirlpool* provides may be more helpful for employees who are not aware of what their rights are under the NLRA, but who now can simply refuse to work and be supported by the courts.

DR. HOLTHOUSER: How does the worker show that he acted in good faith? This is one of the major requirements, and it would appear to be difficult to establish.

MS. ALEXANDER: I believe that the employee must show that he sincerely feared for his life or feared serious injury and was not attempting to avoid work. I think that the fact-finder must decide whether or not it was in good faith.

CHAPTER 4

THE OCCUPATIONAL SAFETY AND HEALTH ADMINISTRATION (OSHA) IN THE COURTROOM: FUTURE BATTLES

Timothy F. Cleary, JD

U.S. Occupational Safety and Health Review Commission
Washington, DC

The Occupational Safety and Health Act (OSHAct) has been in existence ten years and, while it has been the subject of much controversy, I believe that it is working and is here to stay. Conferences such as the Third Annual Park City Environmental Health Conference on Legal and Ethical Issues in Occupational Safety and Health play a very important role in pointing out the importance of education in making OSHAct work in our lifetime. Education has also played an important role in making the Review Commission better understood. Shortly after being named chairman by President Carter in August 1977, I established a series of quarterly seminars around the country, in which all three commissioners, our general counsel and chief administrative law judge participated, explaining the role of the Review Commission in the overall OSHAct scheme. More conferences are planned.

I enjoyed my almost 3½ years as chairman of the Review Commission and look back on the period with a real sense of accomplishment. I know that the Commission has never been more efficiently organized, our judges have never been more productive, and our attorneys at the review level have never worked harder. We have had significant changes in the rules of procedure, including our experiment with simplified proceedings. The reorganization of the Commission's attorney resources resulting in a 50% increase in case production has been a great success, and the role of the Commission as an independent adjudicatory agency has never been better understood by a wider group of people. I walk away from the chairman's seat with no regrets. There are, however, a number of advantages to no longer being the head of the Review Commission. For example, I can avoid to a greater extent the public relations aspects of a presentation such as this and concentrate on making my address to you more educationally substantive. In the final analysis, I can speak more freely. With that in mind, let me turn my remarks to some of our "courtroom battles."

OSHA IN THE COURTROOM

To discuss realistically the topic for my presentation, and to place in proper perspective what the future holds for the OSHAct and those affected by it, it is necessary to review what has transpired in the past.

It can be stated unequivocally that the types of decisions coming from the federal courts on issues affecting the health and safety of workers reflect the need for each participant to take a reasonable approach to the problems.

For example, when the Supreme Court upheld an appeals court ruling in July of last year that set aside the benzene standard (*Industrial Union Department, AFL-CIO v. American Petroleum Institute*), Justice Powell cautioned the Secretary of Labor that, in setting occupational safety and health standards, he must not forget that "the economic health of our highly industrialized society requires a high rate of employment and an adequate response to increasingly vigorous foreign competition." Justice Powell concluded that Congress "intended OSHA to balance reasonably societal interest in health and safety with the often conflicting goal of maintaining a strong national economy." In addition, the plurality stated that OSHAct does not "require employers to provide absolutely risk-free workplaces whenever it is technologically feasible to do so, so long as the cost is not great enough to destroy an entire industry." One would have to ask, what is the court saying to OSHA? Be reasonable, take a balanced approach to rulemaking—do not regulate with blinders on.

A similar message was delivered by the United States Court of Appeals for the District of Columbia in the famous "walkaround pay" case (*Chamber of Commerce v. OSHA*).

In the walkaround pay case, the court took issue with what it perceived to be an extreme position. The extreme position was not walkaround pay itself. Instead, the DC Circuit Court had difficulty with a general *policy* favoring walkaround pay. By means of an "interpretive rule," which was promulgated without public notice or comment, the Secretary of Labor made the failure to pay employees engaged in walkaround inspections an issue of discrimination under Section 11(c) of the OSHAct. The court rejected the Secretary's "interpretive rule," finding it to be "legislative in nature" and, therefore, improperly promulgated because the notice and public comment requirements of the Administrative Procedures Act were ignored.

The court was not delicate in communicating its message. To quote only a small, but instructive, part of what the court said: "Finally, and most important of all, highhanded agency rulemaking is more than just offensive to our basic notions of democratic government; a failure to seek at least the acquiescence of the governed eliminates a vital ingredient for effective administrative action." The court also remarked that "Public participation in a legislative rule's formulation decreases the likelihood that opponents will attempt to sabotage the rule's implementation and enforcement."

By the same token, when the Review Commission has taken what might be characterized as administratively generous positions, our experience mirrors that of other parties — court rejection.

A good example is the Eighth Circuit's decision in *H.S. Holtze Construction Co. v. Secretary of Labor and OSHRC.*

The case involved an alleged violation of OSHA's construction standard requiring the installation of guardrails on every open-sided floor. At the time of OSHA's inspection of the worksite, employees of Holtze were installing preassembled wall sections, which were to be the exterior walls of the building. The Commission majority, of which I was a member, reversed the conclusion of the administrative law judge and found Holtze in violation. We required the employer to install guardrails around the perimeter of the building under construction to protect the employees who were installing the wall sections.

The Eighth Circuit reversed the Commission and was very direct in assessing our decision. The court stated that, while it was mindful of the OSHAct's broad scope and remedial purposes, it was also "of the opinion that some modicum of reasonableness and common sense is implied."

The court stated that "There is a point at which the impracticality of the requirement voids its effectiveness." In the court's opinion, "that point has been reached when to erect an entire wall, a project said to take approximately two hours (Holtze) must begin an endless spiral of tasks consisting of abatement activities which necessitate further protective devices, i.e., guardrail to erect wall, scaffold to erect guardrail, safety devices to erect scaffold, etc." In agreeing with the dissenting Commission, the court said "some demarcation line must be drawn between that which is genuinely aimed at the promotion of safety and health and that which, while directed at such aims, is so imprudent as to be unreasonable."

One of the more controversial areas in recent years has been the treatment by the Commission of settlements submitted by the parties. The Commission ventured out of the Dark Ages (or took its head out of the sand) on this issue with its decision in *Farmers Export.* In *Farmers Export,* the Commission not only reaffirmed its authority to review settlements (and unanimously expressed its disagreement with the Third Circuit's opinion in *Sun Petroleum*), but it also held that the commission would approve settlement agreements containing exculpatory language, even if that exculpatory language could be interpreted to limit the settlement aggreement's future use under the Occupational Safety and Health Act.

In our decision involving Nashua Corporation, the Commission's recent history of acting unanimously in the area of settlement agreements came to an abrupt end.

The settlement agreement in *Nashua* was an interesting document for a number of reasons. For purposes of this discussion, the most interesting aspect of the agreement was the very clear indication that neither the Respondent, Nashua Corporation, nor the Secretary of Labor really knew whether the problem (the noise being generated by Respondent's machinery) could be abated. In short, it was a question of whether *feasible* engineering controls existed to control the noise problem created by Respondent's machinery. Rather than litigate the matter, the parties came to an interesting accommodation. Respondent agreed to experiment with a prototype noise enclosure. If the enclosure was successful, Respondent would install the enclosure on similar machines in its plant. If the prototype enclosure was not successful, it was unwritten but assumed that Respondent would have met its obligation under the citation by its agreement to experiment with a prototype noise enclosure. It was understood, of course, that if no feasible engineering controls existed to control Respondent's noise problem that its employees would be protected by the use of personal protective equipment.

My colleagues found the agreement deficient and specified that the deficiency could be remedied by changing the agreement to include two dates: one date by which the feasibility of the prototype will be evaluated, and one date by which actual abatement will be accomplished if the prototype device is deemed feasible. If you read carefully the majority's opinion, you come away with the clear impression that they did not trust the parties to work out the problem on their own. I dissented from the majority opinion. I viewed the case as one presenting an unusual set of circumstances — a set of circumstances crying out for a more flexible application of our rules of procedure. True, Commission Rule 100 requires that the date of abatement be set forth in all settlement agreements, but this case presented a situation where the blind application of our rules of procedure would be, to say the least, inappropriate.

I concluded that where the Secretary and the employer have reached agreement on a basis to settle their dispute, and the affected employees or their authorized representative have had an opportunity for meaningful participation in the settlement process (as they had in this case), and the employees raise no objection to the settlement (and they raised no objection in the *Nashua* case), the Review Commission should approve the agreement reached by the parties. I stated that the settlement agreement before the Commission was novel, but that it represented both an honest attempt by the parties to cooperate in the name of safety and health, as well as a true exercise of the Secretary's prosecutorial discretion. In my view, there was no dispute before us. The Secretary was satisfied with the settlement; the Respondent was satisfied with the settlement; and the authorized representative for the affected employees was served with a copy of the agreement and offered no objection. In short, I saw no reason to disapprove the settlement agreement.

Recently, there have been two Circuit Court reversals on the issue of a union's objections to settlement agreements and the Secretary's desire to, in the first case, withdraw from the agreement, and in the second, withdraw the citation. The U.S. Court of Appeals for the Third Circuit, last August, rejected the Commission's decision in the case of the *Oil, Chemical and Atomic Workers International Union v. Sun Petroleum Products and the Secretary of Labor.* By a one-to-one vote, the Commission had affirmed a decision of one of its judges, who approved the settlement despite the union's objections. The Secretary attempted to withdraw from the settlement after it was directed for review and after the time allowed for briefing issues on review. The Commission did not consider the Secretary's request to withdraw. The Third Circuit ruled that the Secretary had the power to withdraw from the settlement agreement after the judge approved it and while it was before the three-member Commission for review.

In December 1980 the Sixth Circuit Court of Appeals reversed another decision in a case involving the Secretary's right to withdraw a citation over a union's objection — the *IMC Chemical Group, Inc.* case. The Commission had held, in a two-to-one opinion, that the union representing affected employees had to be heard on its objections to the Secretary's withdrawal of the citation at issue. The majority of the Commission implied but did not hold that when the Secretary decides not to prosecute a citation, affected employees or their union in effect may proceed to prosecute the citation originally issued by the Secretary if the employees elect party status.

The Sixth Circuit issued a decision sharply reversing the Commission's decision. There is a great deal that could be written about the court's decision in *IMC*. Suffice it to say that the court made it clear that affected employees or their representative may not litigate at cross purposes to the Secretary of Labor. In essence, employees may litigate in tandem with the Secretary or not at all. The only exception to this is the issue of the time set for abatement. In that instance, affected employees clearly can disagree with the Secretary's position and litigate the matter fully.

In the area of party status and the individual employee, the Commission, in September 1980, issued a decision in *Babcock and Wilcox* involving the interplay of Commission rules 20(a), 22(c) and 1(g). The question presented in the case was whether an individual who was arguably an affected employee and who was represented by a union can assert party status as an individual employee when his union chooses not to participate in the Commission proceeding.

Commission rule 22(c) provides that "affected employees who are represented by an authorized employee representative may appear only through such authorized employee representative." The Commission interpreted rule 22(c) literally and answered the question before it quite simply — a Commission majority said "no." In other words, if an affected employee is represented by a union, the employee may appear in Commission proceedings only through that union. Obviously, the case has significant labor relations overtones.

In the *Darragh Company* case issued in September 1980, the Commission moved further in the direction of pragmatic decision-making — the direction of the future. The issue tackled involved the exemption for agricultural operations. Darragh, which operated under contract to deliver chickens and feed to poultry farms, was eligible for the agricultural operations exemption in Section 1928.21(b), the unanimous Commission determined. In holding that the exemption applied, the Commission reversed the judge's decision, finding that Darragh allowed its employees to use unsafe ladders at several farms. We examined the specific task that exposed the workers to the alleged hazardous condition and decided that the task was integrally related to the farmers' agricultural operations. We ruled that the task performed by Darragh's employees made it no less a part of an agricultural operation than if it had been performed by the farmers themselves or by the farmers' employees.

One of the more important areas of the law that the Commission considers regularly is section 4(b)(1), exemptions from OSHA jurisdiction.

In this area of the law, the Commission has recently issued two relatively significant decisions. The first is *Northwest Airlines, Inc.* This case involved Northwest flight mechanics at New York's JFK Airport. Specifically, the flight mechanics were performing maintenance on the landing lights of Boeing 747s. In order to perform the maintenance, the leading edge of the wing (the flaps) had to be extended to give access to the lights. These flaps are electropneumatic and, if activated, they retract quickly. If a mechanic is working in the flap cavity on the landing lights and the flaps are activated, he would be crushed.

OSHA issued a citation alleging a violation of Section 5(a)(1) — the general duty clause — for Northwest's failure to provide a means to protect employees from this crushing hazard. In its defense, Northwest claimed that the condition at issue was exempt from OSHA coverage under section 4(b)(1) of the OSHAct because of the regulatory authority of the Federal Aviation Administration (FAA). The administrative law judge assigned to the case vacated the citation against Northwest on the ground that the working condition covered by the citation is exempt from the OSHAct's coverage under Section 4(b)(1). The two issues before the Commission on review were: (1) whether the FAA has the statutory authority to regulate the health and safety of ground maintenance employees, and (2) if the FAA does have the authority, were Northwest's maintenance manual and the rule dealing with the cited condition in that maintenance manual a proper exercise of that authority in order to give rise to a Section 4(b)(1) exemption?

As for resolving the issue of the FAA's authority, the Commission concluded that it should give considerable weight to the FAA's interpretation of its own regulation; indeed, we decided that the FAA's interpretation should be *controlling* as long as it is reasonably supported by its enabling legislation. On the bottom line, we agreed with the FAA that its authority to regulate aircraft safety is broad and that it encompasses the health and safety of maintenance personnel.

As to the second issue — whether the maintenance manual was a sufficient exercise of the FAA's statutory authority to exempt the condition cited by OSHA — the Commission indicated that it would apply a two-part test. First, we would examine whether the rule at issue, in this case the rule in the maintenance manual, has the force and effect of law. And, second, we would examine whether the rule is enforced or, at least, subject to an enforcement mechanism.

The Commission concluded that, because the maintenance manuals are required by the FAA and subject to FAA disapproval, because the maintenance manuals have employee safety and health as one of their goals, and because the rules in the maintenance manuals are subject to enforcement by the FAA, a section 4(b)(1) exemption should be granted for the specific cited condition, since the cited condition was covered in the Northwest maintenance manual.

Before moving to the second case in this subject area, I should mention that, as I view it, there are only two ways to remedy the confusion brought about by Section 4(b)(1). The first is to have, as I call it, one OSHA and one Review Commission. In other words, there should be only one agency dealing with occupational safety and health. The other option is for OSHA to meet with all other agencies having overlapping safety and health jurisdiction to work out jurisdictional agreements, or more commonly referred to as memoranda of understanding. Such agreements certainly are preferable to the case-by-case approach, which is the only approach available to an adjudicatory agency such as the Review Commission. Adjudication is simply not well suited for the resolution of jurisdictional problems with broad implications such as those we see in the Section 4(b)(1) cases brought before the Commission. They provide a great legal exercise but, in my view, do not advance worker safety and create nothing but confusion for employers.

The second and most recent case in this area is *Consolidated Rail Corporation.* This case was actually a consolidation of three separate Consolidated Rail cases. All three cases presented the same legal question: Is a policy statement of the Federal Railroad Administration (FRA) published in the *Federal Register* an exercise of statutory authority within the meaning of Section 4(b)(1) of the OSHAct such that it preempts the OSHAct's application to certain working conditions of Respondent's employees?

The FRA policy statement at the center of this controversy was issued on March 14, 1978. In that statement, the FRA determined that it should not attempt to regulate "in an area covered by regulations issued by the Department of Labor." The FRA, however, specifically addressed the applicability of OSHA's regulations on guarding open pits and ditches in railroad repair facilities. The FRA stated:

> As the agency which has exercised jurisdiction over railroad operations, FRA is responsible for the safe movement of rolling stock through railroad repair shops. OSHA requirements for general industry are in some respects inconsistent with the optimum safety of employees in this unique environment where hazards from moving equipment predominate. Therefore, OSHA regulations on guarding of open pits, ditches, etc., would not apply to inspection pits in locomotive or car repair facilities.

In addition to the FRA's policy statement issued in the *Federal Register,* counsel for Conrail sought advice on the matter from the Administrator of the FRA. The Administrator concluded that the policy statement issued by the FRA was a sufficient exercise of the FRA's statutory authority to preempt OSHA regulation of open pits in locomotive and car servicing areas. The items before the Commission on review in these Conrail cases involved open pits in locomotive and car servicing areas.

The Commission majority held that the FRA's March 1978 policy statement was not a standard or regulation under Section 4(b)(1) of the OSHAct and did not preempt OSHA's authority to issue the citations in these cases. I dissented from the majority's opinion.

Regarding the validity of standards, the Commission issued a decision in late 1980 that is destined to be a very important decision on this topic, the case of *Rockwell International Corporation.* Rockwell represents the first time that the Commission has come to the unanimous conclusion that it has the authority to entertain validity attacks on the standards promulgated by the Secretary of Labor and, if necessary, to declare those standards invalid and unenforceable. For some time, I have taken the position that the Commission is without the power to invalidate standards promulgated by the Secretary. Although I continue to have some reservations, as explained in my separate opinion in the Rockwell case, I agreed generally with the majority.

In *Rockwell* the Commission rejected the Secretary's argument that the mere existence of the preenforcement challenge provision in Section 6(f) of the Act indicates that Congress intended to bar all challenges to the validity of a standard in enforcement proceedings. We noted, however, that Congress' rejection of a bill containing an exclusive preenforcement challenge provision does not indicate necessarily that Congress intended to permit *all* challenges to a standard's validity in enforcement proceedings, *regardless* of the circumstances. However, the Commission did not make a distinction between procedural validity attacks and substantive validity attacks in *Rockwell,* and that is one part of this issue that will certainly be litigated more fully.

On February 23, 1981, the Commission issued its decision in *General Motors Corporation.* The issue in *General Motors* was whether the validity of an occupational safety and health standard may be challenged before the Review Commission on the ground that the established federal standard from which it was derived was invalidly amended before its adoption under the Occupational Safety and Health Act of 1970. Very briefly, the Commission held (once again, unanimously) that the validity of the procedure by which such ancestor standards might be amended may not be challenged in a Commission proceeding. In reaching our conclusion, we made the point that Congress authorized the Secretary to summarily adopt established federal standards and national consensus standards "as soon as practicable" because it (Congress) found a "pressing need for adoption of OSHA standards on an exceedingly broad industrial front without undue delay." We went on in *GM* to note that Congress expected that, despite their defects, the established federal standards and the national consensus standards "would provide a sound foundation for a national safety and health program" and that the summary adoption of established federal standards would be fair to employers because "such standards have already been subjected to the procedural scrutiny mandated by the law under which they were issued." After a great deal of deliberation, it was the Commission's collective opinion that to strike down an occupational safety and health standard at this time because of a procedural misstep involving its ancestor standard adopted under another federal statute would, very simply, upset the Congressional expectation. All of the commissioners also gave considerable weight to the fact that freely permitting challenges to the procedural validity of ancestor standards would substantially undercut the public interests in finality and in avoiding the burden that continuous challenges would impose on the Secretary's enforcement program and the Commission's adjudicative processes.

There are a number of difficulties surrounding enforcement of the standard on protecting workers from damaging exposure to excessive noise. The issue of what constitutes "feasible" engineering controls, in the noise standard in particular and in others that similarly require such controls, continues to demand interpretation. The *Samson Paper Bag Co.* case, decided in June 1980, gave us the opportunity to carefully reexamine the entire issue of feasibility in light of what has transpired since *Continental Can.*

Although I did not agree with what has become known as the "cost-benefit theory," I had to agree that economics must play some role—however limited—in determining whether controls are "feasible." In my separate opinion, I stated

> Factors such as the noise level, the possibility of hearing loss, the cost of abatement, the *possible impact upon the jobs of affected employees,* the state of the technical art, industry practice, and comparable situations in other industries may merit consideration.... One aspect of the problem, however, is clear. Our approach must be reasonable and realistic. No fair reading of the standard should force an employer into an open-ended research and development program in order to abate a noise situation. Further, it should not force an employer into an abatement situation that, from an economic standpoint, is patently unreasonable. On the other hand, there is no reason to use a formal "cost benefit analysis" with all that term implies, in order to make our determination in a particular case.

The topics of future courtroom battles in the job safety and health law field remain varied and complex. Looking harder at the future, I see several additional topics, not confined to cases, that will present other important and difficult challenges in the coming years. Cases arising under the general duty clause, Section 5(a)(1) of the OSHAct, will continue to be of great interest.

The general duty clause was included in the OSHAct because no group of standards can cover every conceivable hazard that may exist in the workplace. It requires employers to furnish employees with workplaces free from recognized hazards causing or likely to cause death or serious physical harm. General duty clause citations usually result from an accident investigation. More and more, however, the Secretary is issuing citations under the general duty clause before an accident occurs, seeking to fulfill the OSHAct's purpose of abating hazardous conditions before employees are killed or injured. At the Commission, we are seeing numerous general duty citations involving alleged health hazards. These new types of general duty cases may well lead to a reexamination of the term "recognized hazard," as well as an even closer look at the "feasible" means of abatement suggested by the Secretary of Labor.

The Commission will be called on to break new ground in the area of using workers' medical records. The medical records issue is but one emanating from the increased number of occupational health cases. We may be asked to determine how freely such records may be used in the discovery process and what steps will be taken to ensure a worker's right to privacy.

In conclusion, excessive zeal is generally counterproductive. It creates an atmosphere that only threatens the OSHAct's legitimacy. Industry, labor and government must work together to dispel the atmosphere of distrust that has unfortunately settled over the Act. Rather than engaging in an ideological tug of war, they should work together and learn to respect each other's interests. Without an end to extremism — unless the virtues of moderation are appreciated — safe and healthful workplaces for all working men and women of the nation will always lie beyond our grasp.

DISCUSSION: CHAPTER 4

Nicholas A. Ashford, Timothy F. Cleary and John J. Mattick

DR. ASHFORD: Do you think that the Commission's decision to allow the attack on the validity of the standards beyond the statutory time provided in the OSHA Act will survive the circuit court review?

MR. CLEARY: I indicated earlier that, initially, I did not concur in this Commission decision. I felt that it was wasteful of time and resources of the Commission. The reviews are excellent legal exercises, but they do tie up the Commission. *Rockwell International* is a good example. The only court decision that was even half-way supportive of my position to not allow attacks on the validity of standards was the *National Industrial Constructors* case, which held that employers are barred from challenging procedures used in adopting safety and health standards in an enforcement proceeding. That decision has been disavowed by other courts.

So, I abandoned my position, because there comes a time when it becomes obvious that we cannot realistically change an emerging appellate court consensus.

DR. ASHFORD: Would you explain to me the legal theory of allowing the validity of challenging beyond the 180-day statutory time?

MR. CLEARY: In a nutshell, the 180 days, or six months, did not allow enough time for the public to examine the copious standards initially adopted by OSHA. In effect, the courts were saying, if there is no review in the enforcement stage on the validity of a standard, there may be no effective review.

The argument has less force in later standards, but I am fearful that long-range precedent has been set.

DR. ASHFORD: I hope you're wrong. Let me ask you a second question. What constitutes a recognized hazard for toxic materials under

5(a)(1)? In your opinion, if a substance is not regulated by six feet of standards, is not on OSHA's "worst 10" list or is not on NIOSH's [National Institute for Occupational Safety and Health] list of 180 carcinogens, would it meet the test of being a recognized hazard?

MR. CLEARY: That is a pretty tough question to answer right off the top of my head.

The answer can be "yes" or "no," depending on how the issue of adequate notice is resolved in particular cases. There is a fundamental due process problem on the notice issue, e.g., what level of exposure to phosgene gas would be a recognized hazard? I would prefer not to discuss 5(a)(1) cases any further because, as you know, there are a lot of them before the Commission.

The Secretary has increasingly used 5(a)(1) when he has been unable to adopt new standards. I have no problem with using 5(a)(1) for a given situation. I was in the majority in the *ASARCO* case.

DR. ASHFORD: I didn't mean to "box you in," but I think your response has been helpful for this audience.

MR. MATTICK: I believe that the U.S. Postal Service has approached OSHA and the U.S. Coast Guard for assistance in developing standards for themselves. In your personal opinion, I would like to know if you would like to see the OSHA standards applied to all industries.

MR. CLEARY: The possibility of that happening is remote. Congress only recently passed the Mine Safety and Health Act, with its own administration and review commission. I would emphasize that all workers are entitled to safe and healthy workplaces and should have relatively uniform protection on their jobs. The accident of agency alignment should not be allowed to impair job safety and health. By the same token, each employer is entitled to know which agency has jurisdiction over his business. As I indicated earlier, there should either be one OSHA and one Review Commission for all industry, or all federal agencies purporting to govern employee safety and health should resolve their jurisdictional differences, so that employers and employees will know which agency governs their workplace.

From reviewing cases at the Commission level, I think there needs to be greater emphasis on cooperation among all agencies, just as I have thought there should be greater cooperation between labor and industry. I think everybody ought to realize that OSHA is here to stay, and it is in everybody's best interest to make it work.

CHAPTER 5

COST-BENEFIT ANALYSIS:
A TECHNIQUE GONE AWRY

John J. Sheehan
United Steelworkers of America
Washington, DC

I should like to quote from two documents that were recently released:

> By the authority invested in me as President, by the Constitution and the laws of the United States of America, and in order to reduce the burdens of existing and future regulations, increase agency accountability for regulatory actions, provide for Presidential oversight of the regulatory process, minimize duplication and conflict of regulatory actions, and insure well-reasoned regulations, it is hereby ordered...

The purpose of this presidential proclamation was, according to the fact sheet which accompanied it, "to impose to the extent permitted by law a requirement that agencies choose regulatory goals, set priorities to maximize benefits to society, and choose the most cost-efficient means among legally available options for securing those relatory goals." Note that the order is designed to impose both *cost-benefit* and *cost-effective* analysis on regulatory actions.

The result, according to the proclamation, is to ensure that "the regulatory actions shall not be undertaken unless the potential benefit to society from the regulations outweigh the potential cost to society."

I have read the introduction of the order primarily because one thing struck me about it. It was the imperial tone that was used—"by order of God and the Constitution I'm now promulgating this order."

Recently my daughter was home from college. One of her books related to Nurenburg in sixteenth-century Germany. There was a description of the Councils of the Nurenburg, which also issued proclamations by the order of the Council and that power invested in them by God. Well, nevertheless, this particular proclamation by the President is to revoke the previous existing Presidential Order covering regulatory actions, in particular, Executive Order 12044. In President Reagan's order, there is a very positive statement that cost-benefit analysis is to be applied "to the extent permitted by law" on the various regulatory agencies. Incidentally, I hope that you notice that it is not an automatic requirement, inasmuch as some of the agencies may not be able to comply because their own legislative mandate requires otherwise. That was dated February 17, 1981.

On March 30, 1981, the present Assistant Secretary of the Occupational Safety and Health Administration (OSHA), Thorne Auchter, requested a withdrawal of the cotton dust standard from review by the Supreme Court. His purpose, as announced in the press statement: "We plan to apply a cost-benefit analysis to the cotton dust standard."

So, now you see the immediate movement to implement the requirement for cost-benefit analysis. Two days later, the *Washington Post* included a statement on the editorial page, in which the writer indicated that there had been a lengthy cost-benefit analysis of the cotton dust standard and conceded that certainly more refinement of the studies was possible.

> But no magic is to be expected. Analysis of this sort is woefully imprecise. To call it a science is to invite gales of laughter from the economics profession that practices it. Any honest analysis will show large, and sometimes huge, ranges in the estimates of even the most tangible sort of cost and benefits.
>
> Even if those uncertainties could be resolved, we would still not know what value to place on reducing pain or the avoidance of premature death. A matter that is essentially one of both policy and politics.
>
> Comparing the relative costs and benefits is no doubt a useful device for deciding among alternative regulations. But in the cotton dust case, and many others like it, no computation, simple or complex, can absolve the Congress of its responsibility for seeing that industries do not pass off the real cost of doing business on either their workers, their communities, or the general tax payer.

It is interesting to note that the forced embrace of cost-benefit analysis technique by a number of agencies in government is occurring at the same time that there is an emergence of supply-side economics. I feel that basically the two are somewhat related. It is just not coincidental that these two theories, or, if you wish, code terms are now bandied about quite frequently.

First of all they both are intrinsically free-market oriented. The theory behind supply-side economics and cost-benefit analysis is that first, the free market will produce economic growth in the country if you let it function unimpeded; and, secondly, the free market will provide occupational and consumer health and will also produce environmental protection. Surely there is a recognition that government may need to intervene, but that the government would intervene only for the purpose of assisting the free-market forces.

Supply-side economics, without going into too much detail, would recognize the need for the antitrust laws and tax incentives, in order to get the free-market moving. But essentially, supply-side is relying on the free-market forces to push the country into prosperity and worker/community protection.

However, on the cost-benefit side, environmental and safety values sometimes are not immediately or easily translated into economic values. The costs of these values are instead externalized and hence, there is no protection of safety and health. There is, therefore, an additional need for government intervention over and above the mere assistance it might give the market forces. It is needed when environmental values are not internalized to pay for the conservation of clean air and workers' health. If the free-market concept of cost-benefit analysis is applied, the basis of the economic balance will be in favor of economic gain and the environmental values will suffer. In other words, the imposition of regulations may not provide an economic advantage and thus the government intervention would be excluded. However, I think it is reasonable to insist that if government intervention is a means to achieve economic gain, it is no less a means to achieve social gain. But the concept of cost-benefit analysis is so purely free-market oriented that it would assure that market forces would override social concerns.

There is another idea that I would like to discuss with regard to cost-benefit. There has emerged an opposition to the redistributive function of government. The government has promoted the redistribution of some of the wealth of the country. Supply-side theories, now being translated into concrete proposals, dramatically detail the elimination of some of those past redistribution programs.

Cost-benefit analysis is also designed to get at those regulatory rules that do, in effect, redistribute part of the resources of our society and redistribute them not according to the free-market.

I would like to quote from Murray Weidenbaum, Council of Economic

Advisors, "In the final analysis, however, the political factors in regulating decision-making cannot be ignored. Many social regulations involve transfer of economic resources from a large number of people to a small group of beneficiaries." Again, the cost-benefit theories designed to attack regulations fit in with the overall political philosophy ascribed to by the Reagan White House.

Basically, the issue with regard to the application of cost-benefit analysis to the Clean Air Act and to OSHA is whether the regulator can produce standards for social purposes and not necessarily weigh them as to their final bottom line economic consequences.

Under OSHA, the mandate is to promulgate at the "lowest level feasible" and to do this with engineering controls. Counterproposed against this obligation is whether the administrator must promulgate at the cost-benefit levels with cost-effective allowance for the use of respirators or other devices rather than the use of engineering controls.

The *Wall Street Journal* in an article entitled "Cost Benefit Study of Health, Safety Rules is Threat to Workers," remarks on the cotton dust standard pending before the Supreme Court:

> However, George Cohen, a lawyer representing the Amalgamated Clothing Workers, contends that Congress made clear that the Labor Department must set standards subject only to whether they are capable of being achieved. He maintains that "if we are right on that, then any less protection in the name of cost-benefit analysis is foreclosed."

Perhaps we might want to back up a little to see why we are now on the front line with cost-benefit terminology, cost-benefit systems. My knowledge of this subject relates primarily to the Occupational Safety and Health Act (OSHA Act), and in particular, to Section 6(b)(5), which has to do with the health standards. There is found the almost innocuous phrase that standards ought to be promulgated which, "to the extent feasible," adequately assures no material impairment of health. What does "feasible" mean? Certainly we all acknowledge that it meant feasible in terms of technical feasibility, but there was no acknowledgment nor acceptance by many of us that this necessarily included economic feasibility. However, after a number of years of court cases, "to the extent feasible" meant to the extent technologically and economically feasible.

When we talk about economic feasibility, we are talking primarily about the ability, the affordability, to bring the hazardous levels down to what was being proposed by the promulgated standard. I think it rather interesting that, as a result of the courts' involvement in this issue, the conclusion evolved that in the feasibility argument OSHA must take affordability into consideration. But it should be noted that we are talking about affordability on the basis of the industry as a whole, and not affordability interpreted at the lowest level in terms of a specific plan. The courts did indeed consider that certain plants might not be able to afford the standards being promulgated, but if, nevertheless, the industry was able to survive, that was to be considered an economically feasible standard.

I must agree with many of the things that Lord (Chapter 2, this volume) said. Certainly the whole issue of jobs is at stake. Every time industry attempted to turn around the labor movement, economic feasibility was being translated as "this plant that is to be shut down." I think that placed overwhelming pressure on people in the labor movement. Nevertheless, the court maintained that (OSHA) could move ahead with economic feasibility based on an industrywide interpretation.

Gradually, the arguments with the question of economic considerations of OSHA standards moved into "inflationary impact statements," "economic impact statements" and now into the "cost-beneficial impact statements." The political philosophy behind this movement was to get the government off the back of producers, which, in turn, would somehow or other result in an economic revival since the regulations were inflationary and created an economic impact on the economy. According to the new terminology, you cannot justify them; they are not cost-beneficial.

Without surely burdening you with all the terminology that is used, the terms cost-benefit analysis, cost-effective analysis, and regulatory budgets are used throughout the debate with regard to economic considerations. The fundamental issue, however, is whether cost-benefit analysis should be a tool for controlling the administrator's discretionary authority in promulgating standards, or, indeed, if it could be an analytical tool for helping to assign priorities to some of the concerns that he might have and the areas of interest in which he might need to go with regard to regulation.

I would like to distinguish between the issue of cost-benefit analysis and cost-effective analysis.

Many of us (in the labor movement) who spoke positively with regard to cost-effective analysis, embraced it on the basis that industry should have the right to choose from various methods, the best method to arrive at a certain standard. And, indeed, that method could be "the least costly," the most cost-effective in those terms. However, I do realize that the words "cost-effective" mean something much broader than that. This expanded theory would assert that if you were to look at a number of toxic substances, preview the economic cost of lowering the levels to protect the workers for each of the substances, and then compare those costs, you could determine which toxic substance to control. The cost to save a life endangered by one substance might be less than the cost to protect a worker exposed to another toxic substance. Cost-effectiveness, in this more global sense, attempts to get the "biggest bang" out of the health dollar that is available to the regulatory process. It is an administrative tool for rule-making.

I suppose we would say that, for a regulator making decisions, it could be a helpful tool. However, the real issue of concern has been whether the regulator should be governed solely by these cost-benefit and cost-effective mechanisms. Most authorities will reiterate the fact that the state-of-the-art in quantifying costs and benefits just simply is not at a stage where one can bring any reliability in giving a cost-benefit analysis ratio. The fundamental criticism of these systems is that the methodology is essentially faulted. For instance, there are a number of economic schemes designed to come up with a dollar value for a life. I thought it very interesting that OSHA, in the promulgation of its carcinogen standard, mentioned one of these economic methods for evaluating a life which is based on a worker's income and on his economic contribution to society. However, OSHA decried this particular method on the basis that it appeared that the government was looking upon the worth of a person only in terms of his contribution to the gross national product.

> Attempts to assign economic weights to human life often assumed a human capital approach, which values an individual's economic contribution to society. Human life valuation is calculated in terms of future income streams (gross output) or the present discounted value of future income streams minus an individual's consumption (net output). This approach assumes that the only goal of economic policy is to maximize Gross National Product. Human life valuations also depend critically on earnings' levels, leading to lower values for the retired, women and disadvantaged low-wage groups [OSHA Carcinogen Document].

Well, be that as it may, the certainty of the quantification methods is in extreme doubt. Hence, its objective value becomes vitiated.

Second, as has been pointed out in literature, when only economic issues are evaluated, equity in using cost-benefit mechanisms is completely ignored. I would like to quote another section from the carcinogen document with regard to the application of a cost-benefit analysis to the promulgation of standards.

> As a matter of policy, decisions based upon efficiency criteria are not appropriate because they ignore equity considerations. The eco-

nomic savings from less protective regulation accrue to industry in the form of higher profits and to consumers through lower prices. But the costs are borne by workers through increased industrial illness and death rates.

So these economic guidelines do not handle that crucial issue of public policy, the equity of the workers involved. I would like to quote further from this document because it makes reference to Dr. Nick Ashford, with relation to the area of equity and social responsibility. The conclusion is that cost-benefit analysis flies in the face of equity. According to Dr. Ashford:

> The most serious limitation of cost-benefit analysis lies in the failure to successfully deal with the fact that the cost and the benefit streams accrue to different parties. One person's benefit cannot be neatly traded off for another person's cost. Maximizing social welfare should not necessarily be equated with optimizing social welfare. Equity and economic efficiency are sometimes conflicting goals. As a decision tool, cost-benefit analysis can be used in identifying the nature of the trade-offs; as a decision rule it is useless. Regulation of toxic substances is an expression of social policy, not of economic policy.

Recently, a presidential commission was established to look into the future. Its report is called the "Agenda for the 80's." The Commission conducted a series of studies. One dealt with regulations in the 1980s. Again, the critical issue of the recognition of equity and not just economic value was addressed. It was treated in a somewhat different context, perhaps, than that desired by legislators, who are all looking for the magic wand that will determine objectivity, thereby lessening political controversy.

No legislator wants to be in a situation in which he has to cast the vote and in which he will end up with people on either side of the issue hounding him. The search for scientific and regulatory objectivity by our public officials, given the kind of political climate we live in today, is intense.

Quoting from the panel, it is the panel's view that

> facts are often assumed to play a too decisive role and that legislators are too prone to believe that the decision can be made based on agreed analysis by experts. Because consensus among experts is unlikely, it may be that legislators rather than experts must provide guidance as to the course of action a regulator should follow when confronted by uncertainty. Questions about matters such as the definition of acceptability of risk, [I might interject "informed consent"] are inherently political and regulators cannot be expected to resolve them through expert analyses and quasi judicial, evidentiary proceedings.

The report indicates that these are "value judgments about which people can legitimately disagree. Such conflicts of values require political resolution and not simply fact-finding investigation."

In conclusion, it seems to me that the injection of economic considerations into rule-making by regulators has resulted in a great deal of detailed content in the standard document primarily because the regulatory agencies have been hauled into court to justify the standard. As different companies exercised their due process in contesting standard rights, they increased their articulation as to what kind of theories could be used to stop a regulator from regulating.

In other words, cost-benefit analysis, up to now, was a technique to stop the administrator from promulgating a rule. Certainly the efforts to inject it into the various regulatory laws, i.e., to require that regulators shall not regulate unless there is a cost-beneficial ratio, is a direct effort to prevent some of the equity considerations from being exercised by the administrator. Regulatory reform bills pending in Congress attempt to require an across-the-board obligation to perform cost-benefit analysis because some acts, like OSHAct, do not provide it. The effort to increase the possibility of winning in court under due process rights has brought the cost-benefit analysis concept to the degree of articulation that it is at right now. But it is interesting to note that we certainly do not anticipate further movement in the promulgation of new standards under the OSH-Act. So maybe, indeed, the need for putting into law cost-benefit formulas may begin to wane because—and this is the interesting point about it all—while you may be able to stop a regulator from regulating, there is no way you can force a regulator to regulate when even public equity is at stake. Maybe that is the real story here. Not that we should have more economic measurements to measure environmental and occupational health regulations, but that we do have so few tools to advance environmental and occupational health.

COST-BENEFIT: WORKERS' COST AND INDUSTRY'S BENEFIT OR VICE VERSA

The first question to be addressed in any discussion of cost-benefit analysis in the occupation safety and health field is "who pays the cost and who receives the benefit?" What might be a safety and health regulatory cost to employers, for example, might also be a lifesaving benefit to workers. On the other hand, what might be of great benefit to business, in the form of less safety and health regulation, might exact substantial costs from workers in terms of increased injuries, disease and death. Moreover, while industry may benefit from certain governmental actions, the societal costs may be unacceptable.

The second question is what are the costs and benefits of a given action and are they comparable? While this at first might seem to be a rather mechanical quantification exercise, such is not always the case, particularly when the issue involves social regulation. Indeed, many would argue that the costs and benefits of occupational safety and health standards, for example, are impossible to quantify with any degree of accuracy or acceptability. Much of the necessary data either do not exist or are highly suspect. Furthermore, there are certain dimensions of the cost-benefit equation, such as the value of human life, which can never be fully quantified. In addition, the techniques for comparing economic costs to social benefits are highly subjective and often arbitrary. Finally, cost-benefit analysis assumes that both the technological state-of-the art and the national economy are static rather than dynamic.

Third, there is the question of the distribution of costs and benefits. Who should pay how much in costs and who should receive how much in benefits, and is the distribution equitable? This is largely a societal question, loaded with value judgments, which can only be resolved through political process.

To adequately address these three questions, one must first consider cost-benefit analysis as an analytical technique to determine whether it, in fact, lends itself to decision-making, especially in the area of social regulation. Second, one must examine the cost side of the equation to determine its parameters, and the same must be done in terms of benefits. Finally, one must make some judgments concerning the social distribution of costs and benefits and the equity of the distribution.

First, the cost-benefit issue should be placed in what I call its political perspective. In other words, what are we talking about when we speak of cost-benefit analysis in the current political climate.

Politics of Cost-Benefit Analysis

In its original form, cost-benefit analysis was merely an analytical tool used by economists to assess the economic viability of a given project. Since it was an economic tool, its use was limited to those factors which could be expressed in dollars. The dollar costs of a project were calculated and compared to the dollar benefits to provide some idea of the economic dimensions of a particular project. Even in this very narrow sense, cost-benefit analysis was never intended to be a rigid decision-making device. It was simply an economic tool to help order public thinking concerning a proposed economic decision.

Cost-benefit analysis has its roots in the nineteenth century, when the federal government used it to evaluate public projects. It came into its own during the 1930s when cost-benefit analysis was a standard tool used by the Army Corps of Engineers and other agencies to assess proposed water projects, dams and other public works projects. The Defense Department has used cost-benefit analysis since the 1960s as one way to evaluate alternative weapons systems. In all of these cases, "the central focus was to improve the efficiency of federal expenditures" [1, p. 3].

Even in this respect, cost-benefit analysis was not intended as a rigid decision-making technique. Indeed, once a cost-benefit analysis has been conducted, other nonmonetary considerations, such as the social good or national security, were factored into the decision-making process. Thus, even though the monetary costs of a proposed dam, for example, might outweigh its monetary benefits, other social factors might dictate that the dam, nevertheless, be constructed. The cost-benefit calculations were merely one of any number of social, economic and political factors which entered into the final decision.

More recently, however, cost-benefit analysis has assumed a political dimension which has changed its original character. Rather than being limited to purely economic issues, it has been expanded to include social issues as well, particularly social regulatory issues. Moreover, some advocates of social cost-benefit analysis have proposed that it be the sole determining factor in regulatory decision-making. Thus, according to a bill sponsored by Senator Paul Laxalt, where the estimated economic costs of a proposed regulation are greater than the estimated benefits, the regulatory agency would be directed either not to issue the rule or to modify it to reduce the costs. Under such a mandate, occupational safety and health standards, for example, would be subject purely to dollar concerns with no room for nonmonetary considerations.

As I shall indicate later in this chapter, we in the labor movement do not believe that it is possible to reduce the costs and benefits of safety and health regulations to dollar terms without ignoring some very important human considerations. Most economists, if questioned carefully, would agree with us on this point. However, more than the technical and theoretical limitations of social cost-benefit analysis, its practical, political implications are such that its unrestricted use in making social regulatory decisions would effectively stop, if not destroy, the social regulatory system. This, I am afraid, is the real driving force behind the current movement to impose formal cost-benefit analysis on the federal regulatory system.

This perversion of cost-benefit analysis was quietly initiated during the Nixon Administration when the Office of Management and Budget (OMB) forced regulatory agencies, such as OSHA, to engage in cost-benefit analysis of proposed rules. While this was an informal exercise, and not nearly as sophisticated as what is now being contemplated, it nevertheless interjected cost-benefit analysis into the social regulatory system.

In 1974 then-President Gerald Ford formalized what had been initiated by his predecessor. In an Executive Order, President Ford directed that all regulatory agencies subject to presidential control be required to evaluate the "inflationary impact" of all proposed major regulations. Any proposed regulation in which the costs exceeded benefits would be considered inflationary and, thereby, subject to modification. These "inflationary impact statements," in effect, required the regulatory agencies to undertake cost-benefit analysis of all major rules. In January 1977 immediately prior to leaving office, Ford reissued the Executive Order, changing only the title of the required statement to "economic impact statement."

President Jimmy Carter continued this general policy approach, although he changed the requirements somewhat. While Carter stopped short of requiring regulatory agencies to undertake cost-benefit analysis of major rules, his Executive Order (issued in March 1978) directed the agencies to consider the costs and benefits of proposed regulations and to seek the least costly alternatives.

It should be noted that neither the Nixon, Ford nor Carter directives imposed a rigid cost-benefit decision-making formula on the regulatory agencies. All that was required was that the agencies conduct an analysis of costs and benefits in an effort to determine less costly regulatory alternatives. While these requirements imposed an additional procedural burden on such agencies as OSHA, and, thereby, slowed the regulatory process, they did not dictate decision-making.

Most recently, however, President Reagan has dropped any pretense of subtlety and has issued an Executive Order that explicitly directed all federal regulatory agencies subject to presidential control to undertake formal cost-benefit analysis of all proposed major rules. Moreover, the Reagan order directed that "regulatory action shall not be undertaken unless the potential benefits to society from the regulation outweigh the potential costs to society" [2]. This is as close as any administration has come to applying formal cost-benefit analysis to regulatory decision-making. While the Reagan order is silent on the matter of quantifying costs and benefits only in dollar terms, the implication is clear.

While it is difficult to measure the impact that the Nixon, Ford, Carter and Reagan directives have had on the social regulatory system, it is clear that they have retarded the promulgation of new social regulations. In the case of occupational safety and health, only one new health standard was issued by OSHA during the Nixon Administration (asbestos); three OSHA health standards were issued under the Ford administration (vinyl chloride, 14 industrial carcinogens and coke oven emissions); and nine were issued by the Carter Administration (access to medical records, acrylonitrile, arsenic, benzene, cancer policy, cotton dust, lead, dibromochloropropane (DBCP) and noise).

For its part, the Reagan Administration, with its new Executive Order, has not only sought to stop any further OSHA health standards, but has launched an effort to review, modify and/or revoke many of the OSHA standards currently on the books. Under the direction of Vice President George Bush, the Administration's "Task Force on Regulatory Relief" has already targeted the OSHA standards for carcinogens, lead, noise, cotton dust, walk-around pay and engineering controls for either revocation or substantial modification. According to Bush, these regulations may not meet the cost-benefit requirements of the new Executive Order. This, of course, is only an initial list of existing OSHA standards to be subjected to formal cost-benefit requirements.

In addition to these purely administrative actions on the part of the White House, the Task Force on Regulatory Relief is also developing "regulatory reform" legislation that, among other things, would statutorily mandate formal cost-benefit analysis as a matter of law.

Thus, except perhaps for the Carter Administration, the whole notion of cost-benefit analysis has, at least in part, been used for the political purpose of retarding the social regulatory process. Indeed, in the hands of the Reagan Administration it is being used to fulfill the President's campaign promise to "get the federal government off our backs." In this respect, cost-benefit analysis has lost its technical and theoretical economic character and has been converted into a political tool to subvert the statutory foundations of the social regulatory system.

Given this subversion of what was originally merely an analytical device to evaluate purely economic considerations, it is very difficult for us in the labor movement to accept cost-benefit analysis as anything more than a political weapon wielded by those who are fundamentally opposed to social regulation. Indeed, cost-benefit requirements are a blatant political effort to return occupational, environmental and consumer safety and health, as well as basic social and civil rights protection, back to the "free" market system. As we all know from past experience, the market system alone is simply incapable of protecting our people from harm and often fails to produce social and economic justice.

Cost-Benefit Analysis as an Analytical Technique

The cost-benefit concept has been loosely applied to a number of similar but different analytical frameworks. In its purest context, formal cost-benefit analysis seeks to quantify all the costs and all the benefits in terms of a common metric (usually dollars) and then compare the two to determine if one outweighs the other. The Reagan Administration's approach is the closest we have come to applying formal cost-benefit analysis to the social regulatory system.

At the other end of the cost-benefit spectrum is the no-risk approach which is based on the idea that society, or certain segments thereof, should not be exposed to any risk. The Delaney Clause in the Food, Drug and Cosmetic Act, which prohibits the addition of any carcinogen in the food supply, is the foremost example of the no-risk concept. In this respect, the no-risk approach mandates, as a matter of public policy, that some hazards are so great that they should be outlawed without regard to costs.

In between the two extremes of formal cost-benefit analysis and the no-risk approach are the following:

1. Comparative risk analysis seeks to balance the risks associated with taking a particular action against the risks of not taking such action. The problem with comparative risk analysis is that it is probably impossible to quantify the various levels of risk associated with a series of potential actions. Even when the numbers are developed, they often vary considerably. For example, the cancer risk assessment of vinyl chloride exposure at 1 ppm varied a millionfold. In other words, the risk of cancer from exposure to 1 ppm of vinyl chloride ranged from zero to one million, a differential which made the comparative risk analysis almost meaningless.

2. Informal cost-benefit analysis seeks to make some very rough and general assessments of costs and benefits without insisting on strict quantification in common values. Under this approach, those factors that can be quantified are qualified by nonnumerical factors and value judgments. This was the approach adopted by the Carter Administration.

3. Cost-effective analysis seeks the least costly alternative for achieving a particular goal. Under this approach, the goal itself is not affected by the analysis, but the manner in which the goal is achieved is subject to cost considerations. Once again, strict quantification is not the exclusive driving force. The Carter Administration combined cost-effective analysis with informal cost-benefit analysis as a technique for achieving the least costly regulatory alternative.

Formal Cost-Benefit Analysis

While all the analytical techniques described above have been used or advocated in some degree, the one that lies at the heart of the controversy is the formal cost-benefit analysis approach of the Reagan administration. In this respect, it is probably wise to spell out briefly what the terms "cost" and "benefits" mean in the current context. As applied to social regulation, costs are considered to be the compliance dollar costs to industry of complying with such social regulations as occupational, environmental and consumer safety and health standards. Regulatory benefits, on the other hand, are the prevention of injuries, disease and death. Thus, costs can be viewed as the compliance dollars expended by the industry, while benefits are the "social savings" realized from preventive social regulation.

Aside from the use of formal cost-benefit analysis as a political tool to retard the social regulatory system, the technique itself has severe limitations. While it can and has been used for purely economic analysis, it is simply inappropriate as a method for assessing social issues.

Perhaps its single greatest analytical shortcoming lies in the quantification of costs and benefits. Central to the whole cost-benefit exercise is the ability to first quantify costs and benefits and then reduce them to a common metric for comparative purposes. As I shall discuss later, the necessary data for accurately quantifying compliance costs are either unavailable or are highly suspect, and the date for quantifying social benefits are largely nonexistent or purely speculative.

Even if one could develop acceptable data for quantifying costs and benefits, there remains the very difficult problem of reducing the data to a common, comparative value. Indeed, unless both costs and benefits are expressed in a common unit of measurement, there is no basis for a comparison. Economists, of course, like to measure things in terms of dollars, and, indeed, traditional economic cost-benefit analysis has always used dollars as the basic unit of measurement. While the use of dollars as the common metric makes sense when the issue under analysis is purely monetary in nature, i.e., when only dollar considerations are at stake, it makes no sense whatsoever when the issue involves other than monetary considerations, e.g., the prevention of injury, disease or death.

While it is at least theoretically possible to quantify compliance costs in dollars, it is virtually impossible—especially in a democratic society—to express fully the benefits of social regulation in monetary terms. Although some aspects of the benefit equation are susceptible to expression in dollars (e.g., lost wages), there are other important variables, such as the value of human life, which defy dollar evaluations.

Even the leading advocates of cost-benefit analysis, when pressed on this point, agree that it is not possible to quantify the dollar benefits of social regulation. For example, Dr. Murray Weidenbaum (the new chairman of the Council of Economic Advisors and a leading proponent of cost-benefit analysis) admitted as much during hearings last year before the House Committee on Interstate and Foreign Commerce. In an exchange between Congressman Albert Gore and Weidenbaum, Gore asked: "Let us assume that one million people will die from workplace exposures to asbestos...you decline to put a dollar value on their lives, is that correct?" Weidenbaum responded: "Yes sir,...I do not think that dollars are nearly as important as people." When pressed on how he could then continue to advocate the use of cost-benefit analysis for determining safety and health standards, Weidenbaum retreated in a vague statement on the need to develop some other metric for measuring benefits [3].

There are, of course, other metrics that can be used to evaluate the benefits of social regulations, e.g., the number of lives saved or disease prevented. However, as long as costs continue to be measured in dollars, benefits must also be reduced to dollars or else there is no basis for a cost and benefit comparison. Comparing dollar costs to the number of lives saved, for example, tells you very little. Indeed, it is comparing "apples with oranges." Thus, when cost-benefit economists try to get around the problem of placing a dollar value on human life or health by suggesting the use of some nondollar measurement, the whole concept of cost-benefit analysis begins to break down.

There are those, however, who have advocated monetizing human life and have developed various methods for so doing. Two such methods currently in vogue are the discounted future earnings (DFE) model and the willingness to pay (WTP) model.

The first thing to note about these methods for monetizing human life is that they ignore such nonquantifiable human factors as physical and emotional pain and suffering. These "human costs," although very real, involve complex moral and ethical factors which have both religious and philosophical dimensions. The value of a father's life to his child, for example, simply cannot be monetized or measured statistically. By not accounting for such human factors—which many would agree are the most important—the human life valuation methods are largely incomplete and often arbitrary.

In addition to the irreconcilable moral and ethical problems associated with human life valuation models, there are also serious methodological shortcomings. Indeed, the disparities that often result from such analysis are so large as to question the use of the technique for anything other than theoretical debate. For example, using a DFE model, the National

Highway Traffic Administration established a $395,000 value for "societal loss" for such traffic fatality. Using the same model, adjusted for individual earning power, two researchers developed a DFE index which valued the life of an 85-year-old black woman at $128. Another researcher, using a WTP survey, set the value of human life at $28,000, while still another WTP analysis valued a life at $5 million. Such variations, of course, in addition to raising questions of social justice, demonstrate the very primitive state-of-the-art for monetizing human life [4].

The final comment I wish to make concerning the concept of formal cost-benefit analysis is that, even as a purely economic tool, it was never intended as a mechanical or automatic decision-making instrument. Assuming that both costs and benefits could be quantified and expressed in a common value, there are other social, moral and ethical considerations which must be factored into the decision-making process. As the Reagan Administration would have it, cost-benefit analysis should be an automatic rigid policy device. Where the quantified dollar costs outweigh the quantified benefits, the regulatory decision should be in favor of costs. This approach would change the basic character of cost-benefit analysis from a largely neutral economic analytical technique to a social policy-making mechanism largely dictated by the quality of available data. Questions of equity, justice and social need would be left out of the policy equation. It is this rigid policymaking dimension – along with the methodological problems – which has been rejected by organized labor.

It would, indeed, be unfortunate if this administration compounded its administrative actions to force cost-benefit analysis on the regulatory agencies by also seeking legislation mandating cost-benefit analysis as a matter of law. The complexity of social protection laws is such that statutory changes requiring cost-benefit analysis would cause havoc on the legal and judicial system. To transform cost-benefit analysis from an administrative tool into a legal principle would seriously compromise the existing body of socially protective law in the United States.

Compliance Costs of Social Regulations

One of the major problems with cost-benefit analysis is accurately estimating the dollar costs of complying with such social regulations as occupational safety and health standards. According to Dr. Nicholas Ashford of the Massachusetts Institute of Technology (MIT):

> With regard to estimating the cost of regulation, it is often assumed that because the cost of complying with regulation can be easily mon-

etized, they [cost-benefit analyses] are reliable estimates of true cost. Unfortunately, there are many instances in which the costs are not only uncertain but unreliable [1, p. 11].

Organized labor's experience with occupational safety and health standards tend to substantiate Dr. Ashford's observation. For example, the plastics industry originally estimated that the OSHA vinyl chloride standard would close down the industry at a cost of $90 billion. However, today—some seven years after the standard was issued—the industry is not only still in business, but thriving. The total compliance costs, according to the plastics industry's own figures, has been about $300 million.

The question arises: why the large disparities between the estimated and real compliance costs? According to Dr. Ashford, there are four basic problems with estimating regulatory compliance costs:

1. Since the source of cost data is largely the regulated industries themselves, there is a built-in bias for inflating potential compliance costs.
2. Compliance cost estimates often fail to consider the economies of scale which result from production increases which result from compliance technology.
3. Compliance cost estimates also fail to consider the ability of regulated industries to learn over time how to comply with regulations in the most cost-efficient manner.
4. Perhaps most importantly, compliance costs based upon present technology ignore the crucial role played by technological innovation which yields productivity benefits for the regulated industry as well as the society at large [1, p. 11].

In addition to these four problems identified by Ashford, there is the additional problem of separating cumulative costs from incremental ones. Weidenbaum, for example, has estimated annual industry safety and health costs at $3.5 billion [4, p. 31]. However, his data source was a 1976 survey by McGraw-Hill Company which noted that the $3.5 billion figure was what business had been spending on safety and health even prior to the passage of the Occupational Safety and Health Act. Thus, Weidenbaum's estimates were gross or cumulative costs rather than the incremental compliance costs solely attributable to specific regulations and are, therefore, overinflated. In seeking to assess the compliance costs of social regulations, only the incremental costs of regulation should be counted rather than the cumulative costs which would have accrued even in the absence of regulation.

Finally, regulatory compliance costs fail to consider the dynamic character of the national economy. Indeed, cost estimates are largely based on a given set of economic assumptions at any one point in time. As we all know, these assumptions keep changing to accommodate national and international economic events. What might be acceptable compliance costs during periods of high economic growth rate might prove to be unacceptable during periods of recession or depression. Thus, while the benefits of social regulation remain constant (e.g., the prevention of injury), the costs of regulation are fluid in a constantly changing economy.

To ignore these economic dynamics, as cost-benefit analysis does, is to subvert the whole notion of increasing society's economic ability to improve the quality of life. I believe that American industry currently has the economic resources to dramatically improve worker safety and health and still make a profit. I also believe that further improvements can be made in the future, and that the compliance costs of social regulation should be viewed as an economic investment rather than a nonproductive burden.

Thus, in viewing the cost side of the cost-benefit formula, a great deal of care must be exercised in determining the true compliance cost of regulation. Unfortunately, in the case of occupational safety and health standards, the compliance costs have been exaggerated by the business community. By not factoring in such considerations as technological innovation, economies of scale, the ability to learn over time and the incremental costs of regulation, industry's figures have tended to be more political than economic. In this sense, the inflated cost estimates are used more to resist regulation than to arrive at an accurate assessment of the actual costs involved. And as long as this situation continues, the cost-benefit exercise will be politically skewed and biased.

Benefits of Regulation

If the direct compliance costs of social regulations are difficult to determine, then the regulatory benefits are nearly impossible to quantify with any degree of accuracy or acceptability. While some of the regulatory benefits may be subject to quantification, others defy statistical measurement. Implicit in the whole benefit analysis exercise are certain moral, ethical, philosophical, religious and social judgments concerning the value of human life and health.

In very broad terms, the benefits of worker safety and health regulation can be defined as the prevention of injury, disease and death. Other related benefits might also include improved worker productivity, more efficient industrial processes, and more responsive public and private institutions [5].

Perhaps the best way to analyze the benefits of injury and disease prevention is from the perspective of costs. In this respect, the question becomes: "what are the nonregulatory, socioeconomic costs of injury and disease which can be reduced or eliminated by the greater regulation of the workplace environment?" Viewed in this manner, the benefits of prevention can be defined as "social savings." Thus, while compliance costs might rise because of increased regulation, the other costs of injury and disease may be reduced and a social savings realized.

Some of the "other costs" of injury and disease are quantifiable and others are not. Among the quantifiable costs are medical care, insurance, lost wages, reduced productivity, workers' compensation benefits, corporate liability claims and social welfare payments. The nonquantifiable costs are those human costs associated with the physical and emotional pain suffered by victims and their families. As indicated above, these human costs, although very real, involve complex moral and ethical factors which have both religious and philosophical overtones. Indeed, the value of a father's life to his children simply cannot be monetized or measured statistically nor can it be expressed in government regulations. What we have, therefore, are at least two important dimensions of the value of human life and health, one which is measurable and the other which is not. To a worker and his family — as well as his immediate community — the second dimension is undoubtedly the more important.

While some theoretical economists continue to tinker with human life valuation models, many have rejected attempts to monetize life and have sought instead to simply quantify the costs that are subject to quantification as a rough indicator of at least some of the social costs of injury and disease. Even in this respect, the data are at best suggestive.

In terms of occupational safety and health, the National Safety Council estimated that in 1971, workplace injuries (exclusive of disease) accounted for about 1% of the Gross National Product, or some $9.3 billion in lost wages, medical care, insurance claims, productivity and lost time of co-workers. The corresponding figure for 1978 was $23 billion. In addition, Dr. Nicholas Ashford had estimated that another $10 billion was lost if occupational disease was responsible for just one day of absence per worker for the 25 million workdays lost in 1972 due to injury. The Department of Labor has stated that if the costs to workers could be fully established for occupational disease and reimbursement paid to cover the costs, the resulting annual sum would be somewhere between $30 billion and $50 billion [1, pp. 76–77].

With respect to cancer, DHEW's Task Force on Environmental Cancer and Heart and Lung Disease has estimated that the annual cost of cancer in the United States is $15.6 billion in medical care, lost wages, and reduced productivity [6]. Dr. Samuel Epstein, a leading environmental health scientist, has adjusted this figure (first estimated by the Government Accounting Office in 1971) to $25 billion in 1978 [7]. How much of this cost can be directly attributed to occupational cancer has not been determined. However, if one accepts the DHEW estimate that at least 20% of all cancer in the United States may be occupationally related, the social cost of occupational cancer may be as high as $5 billion annually. To this, of course, must be added such other costs as product liability claims and social welfare payments. In terms of asbestos alone, there are some $2 billion in outstanding product liability suits pending against the asbestos industry. Moreover, since state workers' compensation programs only compensate a mere 5% of all occupational disease cases, most workers disabled by workplace diseases must rely on social welfare systems, such as Social Security Disability, which in 1978 paid some $13 billion in benefits, of which at least $2.2 billion was directly related to occupational illness [8].

All of the cost figures quoted above are, of course, aggregates that are subject to many of the same methodological and data collection problems associated with estimates of compliance costs. Thus far there have been very few attempts to monetize the benefits that might be realized from specific OSHA standards. Indeed, the only such study that we know of was done by Russell Settle of the University of Delaware [5, p. 32]. By applying a DFE model to the OSHA asbestos standard, Settle concluded that, if the standard saved 2563 lives per year, it would result in an annual social savings of $651.5 million.

Thus, on the benefit side of the cost-benefit equation, what we have are various sets of numbers and estimates which merely establish some rough indicators of the social-dollar costs of workplace injury and disease. Taken together, these figures do not reveal very much because, first, they are not inclusive and, secondly, they cannot account for the human costs. Moreover, the further out one seeks to extend the benefits of regulation, the more speculative the exercise becomes. The state-of-the-art is such that it is extremely difficult to estimate the number of lives saved or disease prevented 20 or 30 years after a regulation has been promulgated. Many occupational cancers, for example, have latency periods of 20, 30 and even 40 years from the time of exposure to the onset of disease. Given this and such other complicated factors as the dynamics of technological innovation, the best that can be done is to recognize that large numbers of workers are exposed to increased workplace hazards and to take preventive regulatory actions which may result in large social savings. That relatively large compliance costs are also involved must also be recognized, but society has made the decision, through the OSHAct, to protect workers as best as possible. While cost-benefit analysis may have some limited use in establishing certain rough parameters of both costs and benefits, as a decision-making tool, its defects are such that it is no magic substitute for informed policy judgments.

Distribution of Costs and Benefits

As we have seen, those costs which can be quantified include the direct costs of regulatory compliance ("costs") and lost wages, reduced productivity, medical care, insurance and social welfare payments ("benefits"). Presently, the distribution of these costs falls disproportionately on individual workers and the public treasury, while industry is bearing only the relatively small costs of complying with safety and health regulations. It has been suggested that because industry is not responsible for paying the full economic costs of occupational disease; for example, there is little economic incentive for employers to clean up the workplace and prevent disease.

While it is difficult to estimate the future compliance costs of OSHA standards, there are economic data on the regulatory costs of those health standards thus far promulgated. Prior to those OSHA standards issued in the closing days of the Carter Administration, OSHA determined that the total compliance costs of the ten health standards issued up to that point included about $2.8 billion in capital expenditures and some $686 million in annual operating costs [9]. Thus, in spite of the large cost estimates advanced by industry, after nine years of regulation the OSHA program had resulted in comparatively modest incremental economic costs for the regulated industries, particularly when one considers that these industries are among the largest in the world, including oil, chemicals, steel, plastics and rubber.

Compared to some of the other costs of occupational injury and disease, the compliance costs are indeed modest. In lost wages alone, workers are forced to bear a much greater burden than their employers. According to a study on occupational disease prepared by the Department of Labor, in 1978 workers lost some $11.4 billion in wages due to occupational disease alone, only about $2.2 billion of which was recovered through social security disability and other welfare benefits [8, p. 3].

When compared to some of the other economic data we have examined, the figures cited above lead to the conclusion that, in general, the costs of occupational injury and disease have not been fully reflected in the marketplace, and both the American business community and American consumers have, in effect, been subsidized by workers and the public treasury. In terms of lost wages due to occupational disease, this subsidy amounts to $11.4 billion a year from workers and another $2.2 billion from the social welfare system. Compared to this $13.6 billion "wage cost," the business community is spending at most $3.5 billion (Weidenbaum's figure) for all safety and health activities, including those required by OSHA regulations. Thus, from a pure cost-benefit approach, the economic costs of compliance are far less than the economic benefits which could be realized by the elimination or marked reduction of occupational disease and injury. On the basis of the rough numbers alone, a cost-benefit advocate would have to conclude that OSHA regulations cannot be considered economically burdensome to industry until compliance costs equal or exceed the individual and social costs of occupational disease.

CONCLUSIONS

As an analytical tool, cost-benefit analysis has some utility when applied to purely economic issues involving dollar considerations. However, as a technique for making social decisions its limitations are such so as to question its use for anything other than providing some very rough and general parameters of costs and benefits.

From a purely analytical perspective, cost-benefit analysis can be helpful in providing a framework for:

1. organizing and structuring available information;
2. making some limited choices concerning priorities;
3. identifying the least desirable alternatives; and
4. identifying those areas of uncertainty requiring additional research.

On the other hand, the analytical weaknesses of cost-benefit analysis include:

1. overemphasis on economic efficiency at the expense of other social and human factors which may be more important;
2. predisposition for translating gains and losses into purely monetary measurements;
3. failure to recognize the dynamics of technological innovation;
4. failure to consider the social distribution of costs and benefits; and
5. dependence on a very limited data base which is often biased and politically skewed.

As a formal decision-making mechanism, cost-benefit analysis is a primary example of an analytical technique gone awry. Even as an economic tool, cost-benefit analysis was never intended as the sole determining force for public decision-making. Other social, economic and political considerations often outweigh the cost-benefit calculations. To now suggest that the cost-benefit numbers alone (which themselves are highly questionable) should dictate decision-making in the area of social regulation is a political perversion of what traditionally has been a very limited technical device for analysis. In this sense, cost-benefit analysis is a pseudoscience intended to provide an air of scientific respectability to purely political motivations. It is backdoor attempt to subvert existing socially protective laws and to return workers and others in our society to the mercy of the market system. Indeed, as *The Washington Post* recently editorialized with respect to the OSHA cotton dust standard:

> Comparing *relative* costs and benefits is, no doubt, a useful device for deciding among alternative regulations. But in the cotton dust case and all the many others like it, no computation, simple or complex, can absolve Congress of its responsibility for seeing that industries do not pass off the real costs of doing business on either their workers, their communities or the general taxpayer.

The same, of course, can be said for the new Reagan Administration which has a constitutional responsibility to uphold and enforce the laws of the land, including those which protect workers, consumers and the environment. Cost-benefit analysis is no magic substitute for sound value judgments and should not be used to reward one segment of society at the expense of others.

REFERENCES

1. U.S. House of Representatives, Interstate and Foreign Commerce Committee, Subcommittee on Oversight and Investigations: Cost/benefit analysis: wonder tool or mirage. 1980.
2. Office of the President of the United States: Executive Order 12291. 1981.
3. U.S. House of Representatives, Committee on Interstate and Foreign Commerce, Subcommittee on Oversight and Investigations and Consumer Protection and Finance: The use of cost/benefit analysis by regulatory agencies. Public hearing, October 24, 1979:101.
4. Corporate Accountability Research Group: Business war on the law: an analysis of the benefits of federal health/safety enforcement. 1979:40–48.
5. Center for Policy Alternatives, Massachusetts Institute of Technology: Benefits of environmental, health and safety regulation. Report prepared for the U.S. Senate, Committee on Governmental Affairs: 1980:7.

6.   U.S. Department of Health, Education and Welfare: Environmental pollution and cancer and heart and lung disease. First Annual Report to the Congress. 1978:13.
7.   Epstein S: OSHA's proposed cancer policy. Statement before the Occupational Safety and Health Administration. 1978.
8.   U.S. Department of Labor: An interim report to Congress on occupational disease. 1980.
9.   Occupational Safety and Health Administration: Safety and health standards. 1979.
10. U.S. Department of Labor: An interim report to Congress on occupational disease. 1980:3.

DISCUSSION: CHAPTER 5

Theodor D. Sterling and John J. Sheehan

DR. STERLING: I'm curious about the failure of representatives of labor, as well as other occupational health professionals, to press for government subsidies to industry to enable them to meet occupational health standards. We subsidize industries in Canada and the United States in many different ways, including direct payoffs, tax rebates and tax write-offs. It would seem to me that, within the tradition of the system involved, there should be nothing contrary to government subsidies in order to meet the OSHA standards.

Have there ever been proposals of this nature advanced either by labor or by people in the occupational health bureau? If not, why not?

MR. SHEEHAN: I have a number of comments to make with regard to your statement.

First, the claims made to support regulations that would cover alleged injuries or illness have been constantly disputed. For example, we claim cancer is a danger to those who work around the coke ovens and that we have serious arsenic problems at some of our smelters. But you find the opposition states there is no problem and the exposure is not serious. Then, rather than concentrating on handling the problem, you are forced into an adversarial position.

Second, under our current laws, there are a number of tax advantages that are available for industries that engage in abatement activity. Investments in control equipment are subject to 10% investment tax credit. Some might say that everyone gets the credit whether they put up a hot-dog stand or a bag house. Therefore, maybe we ought to offer more than the 10%.

There is an additional depreciation schedule available to those industries involved in environmental abatement under the Clean Air and Water Acts. Industrial revenue bonds are also available for environmental abatement activities. Some of these investments may be written off entirely by virtue of the industrial revenue bonds' low interest rate.

Third, there may be an issue of some fundamental philosophy here. Much of the concern relates to the "free market." Are you externalizing or internalizing these costs? If, indeed, you have government aiding and assisting industries, then you might say you are defeating the issue or the question of internalizing costs. Many people believe that the destruction of the air or the destruction of a person's body should be paid as a cost of running that business.

I must readily admit, however, that in the United States we have only in the last ten years come to recognize what we have been doing to our environment and our people. We have a tremendous retrofit problem — old plants, old facilities. Hopefully, if we initiate a modernization program, we can clean out these facilities — certainly at least the steel, copper and various oil facilities. Perhaps legislation will be required for those that remain standing and needing heavy retrofitting. But it certainly will not happen in this Congress. The effort is to clean out the federal budget. Therefore, I think that, for the present, it is simply impractical.

CHAPTER 6

USE AND ABUSE OF THE OCCUPATIONAL SAFETY AND HEALTH ACT GENERAL DUTY CLAUSE

Richard F. Boggs, PhD

Organization Resources Counselors, Inc.
Washington, DC

On December 29, 1970, President Richard Nixon signed into law the Occupational Safety and Health Act (OSHAct) of 1970 (PL 91-596), "to assure safe and healthful working conditions for working men and women; by authorizing enforcement of the standards developed under the Act; by assisting and encouraging the states in their effort to assure safe and healthful working conditions; by providing for research, information, education, and training in the field of occupational safety and health; and for other purposes." At the signing ceremony the new legislation was proclaimed to be one of the most significant steps taken by the United States to assure worker protection.

Section 5(a)(1) 29 U.S.C. 654(a)(1) of the act states, "(a) Each employer...(1) shall furnish to each of his employees employment and a place of employment which are free from recognized hazards that are causing or are likely to cause death or serious physical harm to his employees." As interpreted and enforced by the Occupational Safety and Health Administration (OSHA), this section, better known as the General Duty Clause, has been the subject of continued and intense controversy. Although the clause originally was intended to illustrate the breadth of an employer's safety and health responsibility, it has lately become a greatly abused OSHA enforcement tool.

Occupational Safety and Health Review Commission Chief Paul Tenney has observed that the Secretary of Labor is trying to accomplish through the General Duty Clause, "what he has been unable to do through rulemaking proceedings" [1]. A recent statistical study conducted by the Commission showed that 25% of final orders involved alleged violations of Section 5(a)(1). Whether or not the congressional intent of this Section will continue to be abused under the new administration is as yet unclear.

Examples of abuse include:

- A serious citation and a $420 penalty for Section 5(a)(1) to a small firm in Bridgewater, Massachusetts, for failure to train employees in procedures to be followed in the event of a cyclone;
- a serious citation and a $1260 penalty for Section 5(a)(1) for failure to protect employees against being struck by lightning when the antennas on vans were elevated during electrical storms; and for failure to provide devices to insure that the antennas were in lowered positions before moving the vans;
- a serious citation and a $350 penalty for Section 5(a)(1) for failure to ensure that employees were not exposed to vehicular traffic when stepping out of toilets; and
- a willful citation and a $4800 penalty for Section 5(a)(1) for failure to ensure that front-end loaders were not used to topple a water tower.

Although there is a need for the General Duty Clause, actions such as these undermine the seriousness of the effort made by our elected representatives to provide a realistically safe environment for workers.

HISTORICAL REVIEW

Members of Congress who were instrumental in the inclusion of the General Duty Clause recognized that its language was broad and, therefore, attempted to ensure that its intent would be narrowly construed.

The Senate Committee report on the Occupational Safety and Health Act (Senate Report 91-1282) stated:

> The general duty clause in this bill would not be a general substitute for reliance on standards, but would simply allow the Secretary of Labor to insure the protection of employees who are working under special circumstances for which no standard has yet been adopted. Morever, the clause merely requires an employer to correct *recognized hazards* that have been discovered on inspection and made the subject of an abatement order. There is *no penalty for vio-*

lation of the general duty clause. It is only if the employer refuses to correct the unsafe condition after it has been called to his attention and made the subject of an abatement order that a penalty can be imposed (emphasis added).

Compromise language was agreed on during a House-Senate conference. In urging adoption of the compromise language, Congressman William Steiger stated on the floor of the House on December 17, 1970:

> The conference bill takes the approach of this House to the general duty requirement that an employer maintain a safe and healthful working environment. The conference-reported bill recognizes the need for such a provision where there is no existing specific standard applicable to a given situation. However, this requirement is made realistic by its application only to situations where there are "recognized hazards" which are likely to cause or are causing serious injury or death. Such hazards are the type that can readily be detected on the basis of the basic human senses. Hazards which require technical or testing devices are not intended to be within the scope of the general duty requirements. It is also clear that the general duty requirement should not be used to set ad hoc standards. It is expected that the general duty requirement will be relied upon infrequently and that primary reliance will be placed on specific standards which will be promulgated under the Act.

Congressional concern about the use of general duty clauses was clearly expressed during the debate on passage of the Federal Mine Safety and Health Act of 1977. The Senate bill, S717, contained a general duty clause, whereas H.R. 4287, the House substitute, contained no such proviso. Congressman Sarasin twice pointed out the lack of a general duty clause. In response, Congressman Gaydos, floor manager of the bill, stated:

> The general duty clause is a catchall. We have been using every effort we can to provide specific standards so that if they violate standards they are fined. That is fair. That is what we are talking about. The general duty clause catches them for a violation of practically anything an inspector may consider a serious violation. So in responding to individuals who have been critical of standards and how they are met and enforced, we say specifically, "Forget about the general duty clause."

The House-Senate conference compromise language conformed to the House amendment.

COURT INTERPRETATIONS

Despite OSHA's continued effort to broaden its application, the courts have, with few exceptions, held to Congress' narrow interpretation of the General Duty Clause. The leading case on Section 5(a)(1) is *National Realty and Construction Co., Inc. v. OSAHRC* (489 F.2d 1257). In that case it was held that the employer was subject to the general duty clause "where no promulgated standards apply." The Court went on to state that a violation of Section 5(a)(1) could be sustained only where OSHA proved:

> (1) That the employer failed to render its workplace "free" of a hazard which was (2) "recognized" and (3) "causing or likely to cause death or serious physical harm."

In *Babcock & Wilcox Company v. OSAHRC* (8 OSAHRC 1317), a more recent case, the U.S. Court of Appeals generally supports these criteria and emphasizes that the hazard must be reasonably foreseeable.

On February 18, 1981, in *Secretary of Labor v. Kastalon, Inc.* (OSAHRC Docket No. 79-3561), an administrative law judge stated:

> The Secretary has the burden of proof to establish that a "recognized hazard" exists which is "likely to cause" death or serious harm to respondent's employees under Section 5(a)(1) of the Act. The Secretary must quantify by a preponderance of evidence the exposure levels which will establish these elements of a violation and that these exposure levels were exceeded in the respondent's workplace.

These court-imposed requirements are in addition to other elements of proof that OSHA must establish in order to demonstrate a violation of Section 5(a)(1) of the Act, including:

> (1) noncompliance with an OSHA standard, (2) employee endangerment as a result thereof, (3) employer knowledge of the cited condition or practice, (4) the existence of effective and feasible measures the employer could have taken to avoid citation [2].

OSHA REQUIREMENTS FOR ENFORCEMENT

Much of the abuse of the General Duty Clause could be eliminated if OSHA inspectors would adhere to the instructions provided in Chapter 8 of the OSHA "Field Operations Manual":

A hazard is *recognized* if it is a condition that is (a) of a *common knowledge* or general recognition in the particular industry in which it occurred, and (b) detectable (1) by means of the senses (sight, smell, touch, and hearing) or (2) is of such wide, general recognition as a hazard in the industry that, even if it is not detectable by means of the senses, there are generally known and accepted tests for its existence which would make its presence known to the employer. For example, excessive concentrations of a toxic substance in the air would be a *recognized hazard* if they could be detected through the use of measuring devices. However, if a specific standard covered the hazard, the General Duty Clause would not be applicable.

All too often, OSHA inspectors have resorted to the General Duty Clause when they were unable to find a reason for citing under any other provision of the OSHAct. In fact, the General Duty Clause is used as a device for penalizing employers when a serious accident occurs, rather than as an avenue for correcting conditions which may cause an accident.

CONCLUSION

The General Duty Clause should be used as it was intended by the Congress of the United States: to correct recognized hazards not identified by a specific OSHA standard, particularly those likely to cause death or serious physical harm to workers.

The General Duty Clause should *not* be used:

- to circumvent or subvert established rulemaking procedures;
- as a substitute for reliance on standards;
- to cite for stricter requirements than have been established in specific standards;
- to address chronic hazards presenting no immediate danger;
- to set ad hoc standards; or
- as a mechanism for citations based on documents developed for research or administrative purposes which do not meet the criteria previously described.

REFERENCES

1. Occupational Safety Health Reporter (October 25, 1979):495.
2. Moran: The developing law of occupational safety and health. Oklahoma Law Rev. 30:354, 1977.

DISCUSSION: CHAPTER 6

William L. Wagner, Richard F. Boggs and Nicholas A. Ashford

MR. WAGNER: You referred to the Federal Mine Safety Health Act of 1977. It is my understanding that the General Duty clause was eliminated from the Act because the Act includes a clause that allows a mine inspector to close a mine if he sees an imminent danger situation. It is my recollection that at the time the Act was being constructed, it was felt that this clause would suffice for the purpose of the General Duty clause.

DR. BOGGS: There is such a clause in the Act, and mine inspectors can shut down the mine for an imminent hazard.

DR. ASHFORD: I'd like to make some comments that go to the tone rather than the letter of your remarks. First, I would hope that finally in 1981 we can remove all scatological references to OSHA standards and activities. Let me address one and then hope to lay it to rest forever: the split toilet seat standard was used to poke fun at OSHA. It turned out to be an industry standard. Reducing OSHA concern to scatological health hazards is an injustice. Citing those few examples, which you pulled out of some obscure hearing cases, as exemplifying the use of the General Duty Clause is unfair. That is not what the General Duty Clause is all about—this audience knows better. Let's hope to lay that issue to rest and really deal with the serious issues.

What do you do with a clause that was put in at a time when the Act was passed, when we were not as aware as we are today of the occupational health hazards? It is fair to say that this Act was passed on the heels of the Coal Mine Safety and Health Act when we were concerned with the industrial accident rate—when our consciousness of industrial disease was not as broad.

The General Duty Clause is properly limited to those hazards which are likely to cause serious bodily harm, not skin irritations, which lead to nothing. The General Duty Clause is not, in fact, used to replace standards. We see a situation or workplace found to expose workers to substances like Kepone, for which there was no standard. One company made the material. Then there is the case of DBCP [dibromochloropropane] where 50% of the workers were sterilized. When the Secretary of Labor enters those workplaces, you want to have a full-fledged authority of the Act to indicate that when you have 50% of your workers sterilized, you don't need to do a hygiene workup in order to understand that there is a violation of the statute's general duty to provide a safe workplace. This duty also existed in common law before the passage of the OSHAct.

A very clear question to ask is: would you let your daughter work in a vinyl bromide plant knowing full well that OSHA hadn't gotten around to setting standards for vinyl chloride's first cousin? I presume the answer would be "no." Well, if you don't want your daughter to work in such a plant, then I suggest that you shouldn't want anybody else's daughter or son to work in those plants either. And without the General Duty Clause we would not be able to protect workers from many harmful substances.

The Act is clear on its face. We are not talking about an OSHA inspector saying you've exceeded 10 parts-per-million for a substance we never set a standard for. We are talking about a substance that causes great illness for which there are recognized avenues of control.

For example, one of the carcinogens on OSHA's top 10 list, even though we have no standard, would be considered a recognized hazard. There are data to force the employer to provide some protection.

Now, I can agree with perhaps the letter of what you said, that is, we ought not to abuse the General Duty Clause, but we ought not create a straw man to imply the General Duty Clause is abused or was subverted from its original intent.

DR. BOGGS: I appreciate your comments, I was hoping for a chance to debate with somebody. I did clearly state that I think there is a strong need for a General Duty clause. In fact, I am disappointed it was not included in the Mining Act, but we need to attempt to minimize abuse of it. And that is easier said than done.

CHAPTER 7

THE STANDARDS-RECOMMENDING PROGRAM OF THE NATIONAL INSTITUTE FOR OCCUPATIONAL SAFETY AND HEALTH

William F. Martin, PE and Richard A. Lemen, MS
Division of Criteria Documentation
and Standards Development
National Institute for Occupational Safety and Health
Rockville, Maryland

Occupational safety and health issues and government regulatory activities are current topics in professional journals, political speeches and news media. As a result, words and phrases such as hazardous waste, cancer, mutagenesis, toxic substances, occupational health and safety have become terms commonly used by the public. Because of the political debate and controversy surrounding occupational regulations, it is difficult for a federal agency to connect these terms to a proposed government action involving regulations or standards.

As public health officials, we are constantly looking for ways to improve the nation's health. With this in mind, we will review the status of the National Institute for Occupational Safety and Health (NIOSH) development of new and updating of existing occupational safety and health recommendations for standards.

Criteria documents have been the primary mode for transmitting standards recommendations to the U.S. Department of Labor for both the Occupational Safety and Health Administration (OSHA) and the Mine Safety and Health Administration (MSHA). This process was established under the Occupational Safety and Health Act (OSHAct), PL 91-596, in 1970, and the federal Mine Safety and Health Amendments Act of 1977, respectively. More than 100 criteria documents have been developed and transmitted to the Department of Labor during the past ten years.

The standards-setting process has been applauded and criticized throughout its federal history. There was a great deal of controversy surrounding the use of law vs education to obtain better occupational safety and health conditions. A major criticism of the past ten-year process was the low conversion rate of NIOSH recommendations to OSHA standards. The effectiveness of federal expenditures was questioned. Currently, NIOSH and OSHA exchange priorities and interact through a work group and a national planning group. This is an ongoing process to rank and identify needed occupational safety and health standards.

Several years ago, NIOSH hired an outside consultant to evaluate the document process [1]. More than 90 interested occupational safety and health professionals were interviewed and participated in the evaluation. That study recommended several major changes in the criteria document and standards recommendation process. In 1979–1980 a thorough reevaluation of NIOSH's total information dissemination procedure was undertaken in preparation for an institute change. A major portion of the recommendations from the consultant's report was included [1]. Their contributions have had considerable influence on the changes NIOSH is experiencing.

Let me review a few major changes that NIOSH is attempting to implement.

One of the first recommendations was to evaluate critically the scientific validity and scientific assessment of information supporting the recommended standards. This includes peer review by experts outside of NIOSH, using the most experienced professionals in the development and review of the documents, and openly identifying the weaknesses and gaps in the supporting data. NIOSH has been following this recommendation.

A second major issue was to remove NIOSH from a committed, fixed number of criteria documents to be delivered annually, and to concentrate more on the development of a few well supported, well prepared criteria documents.

A third major recommendation was to develop and encourage a more compatible and efficient relationship between OSHA and NIOSH. A very small percentage of the NIOSH criteria documents actually resulted in the establishment of new OSHA standards. The improved communications between the Department of Labor and NIOSH and the joint participation in setting standards priorities should make the entire process more efficient. NIOSH will apply additional resources to OSHA's highest priority topics by updating past criteria documents and developing new recommendations. In addition, NIOSH will develop documents on other emerging occupational health problems for direct dissemination.

A fourth issue raised was the quality and cost of outside consultant services in developing criteria documents. To overcome this, NIOSH professionals are now more actively involved with existing contractors, and a higher percentage of documents are being prepared within the institute.

A fifth concern of the consultants, with strong agreement from the NIOSH staff, was the need for broader utilization of the criteria documents. It was felt that, in addition to their applicability to the Department of Labor standards setting process, the documents could be resources for industry and workers. This recommendation has been studied in considerable detail. NIOSH is attempting to develop a companion document to the criteria document in a form that managers, supervisors and workers can utilize to protect the worker's health.

A sixth area that raised considerable discussion has to do with the relationship between the Department of Labor and NIOSH as to who will identify the necessary engineering and control technology, conduct the economic analyses and determine feasibility. This debate continues today with no resolution as to the best way to deal with these areas. Also in question is the role that each agency should play in the actual limit on exposure and in determining if that limit should be contained in the NIOSH recommendations. This area will continue to be negotiated between the Department of Labor and NIOSH. It will most likely remain a dynamic area, depending on the topic, the political and judicial climate of the time, and the agency's available resources.

As of this date, NIOSH has transferred to OSHA 107 criteria documents, of which 10% have been updated, and another 10% are in the process of being updated and should be completed in fiscal year 1981-1982. The current schedule calls for four to six criteria documents to be transmitted to the Department of Labor each fiscal year. The priority list for fiscal 1983-1984 is currently being developed. Toxic agents, physical agents, processes and generic industries are selected for criteria documents based on a priority rating system used by NIOSH that includes input from OSHA, external consultants and NIOSH professionals.

We feel progress has been made toward improving NIOSH's performance in criteria document development. In April 1979 the previous Director of Criteria Documents Division, Dr. Vernon Rose, presented an overview of NIOSH's five-year program for criteria document prioritization and production at a meeting of the American Public Health

Association [2]. Many of the issues concerning NIOSH and DOL during 1979 were well presented by Whorton and Wegman at that same APHA meeting [3]. (A perspective of that first five years can be obtained by reviewing these articles.) It was shortly after the 1979 presentations that the contract Dr. Rose had initiated on criteria document process evaluation was completed. The projected FY 1980–1981 list of NIOSH criteria documents, extracted from Dr. Rose's presentation, is included in Tables I and II. A comparison of that list to the latest NIOSH list (Table III) of

Table I.   Criteria Documents and Recommended Standards
Transmitted to the U.S. Department of Labor [2]

Document Name	Document Name
Acetylene	
Acrylamide	
Acrylonitrile (emergency hazard review)	
Aldrin/Dieldrin (special occupational hazard review)	
N-Alkane Monothiols, Cyclohexanethiol and Benzenethiol	
Alkanes	
Allyl Chloride	
Ammonia	
Anesthetic Gases	
Antimony	
Arsenic	
Arsenic (revised)	
Asbestos	
Asbestos (revised)	
Asphalt Fumes	
Benzene	
Benzene (revised)	
Benzoyl Peroxide	
Benzyl Chloride	
Beryllium	
Boron Trifluoride	
Cadmium	
Carbaryl	
Carbon Black	
Carbon Dioxide	
Carbon Disulfide	
Carbon Monoxide	
Carbon Tetrachloride	
Carbon Tetrachloride (revised)	
Chlorine	
Chloroform	

Table I, continued

Document Name	Document Name

Chloroform (revised)
Chloroprene
Chromic Acid
Chromium(VI)
Chrysene (special occupational hazard
   review)
Coal Gasification
Coal Tar Products
Coke Oven Emissions
Confined Spaces
Cotton Dust
Cresol and Cresylic Acid
Decomposition Products of Fluoro-
   carbon Polymers
Dibromochloropropane (emergency
   hazard review)
Dinitro-*o*-cresol
Diisocyannates
Dioxane
DDT (special occupational hazard
   review)
Epichlorohydrin
Ethylene Dibromide
Ethylene Dichloride
Ethylene Oxide (special occupational
   hazard review)
Fibrous Glass
Fluorides
Formaldehyde
Glycidyl Ethers
Hazardous Materials
Hot Environments
Hydrazines
Hydrogen Fluoride
Hydrogen Sulfide
Hydroquinone
Inorganic Cyanides
Isopropyl Alcohol
Kepone (emergency hazard review)
Ketones
Lead, Inorganic
Lead, Inorganic (revised)
Logging
Malathion
Mercury, Inorganic
Methyl Alcohol

Table I, continued

Document Name	Document Name

4,4′-Methylene-*bis* (2-chloroaniline)
  (special occupational hazard review)
Methyl Parathion
Methylene Chloride
Nickel Carbonyl (special occupational
  hazard review)
Nickel, Inorganic
Nitric Acid
Nitriles
Nitrogen Oxides
Nitroglycerine and EGD
Noise
Organoisocyanates
Organotin Compounds
o-Tolidine
Parathion
Pesticides, Manufacture and
  Formulation of
Phenol
Phosgene
Polychlorinated Biphenyls
Refined Petroleum Solvents
Silica
Sodium Hydroxide
Sulfur Dioxide
Sulfuric Acid
1,1,2,2-Tetrachloroethane
Tetrachloroethylene
Toluene
Toluene Diisocyanate
1,1,1-Trichloroethane
Trichloroethylene
Trichloroethylene (special occupational
  hazard review)
Tungsten Compounds
Ultraviolet Radiation
Vanadium Compounds
Vinyl Acetate
Vinyl Chloride (emergency hazard
  review)
Vinyl Halides
Xylene
Zinc Oxide

Table II.   Criteria Documents and Recommended Standards
to be Transmitted to the U.S. Department of Labor
as of May 1980 [2]

Document Name	Document Name

FY 1979

Aliphatic Primary Monoamines
Brominated Aromatics
Chlorinated Benzenes
Coal Liquefaction
Cobalt Compounds
Fluorocarbons
Foundries
Furfuryl Alcohol
Methyl Chloride
Nitrobenzenes
Nitrotoluenes
Oil Mists
Oxalic Acid
Paint and Allied Products
   Manufacturing
Plastic and Resin Manufacturing
Printing Industry
Radiofrequency and Microwave
   Radiation
Roofing Industry
Slaughtering and Rendering Plants
Styrene
Synthetic Rubber Manufacturing
Talc
Welding and Brazing Wood Dust

FY 1980

Aliphatic di- and polyamines
Aromatic Amines
Brominated Aliphatics
Dichloropropane
Diesel Emissions
Dying and Finishing Textiles
Hexachlorobutadiene
Hexachloroethane
Infrared Radiation
Manufacture of Nonmetallic Pigments
   and Dyes
Monochloroacetic Acid
Monochloroethane
Pentachloroethane
Pulp and Paper Mills
Secondary Aliphatic Monoamines
Tertiary Aliphatic Monoamines

Table II, continued

Document Name	Document Name
Tetrahydrofuran	
Trichloropropane	
Ultrasonics	
Vibrations, Whole Body	
Wood Preserving	

Table III.   Division of Criteria Documentation and
Standards Development: Current Doccumentation Projects
for FY 1981–1982

Document Type	Document Name
Criteria Document	Radiofrequency and Microwave Radiation; Talc
Criteria Document/Occupational Hazard Assessment	Welding, Brazing and Thermal Cutting
Occupational Health Assessment	Acrylic Acid and Esters; Aliphatic Poly- and Diamines; Coal Liquefaction; Cobalt; Dyeing and Finishing Textiles; Foundries Oil Mists/ Cutting Fluids; Field Sanitation; Methyl Halides; Printing; Pulp and Paper Mills; Rendering Processes; Slaughtering Processes; Styrene; Ultrasonics; Wood Preserving
Comprehensive Standard Package	Asbestos; Diesel; Silica; Radiation
Current Intelligence Bulletin	Ethylene Oxide; Formaldehyde; Nitros- amines; Silica Flour; Toxaphene
Special Hazard Assessment	Trinitrotoluenes
Special Hazard Review	Chlorinated Benzenes' Dinitrotoluenes; Mononitrotoluene; DOP/DEHP; Primary, Secondary and Tertiary Aliphatic Mono- amines; Mill Reagents; Sanitation
Bulletin	Benzidine-Based Dye; Reproductive Hazards
Alert	Alkaline Dust; DDM-4,4'-Diaminodiphenyl- methane; Methyl Ethyl Peroxide
Profile	Oxalic Acid

documents currently under development will point out the direction in which NIOSH is moving, how fast and what topics have been given top priority for the next several years. During the last two years, many of these topics have been addressed with informational publications other than criteria documents, while others have been dropped from the list due to low priority and limited resources.

The Division of Criteria Documentation Standards Development welcomes written comments, including constructive criticism on the utility of the documents we have produced. User feedback aids the division greatly in improving NIOSH products and services. The division is thoroughly committed to developing top-quality criteria documents and other publications that will improve the health of workers. Support to OSHA in standards development is a high priority within NIOSH, and of equal importance is the development and dissemination of occupational safety and health information that will allow plant operators, supervisors, and employees unilaterally to protect and improve the health of workers. The reorganizational efforts within NIOSH during the past year have sharpened the focus on the agency's mission and will aid in producing a steady flow of high-quality documents for standards recommendations and occupational safety and health information for direct dissemination. The incorporation of Technical Information Dissemination staff into the Document Development Division should provide better communication for the Institute and the outside users.

The Division is very interested in identifying new ways to update past documents and to continue to investigate newly identified occupational safety and health problems within the constraints of staff, money, and policies. Working with other agencies, we and the private sector will be able to continue our nation's progress in protecting the health of everyone.

REFERENCES

1. Policy Research Incorporated. "Evaluation of the NIOSH Criteria Documentation Program," Contract No. 210-78-0048.
2. Perkins JL, Rose VE: Occupational health priorities for health standards: the current NIOSH approach. Am J Public Health 69(5):444-448, 1979.
3. Whorton MD, Wegman DH: Occupational health standards: What are the priorities? Am J Public Health 69(5):433-434, 1979.

DISCUSSION: CHAPTER 7

W. Clark Cooper, William F. Martin, J. S. Lee and Mary Alexander

DR. COOPER: I have two questions or comments. First, is there not some concern that, by expanding into the area of economic considerations, NIOSH will increasingly preempt OSHA's decision-making process by boxing them into a position that is very hard to alter after the health-concerned agency has made a strong and specific recommendation?

The second question relates to the medical surveillance portions of the documents. Is there any group within NIOSH, including NIOSH consultants, that is looking critically at the whole matter of medical surveillance, its value and its return, and that is trying to find some kind of internal consistency in recommendations for medical surveillance, rather than depending on blind faith in their adequacy?

MR. MARTIN: Much of our movement into economic and technical feasibility was at the request of OSHA. We were somewhat reluctant to do this, because as a public health organization we should try and identify the health problems and present the best supporting information that we can obtain.

We are cooperating with OSHA in hopes that we can improve our working relationship. If we were being criticized about being too independent, it was only because we were concerned about keeping the research activities independent and concentrating on the worker health problems.

In regard to your second question, medical surveillance has been a major concern to NIOSH for a long time. A number of people within NIOSH are involved with the problem of medical surveillance, including a group of physicians in Dr. Landrigan's Division of Surveillance, Hazard Evaluations and Field Studies in Cincinnati, Ohio, and Dr. Merchant's Division of Respiratory Disease Studies in Morgantown, West Virginia.

94

These individuals have put forth considerable efforts in making the medical surveillance programs workable. This has included eliminating tests or examinations that were needless and, at the same time, coming up with a standardized comprehensive medical surveillance program that both large and small companies could use.

DR. LEE: There has been a lot of talk in the last few years about moving away from substance-by-substance approach to standard setting toward a generic approach. In addition, the joint OSHA and NIOSH standards completion project produced decision logic for specific requirements for standards.

Has NIOSH given any thought to recommending generic standards in such areas as monitoring, medical requirements, and so forth, rather than the substance-by-substance approach?

MR. MARTIN: In the past years, there have been requests to produce generic (industry-type) documents. It was felt that these documents would ease the load on the regulatory agencies. We currently have a number of generic-type documents in the developmental stages.

As far as the decision on how we identify and set our priorities, we still do not have a completely workable system. It has been under study for a number of years, and OSHA is currently working on solving the problem. You might note that we have interacted with the EPA and the National Institute for Environmental Health Sciences group in trying to identify the most pertinent types of documents. Likewise, we have communicated with the various worker groups in an attempt to keep them involved in the decision-making processes.

MS. ALEXANDER: Could you tell us why you are not distributing the NIOSH criteria document for formaldehyde?

MR. MARTIN: The document was a joint project between OSHA and NIOSH. After the document was released to the public, it was requested that we put it on hold until a determination could be made as to whether or not it was a regulatory document.

There is still a continuing dialog, and we had hoped that the decision would be made this past week as to whether or not NIOSH would submit the document as a NIOSH publication, or if OSHA would decide that it would be a joint publication. The decision is still pending, but NIOSH feels that we have finished our input and are satisfied with the information it contains and that it is ready for publication.

CHAPTER 8

THE BENZENE DECISION: LEGAL AND SCIENTIFIC UNCERTAINTIES COMPOUNDED

Douglas G. Mortensen, JD
    Clyde, Pratt, Gibbs, and Cahoon
    Salt Lake City, Utah

Frederick R. Anderson, Jr., JD
    College of Law
    University of Utah
    Salt Lake City, Utah

Chief Justice Burger opened his concurring opinion in *Industrial Union Department, AFL-CIO v. American Petroleum Institute et al.* ("the benzene case") by observing that the case involved the Court in "difficult unanswered questions on the frontiers of science and medicine" (488 U.S. 607, 100 S.Ct. 2844, L. Ed. 2d, 1980). He then added that the statute and the legislative history gave "ambiguous signals" to the Secretary of Labor about how he was to go about answering these difficult questions. The Chief Justice concluded, in something of an understatement, that the Court "takes on a difficult task to decode the message of the statute as to guidelines for Administrative action."

So taxed was the Court by the difficulty of "decoding" Congress' approach to the pervasive scientific uncertainty surrounding the risk of benzene exposure that it was unable to render a majority opinion in the case. Although five justices did concur in the conclusion that the benzene standard was invalid, they did so by means of four separate opinions. The remaining four "minority" justices dissented through a fifth opinion. These five opinions, totalling 120 pages, raise as many questions as they answer. Not to be outdone either by the failure of science and medicine

97

to provide clear answers or by Congress' failure to provide clear guidelines, the Supreme Court returned to the litigants, Congress and critical legal opinion a plethora of unanswered questions on the frontiers of the administrative law of health protection.

After previewing the basic problem presented to the Court and describing the circumstances under which it accepted review of the case, we shall attempt to describe what the Supreme Court decided, as well as what it declined or otherwise failed to decide. In doing so, we shall comment briefly and somewhat disappointedly on the low predictive value of the Court's decision, adding our view that the case has greater value as an illustration of the pitfalls that may ensnare scientists, bureaucrats, legislators and judges in their efforts to perform their respective roles in our "administered" modern American society in the face of scientific uncertainty.

The basic issue posed by the case concerns how far a regulatory agency may go in regulating the workplace exposure to low ambient levels of benzene. Without question, brief exposure to high airborne concentrations of benzene and extended exposure to moderate levels of benzene present significant, unreasonable risks of harm. The Supreme Court's plurality opinion reports that inhalation of concentrations of 20,000 ppm can be fatal within minutes. Exposures ranging from 250 to 500 ppm can cause symptoms of mild poisoning, including vertigo and nausea. Persistent exposures at levels above 25-40 ppm may lead to blood deficiencies and diseases of the blood-forming organs, including aplastic anemia, which is generally fatal (43 Fed. Reg. at 5921). Benzene, like many other widely used substances, both natural and synthetic, is a carcinogen; exposure to high concentrations of benzene causes leukemia.

What is *not* known is the threshold level, if any such threshold exists, at which benzene exposure becomes carcinogenic or otherwise unsafe. Attempts to regulate benzene exposure in the face of this unknown collide with the ubiquity and utility of benzene to modern society. Prohibition of its use or promulgation of a rule requiring zero exposure are not realistic alternatives. Benzene is not an unfamiliar nor an unimportant commodity. Eleven billion pounds of it were produced in the United States in 1976. Benzene is used in manufacturing a variety of products, including motor fuels (containing as much as 2% benzene), solvents, pesticides, detergents and other organic chemicals (43 Fed. Reg. at 5918). The entire population of the United States is exposed to small quantities of benzene, ranging from a few parts per billion to 0.05 ppm, in the ambient air. More than one million workers are subject to additional low-level exposures incident to their employment (43 Fed. Reg. at 5935).

The applicable congressional directives that the Chief Justice described as "ambiguous" and "difficult...to decode" are found in the provisions of the Occupational Safety and Health Act of 1970. Section 3(8) defines an "occupational safety and health standard" as one "reasonably necessary or appropriate to provide safe and healthful employment. . . ." Although the justices' opinions quibble mildly over the meaning of the word "safe," the phrase "reasonably necessary or appropriate" triggered far greater difficulty. Members of the Court found the phrase laden with a wide variety of policy directives.

Section 6(b)(5) directs the Secretary "in promulgating standards dealing with toxic materials or harmful physical agents" to

> (S)et the standard which most adequately assures, to the extent feasible, on the basis of the best available evidence, that no employee will suffer material impairment of health or functional capacity.

The troublesome word in this directive is "feasible." To say there is an absence of consensus among members of the Supreme Court as to its intended meaning is perhaps an understatement. At least one member of the Court was convinced that Congress itself did not know what it meant when it used the word.

The origin of the proposed benzene standard is by now well known. For several years industrial health experts have been aware that exposure to benzene may lead to various types of nonmalignant diseases. The evidence connecting high levels of benzene to serious blood disorders had become so strong by 1948 that the Commonwealth of Massachusetts imposed a 35-ppm limitation on workplaces within its jurisdiction. In 1969 the American National Standards Institute adopted a national consensus standard of 10 ppm averaged over an 8-hr period with a 25-ppm ceiling concentration for 10-min periods, or a 50-ppm maximum peak concentration (43 Fed. Reg. at 5919). In 1971, after the Occupational Safety and Health Act (OSHAct) was passed, the Secretary adopted this consensus standard as the federal standard, pursuant to 29 U.S.C. §655(a).

In a 1974 report recommending a permanent standard for benzene, the National Institute for Occupational Safety and Health (NIOSH) noted that epidemiological studies published in the late 1960s and early 1970s raised the "distinct possibility" that benzene caused leukemia. Noting, however, that all known cases had occurred at very high exposure levels, NIOSH refused to recommend a change in the existing 10-ppm consensus standard and suggested further study to determine conclusively whether there was a link between benzene and leukemia and, if so, what the dose-response relationship was.

Additional studies published between 1974 and 1976, which confirmed the causal connection between leukemia and benzene, caused NIOSH to revise its earlier recommendation. It recommended an emergency temporary standard of 1 ppm exposure limit for airborne benzene. None of these studies had provided the dose-response data that NIOSH found lacking two years earlier.

In the spring of 1977, the Occupational Safety and Health Administration (OSHA) formally proposed the emergency standard that NIOSH had suggested. However, two days before it was to take effect, the Fifth Circuit Court of Appeals issued a temporary restraining order delaying its implementation. Rather than litigate the legality of its proposed emergency standard, OSHA immediately began the administrative process for adopting a new permanent standard equal to the aborted emergency standard.

In publishing notice of the proposed permanent standard, OSHA did not ask for comments as to whether or not benzene presented a significant health risk at exposures of 10 ppm or less. Instead, it asked for comments as to whether 1 ppm was the minimum feasible exposure limit. This formulation of the issue was consistent with the policy OSHA had developed for dealing with carcinogens. Whenever a carcinogen is involved, OSHA will presume that no safe exposure level exists in the absence of clear proof establishing such a level and will accordingly set the exposure limit at the lowest level feasible.

The final standard, which did in fact reduce the permissible exposure limit to benzene from 10 to 1 ppm, was issued in February 1978. On preenforcement review, the Fifth Circuit Court of Appeals invalidated this standard, finding that it was based on findings unsupported by the administrative record. The Fifth Circuit Court concluded that OSHA had exceeded its standard-setting authority because there had been no showing that the 1 ppm exposure limit was "reasonably necessary or appropriate to provide safe and healthful employment" as required by §3(8), and that §6(b)(5) did not give OSHA the unbridled discretion to adopt standards designed to create absolutely risk-free workplaces regardless of cost.

The case was appealed and, as we know, the Supreme Court agreed to review it. Regulatory reform buffs waited with intense anticipation to find out whether, in the conclusive and binding opinion of the nation's highest court, OSHAct requires OSHA to consider the disproportionateness of benefits to expenditures in issuing standards relating to toxic substances (does OSHAct mandate cost-benefit analysis?); and whether OSHA may continue to rely on a general carcinogen policy that presumes in the absence of clear evidence to the contrary that there is no safe exposure level for substances found to be carcinogenic.

The high hopes for what attorneys and journalists are pleased to call a landmark decision were not nearly realized. Far from landmark, the Court's three-one-one-four decision is more of a dust storm obscuring, not guiding, vision. Benzene case watchers wanted to know whether a requirement that economic effects bear a reasonable relationship to expected benefits was implicit in the OSHAct standard-setting directives, not being explicitly placed there by Congress. On this issue, four justices believed that cost-benefit analysis is not required (the four dissenters), one agreed with the Fifth Circuit Court that it is required, and four declined to say what they believed, finding it unnecessary, for one reason or another, to reach the issue. Remarkably, the four "dissenters" were by far the strongest bloc represented on the issue.

On the propriety of OSHA's resort to a general presumption that there is no safe exposure level for carcinogens (otherwise put: is OSHA's generic cancer policy an acceptable administrative response to scientific uncertainty?), four justices voted yes, three voted no, one voted maybe and one abstained. This lack of clarity produced an interesting response from OSHA. In the *Federal Register* (January 21 and 23, 1981) the agency set forth the modifications it felt necessary to conform its generic cancer policy to the Supreme Court's decisions. These modifications remind one of Tallulah Bankhead's comment about another matter: there's less to this than meets the eye. OSHA's changes were lightly cosmetic.

The holding of the plurality opinion—the opinion that supposedly "counts"—is that before OSHA may issue a workplace health standard, it must find that the condition addressed presents a "significant risk of material health impairment." That is, OSHA may not attempt to make workplaces safer without first determining that they are unsafe. This plurality opinion represents the thinking of only four Justices, two of whom were troubled enough by the *ratio decidendi* that they found it necessary to write their own opinions. The remaining concurring opinion (Justice Rehnquist's), supports only the plurality's result (invalidation of the standard), not its rationale. There are four dissenters who vehemently believe, for the same reasons, that such threshold finding is not required. Under any rules but those of Supreme Court procedure, one might be persuaded that the prevailing logic, if not the votes, supported the opposite result than that which the Court's judgment produced.

What then does the benzene decision stand for? This much only is clear: by five votes to four, the Supreme Court affirmed the Fifth Circuit ruling that the benzene exposure standard promulgated by OSHA in early 1978 is invalid because it is not supported by appropriate findings.

If, as one jurist suggests, law is the prediction of what a court will do, the Supreme Court's 120-page work product provides precious little law, as we have seen. Split decisions on all the major issues presented scatter clues in all directions, but do not amount to the type of predictive guidance one expects from the land's highest court.

Nevertheless, the case contains lessons that touch on science, law and government. As badly fragmented as the Court's reasoning is, the separate opinions give us important insights about the extent to which the governance of risk-laden conditions is still problematic in American law. The separate opinions reveal how we have not satisfactorily resolved how to bring either good science or social consensus about the degree of risk we are collectively willing to tolerate, to bear on administrative regulation of health risks.

The problem cannot be dismissed as one more example of the recent inefficacy of Supreme Court decision-making where environmental, health, and safety issues are concerned. The failure of the Court throughout the 1970s to issue clear interpretations of law regarding the standing to sue of recreational, conservational and esthetic interests, the meaning of the National Environmental Policy Act, and the federal common law of environmental nuisance presents a serious issue that constitutional scholars must address, but the difficulties of risk management (as the problem that the benzene case raises is often labeled) cannot be resolved by the courts alone, nor are they its exclusive product. Congress, agency policies and the scientific community have made their own contributions to the risk management imbroglio.

Although OSHAct obligates a reviewing court to leave OSHA's new rules undisturbed if they are supported by "substantial evidence in the record considered as a whole" [29 U.S.C. §655(f)]; under this broad directive there is leeway for a reviewing court to give widely varying degrees of deference to the agency's findings of fact. Some suggest that "whether the court will dig deeply or bow cursorily depends exclusively on whether the judge agrees with the result of the administrative decision" [1]. Justices Stevens, Stewart, Powell and Chief Justice Burger appear to have identified the benzene case for "sharp scrutiny" rather than "mild deference" [1].

Sprinkled throughout the plurality opinion and its footnotes (notably footnote 16) are examples of OSHA's reliance on erroneous or misleading reports, suggesting either prejudgment or sloppiness on the part of OSHA's experts. While one may not expect OSHA's experts to be neutral informants rather than partisan participants, the plurality opinion exposes a surprising array of inconsistencies and weaknesses in OSHA's fact finding. If these judicial observations are accepted at face value, then, clearly, precision has yet to be insisted on uniformity in agency risk management. Developing a professional, scientifically defensible methodology apparently remains problematic, even near the end of the 1970s. Not at all clear, however, is whether congressional (or judicial) insistence on quantification, rather than a more subjective "policy" approach to setting health-based standards, would cure this type of defect in current risk management, making it harder for judges to apply vigorous evidentiary standards in order to supplant an agency determination with which they disagree.

Other perceived shortcomings in OSHA's approach make even clearer the plurality's views and their underlying uncertainties about the correct path for risk management to take. First, the plurality felt that OSHA had misinterpreted its mission. The plurality reasoned that, although it is not explicit in OSHA's authority to promulgate standards "reasonably necessary or appropriate to provide safe and healthful employment and places of employment," nevertheless an implicit requirement rested on OSHA to establish the *need* for a more stringent standard before acting. That is, before promulgating any standard, OSHA must make a finding that the workplace in question is *not* safe. "Significant" risks must be found to be present, and they must be capable of being eliminated or lessened by a change in practices. The plurality opinion found that OSHA did not even attempt to carry this court-manufactured burden.

The origin of the requirement for a threshold finding of "significant" risk must be in the plurality's conception of the logical requirements of rational regulatory policy, because neither the statute nor its legislative history provide even minimally acceptable direction on the issue. Absent congressional guidance, despite a factual vacuum, regulations may be tightened further to reduce possible health risks. The plurality supplied what it viewed as a reasonable (but, we think, decidedly nonstatutory) policy.

The plurality opinion also suggests that OSHA's policymakers erred in construing the meaning of "safe" workplaces. OSHA should not have assumed that when Congress directed it to try to make workplaces "safe," it meant that workplaces should be made entirely risk-free. If OSHA's policymakers had properly studied the legislative history of OSHAct, argues the plurality opinion, they would have realized that Congress was concerned, not with absolute safety, but with the elimination of significant harms only. Again, while the Court may be quite correct that Congress *should have* confronted the familiar conundrum raised by the inescapable trade-off of health and safety for convenience and economic gain (and given the same answer as did the Court), nevertheless it seems clear that Congress did not in fact face the issue squarely. Nor should we be surprised; as legislators, Congress frequently tries to give the electorate what it perhaps illogically asks for: full economic benefits in a risk-free environment. The policy failure here may well be Congress'. The Court should not rescue Congress by doing its thinking for it, as at least one justice — Rehnquist — reasoned.

OSHA's policymakers also erred, according to the plurality opinion, in attempting to shift to industry the burden of proving that a safe level of exposure existed. The plurality pointed out that ordinarily the proponent of a rule has the burden of proof in administrative proceedings. Although the Court noted that, in some statutes regulating toxic substances, Congress has shifted the burden of proving that a particular substance is safe onto the party opposing the proposed rule (for example, in the federal Insecticide, Fungicide and Rodenticide Act), Congress did not follow this course in enacting OSHAct. In attempting to make the shift in the absence of express authorization, OSHA exceeded its power.

Yet it is not entirely clear, in the absence of an explicit congressional directive, that OSHA could not reasonably implement the unique and very high safety standards contained in the Occupational Safety and Health Act by reversing the traditional burden. Obliged as it is by statute to provide "safe" workplaces in which *no* employee suffers material impairment of health, the agency might well adopt a policy of regulating risks to the utmost, up to the point where the employer can show that *no* employee will be materially injured. Again, Congress simply did not think the matter through, or if it did, it did not reveal its insights for the guidance of agency rule-making.

Finally, the plurality opinion accused OSHA of failing to use common sense. After noting that OSHA's proposed permanent standard expressly excluded from its protection more than 795,000 gas station employees whose exposure to benzene is probably greater than any other group of workers, and after listing the detailed requirements of the proposed standard, including its preference for engineering modifications over personal protective equipment, the plurality opinion concluded that the benzene standard was an expensive way of providing additional protection for a relatively small number of employees. Again, however, American risk management has not progressed to the point that statutes require agencies to regulate all similar risks comprehensively, or in a cost-effective sequence. Common sense notwithstanding, folkways do not become law-ways unless the legislature acts.

Although the plurality opinion expressly declined to decide whether OSHA must compare costs to benefits in selecting its standards, the opinion undertakes to demonstrate, using OSHA's estimates and figures, that the costs of implementing the proposed benzene standard would be high and definite, while adverse effects (even at the existing 10 ppm standard) were inadequately substantiated and apparently low. The plurality opinion chastised OSHA for relying on a series of assumptions concerning the carcinogenicity of benzene, with little regard for the factual basis for the assumptions. OSHA's presumptions were found particularly weak as applied to the proposed standard's absolute ban on dermal contact. The dermal contact ban was based, not only on the assumption that benzene is a carcinogen in small doses, but on the further assumption that benzene can be absorbed through the skin in sufficient amounts to present a carcinogenic risk. The plurality opinion found such assumptions an improper substitute for the findings of significant risk of harm required by OSHAct.

Modern regulatory agencies charged with setting standards to control the potential harms which may result from as yet incompletely studied substances have no choice but to make factual assumptions, extrapolate from existing data, and set generous margins of safety against the possibility that impact may be worse than first anticipated. Congress at best occasionally sanctions this approach (e.g., in the Clean Air Act). Usually it is characteristically silent, leaving the agency to chart its own course.

Chief Justice Burger agreed with the plurality opinion's assessment of OSHA's performance and its misconstruing of the Occupational Safety and Health Act, but considered it necessary to add a few words of admonition to each of the three branches of government relative to their differing functions with respect to health and safety regulation. In so doing, he failed to perceive the conceptual difficulty of the task at hand, instead preferring to convert a major challenge into a minor quibble over what he perceived to be unnecessary regulation of minor harms. Further, while he correctly perceived that Congress is particularly responsible as "ultimate regulator," and that the courts' function is "narrow," he does not castigate the ultimate regulator for its many failures to define policy, nor the courts (including the plurality in this case) for overstepping their narrow functions and attempting to supply the guidance not provided by Congress.

> The Congress is the ultimate regulator and the narrow function of the courts is to discern the meaning of the statute and the implementing regulations with the objective of ensuring that in promulgating health and safety standards the Secretary has given reasoned consideration to each of the pertinent factors and has complied with statutory commands.
>
> Nevertheless, when discharging his duties under the statute, the Secretary is well admonished to remember that a heavy responsibility burdens his authority. Inherent in this statutory scheme is authority to refrain from regulation of insignificant or de minimis risks.... When the administrative record reveals only scant or minimal risk of material health impairment, responsible administration calls for avoidance of extravagant, comprehensive regulation. Perfect safety is a chimera; regulation must not strangle human activity in the search for the impossible.

Justice Marshall and his three fellow dissenters, Justices Brennan, White and Blackman, aver that the plurality opinion's discussion of the agency record is "extraordinarily arrogant and extraordinarily unfair" and that its interpretation of the statute is "a fabrication bearing no connection with the acts or intentions of Congress." They accuse the plurality of being "obviously more interested in the consequences of its decision than in discerning the intention of Congress." The dissenting opinion further finds the plurality opinion's conclusion that the Secretary acted beyond the scope of his statutory authority supported by "reasoning that may charitably be described as obscure."

Citing the technical complexity of the subject matter, the need for agency discretion in risk management, and the desirability of deferring to agency judgment on matters not susceptible to factual proof, the dissenting opinion attributes to Congress all the intentions necessary to support the rationale and result reached by OSHA in promulgating the 1 ppm standard. The difficulty with the dissent, however, despite the unanimity of opinion expressed and the passion and cogency with which it is put forward, is that it must implicitly deny that OSHA faced enormous ambiguities and uncertainties in applying this statute. In many respects Congress left the agency on its own. It may be, as the dissent argues, that the proper predominating concern of both Congress and OSHA is to protect American workers from avoidable dangers, that the administrative agency acted rationally on the basis of the best available information, and that in areas of scientific uncertainty an agency must be given wide leeway to make broad assumptions that tend to minimize the risks. But again, the prescription for the cure to the maladies of risk management is effectively mixed one part legislative to ten parts judicial.

Justice Powell agreed with the plurality that OSHA failed to make threshold findings that current permissible exposure levels created a significant risk of material health impairment and that reduction of those levels would significantly reduce the hazard. However, he disagreed with the plurality's conclusion that OSHA did not even attempt to meet its burden of demonstrating that the existing permissible level of 10 ppm presented a significant risk to human health. He conceded that the question was close, and although he did not disagree with the plurality's view that OSHA failed, he preferred to assume, as the dissenters asserted, that OSHA met its burden. This preference allowed him to conclude that "the statute also requires the agency to determine that the economic effects of its standard bear a reasonable relationship to the expected benefits." An occupational health standard is neither "reasonably necessary" nor "feasible," he argues, if it calls for expenditures wholly disproportionate to the expected health and safety benefits.

Justice Powell believes OSHA policymakers were wrong in their contention that §6(b)(5) not only permits but actually requires the promulgation of standards to reduce health risks without regard to economic
effects, unless and only unless those effects would cause widespread dis-
location throughout an entire industry. He stated:

> Thousands of toxic substances present risks that fairly could be
> characterized as "significant"... even if OSHA succeeded in selecting
> the gravest risks for earliest regulation, a standard-setting process
> that ignored economic considerations would result in a serious mis-
> allocation of resources and a lower effective level of safety than
> could be achieved under standards set with reference to the compara-
> tive benefits available at a lower cost. I would not attribute such an
> irrational intention to Congress.

Not finding in the record any documentation supporting OSHA's finding that the substantial economic costs required by the benzene regulations were justified, or of evidence that OSHA even weighed the relevant
considerations, Justice Powell concluded:

> No rational system of regulation can permit its administrators to
> make policy judgments without explaining how their decisions effec-
> tuate the purposes of the governing law, and nothing in the statute
> authorizes such laxity in this case.

The difficulty with Justice Powell's view is again that traditional
notions of "rationality" may be difficult to apply in the new field of precautionary regulation, where Congress may just as well have "rationally"
intended to exact a high known economic cost in return for a greater but
far less determinate measure of safety. This rationale, in fact, does
appear more clearly in other health and safety legislation. Notwithstanding the plausibility of this alternative view and its adoption in other
statutes, however, a more likely explanation for Congress' failure to
make explicit Justice Powell's "rational" policy is that it simply did not
focus clearly on the issue.

Justice Rehnquist, concurring in the judgment, but not in the reasons
supporting it, agrees that the benzene standard was invalidly promulgated, but he does not blame OSHA. Instead, he blames Congress for
unlawfully attempting to pass on to OSHA too much of Congress'
policymaking authority.

> This case presents the Court with what has to be one of the most
> difficult issues that confront a decision maker: Whether the statis-

tical possibility of future deaths should ever be disregarded in light of the economic costs of preventing those deaths. I would...suggest that the widely varying positions advanced in the briefs of the parties, and in the opinions of Mr. Justice Stevens, the Chief Justice, Mr. Justice Powell, and Mr. Justice Marshall demonstrate, perhaps better than any other fact, that Congress, the governmental body best suited and most obligated to make the choice confronting us in this case, has improperly delegated that choice to the Secretary of Labor, and, derivatively, to this Court.

Justice Rehnquist chides the Congress, not for failing to weigh and consider the relevant factors bearing on such a decision, but for failing to reach *any* discernible conclusion. Like the authors of the other four opinions, Justice Rehnquist worked hard over the legislative history of OSHAct, attempting to find in it a coherent risk-management policy. But, unlike his brethren's, his scrutiny disclosed that the legislators themselves failed to reach any consensus as to "where on the continuum of relative safety [OSHA] should set the standard."

Justice Rehnquist found that the feasibility requirement appearing in §6(b)(5) "is a legislative mirage, appearing to some members but not to others, and assuming any form desired by the beholder."

> In drafting Section 6(b)(5), Congress was faced with a clear, if difficult, choice between balancing statistical lives and industrial resources, or authorizing the Secretary to elevate human life above all concerns, save massive dislocation, in an affected industry.... That Congress chose, intentionally or unintentionally, to pass this difficult choice on to the Secretary is evident from the spectral quality of the standard it selected....

Justice Rehnquist's view that Congress' attempt to pass the hardest question on to OSHA for resolution was unconstitutional rests on the nondelegation doctrine, last invoked by a majority of the Court in 1935. The gist of this doctrine is that the legislative branch of government cannot delegate to the executive branch functions that are essentially legislative in character. The nondelegation doctrine, according to Justice Rehnquist, serves three important functions:

> First,...it ensures...that important choices of social policy are made by Congress, the branch of our government most responsive to the popular will.... Second, the doctrine guarantees that, to the extent Congress finds it necessary to delegate authority, it provides the recipient of that authority with an "intelligible principle" to guide the exercise of the delegated discretion.... Third,...the doctrine

ensures that courts charged with reviewing the exercise of delegated legislative discretion will be able to test that exercise against ascertainable standards.

The Occupational Safety and Health Act of 1970, in the justice's opinion, failed on all three counts. In effect, Justice Rehnquist has discerned in Congress' impermissibly broad delegation an abdication of constitutional responsibility. Administrative agencies may not be asked to make difficult risk-control decisions where Congress itself has failed to develop a coherent policy.

> It is the hard choices, and not the filling in of the blanks, which must be made by the elected representatives of the people. When fundamental policy decisions underlying important legislation about to be enacted are to be made, the buck stops with Congress....

Justice Rehnquist's opinion at least offers a plausible diagnosis of the risk-management dilemma, whatever modern constitutional lawyers may think of the cure. Putting aside the novelty of invoking an apparently outmoded doctrine from the tumultuous era of the early New Deal decisions, Justice Rehnquist's candor about Congress' failure to appreciate the complexity of the task facing the agencies engaging in precautionary regulation may both stimulate a better legislative effort from Congress in the future and restrain the more visionary of his colleagues on the bench who believe they understand what Congress "must" have meant despite its refusal to endorse explicitly their judicial notions of "rational" precautionary regulation.

REFERENCES

1. Rogers WH: Judicial review of risk assessments: the role of decision theory in unscrambling the benzene decision. Environ Law 11:309, 1980.

CHAPTER 9

INTERPRETATION OF OCCUPATIONAL HEALTH RESEARCH FINDINGS, ESPECIALLY WITH REGARD TO CARCINOGENS

Charles H. Hine, MD

Division of Occupational Medicine
School of Medicine
University of California at San Francisco
San Francisco, California

Occupational medical research may be defined as careful and critical inquiry or examination in seeking facts and deriving principles related to those data that may impinge on the practice of occupational medicine. Such research will include basic clinical and epidemiological studies and will draw on a multitude of scientific disciplines. These include such basic sciences as biochemistry, physiology, pharmacology, pathology and toxicology; clinical medical sciences such as clinical pathology, dermatology, oncology and internal medicine; and related applied sciences, including biostatistics, behavioral science, demography and epidemiology.

Such research is required because man's varied activities have used, produced and disposed a wide variety of natural and synthetic materials. In all phases of these activities, people are exposed in different ways and to varying degrees. Those exposed include the employee, the consumer and the general public.

Among the many chemical substances and mixtures that are used, developed and produced are those that possess a significant risk of injury to health and to the environment. While the results of some of these activities are immediate and easy to recognize, others produce effects that may not be apparent for years. The lag time between cause and effect is a frequently confounding factor to the growth of our knowledge and understanding and the degree of risk. Of special concern, therefore, are the products and processes leading to carcinogenesis, mutagenesis, teratogenesis and slowly developing dysfunction of critical organs.

NATURE OF RESEARCH

Research may be basic or applied. Basic research may be defined as the inquiry after knowledge for its own sake, without consideration or planning to solve a particular problem. Such scientific inquiry, which is carried out to broaden the base of knowledge, is almost entirely done in academic institutions. Those engaged in performing the work or those sponsoring it may have no intent of developing data applicable to the solution of an occupational medical problem. Nevertheless, these data are frequently of the greatest importance to the final resolution of such problems, i.e., study of the biology of isolated tissues in vitro.

Applied research is the systematic, intelligent treatment of problems for which the data or methods needed for solution are either unsatisfactory or lacking. Such a systematic inquiry can be carried on in any field related to occupational medicine, from basic science to clinical medicine. The distinguishing point of this type of inquiry is that it is goal-oriented. Such research may commence at a very fundamental level, i.e., in vitro study of effects on microorganisms, fungi or other species that have no direct relationship to man. Such investigations are needed, since these systems make possible a study of the effect of chemicals and processes on multiple generations within practical time periods.

RESEARCH GROUPS

Occupational medical research is conducted by a variety of groups. These include academic and nonprofit institutions, government agencies, industry (in-house) research teams and research organizations. The organization of the research effort should involve multiple participants, not only the researchers, but also those who are affected by the outcome of the research and the decision-makers.

Planning for the implementation of research results begins with design of the research project. It is unwise and unprofitable to wait until the results are obtained; specifically, the technical ability of those who will use the results should be taken into account in determining the form and nature of the results that are taught. The extent to which applied research results are usable at all will depend critically on the extent to which those who will use the results understand them and carry them out. Therefore, it is important to formulate the approach to the study of the problems in such ways as to yield a solution that is operationally—and not merely logically—feasible.

Most pure science research is still being conducted by individuals who work within a single scientific discipline, but research teams are being used increasingly (especially in industry) in the solution of occupational medical problems, particularly when these involve several scientific disciplines. It is important, therefore, to compose a team in such a way as to complement one person's weakness with another person's strength. The team performance is usually influenced by the degree of familiarity with the area in which the problem exists. A person who is thoroughly familiar with the problem area is frequently restricted by preconceived assumptions concerning the area of which he is aware. Such a scientist is often unable to see alternative solutions or is too willing to discard suggestions or new alternatives on the basis that they have been tried and have failed. On the other hand, the person who is unfamiliar with the area is likely to think of alternatives but, perhaps, not be in a good position to evaluate them realistically.

All of this is to state that there may be a bias in the research approach to the solution of a problem and, frequently, in the interpretation of results, since interpretation is made using not only the scientific rationale and discipline in which the person has expertise, but also the investigator's background, school of thought and personal experience. This bias can be lessened, however, by a multidisciplinary approach.

TYPES OF INVESTIGATION

Most investigations, including occupational medical research, have four component parts: (1) accumulation of data, (2) evaluation of data, (3) dissemination of information, and (4) implementation of a course of action.

Each of these functions will be addressed in the following paragraphs. The study of occupational medical problems is made through essentially three different approaches: manipulative procedures with living animals, clinical investigations and epidemiological inquiries.

Animal Studies

Procedures such as the Ames test have greatly enhanced the ability to assess the mutagenicity of a substance in a rapid and inexpensive fashion. Similarly, in vitro testing of tissue cultures has assisted in carcinogenicity assessments. Most data generated from toxicological tests, however, depend on the use of mammalian species, including the mouse, rat, rabbit, dog and monkey.

Toxicology is an evolutionary science, and there have been a number of changes in experimental designs utilized in evaluating long-term effects of chemicals. Critics of data reported in the past have suggested that the defects in experimental design are due to the following:

1. poor care of animals, resulting in infection, sickness, and premature death;
2. failure to follow the treatment protocol;
3. use of too few animals;
4. doses that are too high or too low;
5. premature termination of the tests;
6. inadequate autopsies and examination of tissues; and
7. poor record-keeping.

Recommendations have been made by several groups regarding the design, conduct and interpretation of animal bioassays. The minimum requirements for such tests in evaluating the toxicity of chemical agents have been set out in the "Health Effects Test Standard – Toxic Substances Control Act Test Rules" (*Federal Register* 1979). Under the requirements of the Toxic Substances Control Act (TSCA), the sponsor or other persons conducting tests must test the chemical identified according to the Good Laboratory Practices Standard.

To facilitate the government's evaluation of data for regulatory decision-making duties, it has been recommended that uniform testing guidelines and/or protocols should be adopted [1]. The purpose of this is to maximize the value of animal test data, based on the understanding that the best scientific data are of greatest value in the regulatory process.

Uniform criteria for the development of data to be used in regulatory decision-making should include the following:

1. appropriate species, strains, and sexes of laboratory animals;
2. minimum number of animals in control and experimental groups;
3. types of controls needed;
4. appropriate routes of exposure;
5. appropriate range and level of doses;

6. purity of test chemicals;
7. adequacy of animal housing and care;
8. minimum observation times;
9. types of histologic lesions considered; and
10. proper statistical evaluation of the results.

Clinical Studies

The observations of persons at work carried out by occupational health professionals should generate a database that would be useful in the prediction of the effect of the work environment on longevity and the appearance of specific diseases. There have been, and still remain, obstacles to this approach.

Until relatively recently, there have not been the numbers of trained professionals or techniques available to develop meaningful data. Only a handful of the larger corporations contain within their organizational structure a medical research unit whose function it is to develop data from human exposure experiences. While these units are now developing with a greater frequency in the chemical industry, most industries and all of the smaller producers and users of chemicals have failed to devote any significant effort to this problem. Most physicians and health professionals who serve industry have demands on their time that leave little opportunity for clinical research, even if this were an activity that was encouraged by the management of the particular enterprise. In addition, their professional competence generally does not include expertise in research methods. This writer has long been an advocate of more attention being paid to the development of data in the occupational setting; there is some evidence that attitudes are changing in this regard.

To develop this type of meaningful data, commitment has to be made to systems that will more accurately define, both qualitatively and quantitatively, the industrial environment and institute systems for collecting data from quantitative measurements of biochemical constituents, function and behavior, and the longitudinal followup of employees.

There are inherent difficulties in such studies, based on the fact that very few persons are exposed to single chemicals in their environment—there are multiple processes with different degrees of exposure. Further, record-keeping as to the hours spent on a particular process is frequently haphazard. A strong recommendation is made for the expansion of this type of activity and for the medical evaluation of employees, not only during the course of their employment, but also during their retirement.

In keeping with our definition of research, that this is a planned activity, data obtained from accidental episodic exposure should not be recognized as research data. Nevertheless, such data do contribute, on careful study of the case, to knowledge regarding the residue from these episodes and are also useful in predicting the probability of a long-term undesirable effect. For these reasons, careful documentation and dissemination of these data are desirable.

Planned Clinical Research

It has only been in recent years that the morality and legality of carrying out planned experiments on persons who would not immediately benefit from the results was decided. It is now the opinion of jurists who have studied this concept that it is of benefit to society in general and that such procedures, if the experiment does not involve any unnecessary risks, are moral and ethical [2]. Therefore, studies that define the noxiousness, threshold of recognition, pharmacodynamics of absorption, distribution and excretion, and effects on motor skills, coordination and cognition have been conducted on volunteers.

It is a basic premise of any such study that no coercion whatsoever be exerted for participation, that the participant understands the purposes and risks thereof, and that the participant be allowed to disengage himself from the experiment at any time. Prior to initiating such studies, a disinterested party, e.g., the Human Subjects Committee, must review all aspects of the study, especially the protocol and the informed consent form, which the participants in the study should read and sign.

In investigations of this type, it is better to use nonworkers than employees as volunteers, unless there is some special reason for studying the latter, e.g., evaluation of the effects of past and present exposures.

In such studies, the greatest care must be exercised in assuring that there is no coercion whatsoever to participate. The author recognizes that this may not be truly obtainable (as in the use of prisoners), but it is his belief that it is possible if correctly and tactfully handled.

Epidemiological Studies

Epidemiological studies are frequently difficult to design, due to uncontrolled parameters such as change in populations, limits to longitudinal data acquisition, identification of valid control groups, other threats to the population and the basic random variation in measurements. Constraints of time and money frequently restrict the size of the experiment and scope of the data obtained. Controlling biases in the population is also difficult. Nevertheless, this approach is sometimes the only one that yields satisfactory data.

Epidemiological studies have been carried out on industrial groups, with subsequent development of data that were not available from other sources. These data are most useful when they confirm and verify data obtained in the more rigorously manipulatable laboratory observations, when they are applied to studies for which there are no suitable animal experimental models. The majority of studies have been retrospective. While qualitatively identifying the problem, frequently the quantitative aspects are not ascertainable. Studies of populations under conditions where there is no reasonable approximation of the degree of exposure are unsatisfactory. Further, projections as to the degree of risk under present situations of exposure may be erroneous.

In the past, prospective studies have not been utilized to any extent by organized research units within companies and only infrequently by trade associations, which have the advantage of presenting a larger population base. The value of prospective studies in this respect was recently emphasized by Sir Richard Doll in an address to the joint meeting of the Academies of Occupational Medicine and Industrial Hygiene in San Francisco, October 1980.

Government research groups have been hindered in their access to data contained in the medical files kept on employees. Recent judicial decisions regarding the right of access, however, may facilitate such studies. It is the opinion of some (including the author) that greater progress would be made by encouraging such studies to be made by the research groups within the enterprise, since they are more apt to be familiar with the nuances of the industrial exposures and are more intimately familiar with the workings of the industry.

Attention is paid by many industries to the maintenance of better medical records regarding both those data that quantify the risk of the degree and frequency of the exposure. Better quantification of these factors in epidemiological studies will, hopefully, develop a greater amount of useful data to assist in the evaluation of safety of products and work processes in the future. The point has been made, however, that human epidemiological studies should not be used for prediction of safety for both ethical and practical reasons. These include:

1. the inability of such a study to predict the effect that a material will have until after the person has been exposed (moral judgment);

2. exposures are ubiquitous and may not be limited to the material in question;
3. estimation of human exposure doses is most difficult;
4. unless populations are large enough, small changes cannot be identified; and
5. interactions frequently cannot be controlled.

EVALUATION OF DATA

Review of occupational health research data requires multiple evaluations by scientists with different backgrounds. These should include, as a minimum, physical scientists, toxicologists, pathologists, industrial hygienists, biostatisticians, epidemiologists, physicians and policymakers. The purpose of this review is to arrive at two conclusions, the first as to the scientific validity of the data, and the second as to the significance of the observations to those concerned. Evaluation of these research data then involves two broad categories of analysis: mathematical and statistical. The first is entirely scientific; the second "subjective" in nature. The first seeks to evaluate the data as a quantitative expression of the interrelationships that are factual in the sense that, if the study is repeated, the results conform to a similar descriptive statement. The second is, to a greater extent, the product of mental and behavioral mechanisms that are not yet sufficiently understood for systematic treatment with comparable predictive accuracy.

Occupational medical research is replete with situations in which the things that are known are not sufficient to determine the outcome exactly. Because of the constraints of only finite quantities of data, probabilility analysis is employed to define more clearly situations of uncertainty and incomplete information. Evaluation of the data derived from occupational medical research requires assessments of the following:

Experimental Design

Was the experiment designed with sufficient forethought to answer the question to which it addresses itself? Does the assay properly account for all essential pharmacokinetic factors relevant to human exposures? Is the assay applicable to the type of exposures that are of the greatest presumed hazard to man?

True Control Procedures

Are adequate positive and negative controls utilized so that the sensitivity of the system can be evaluated and the randomness of the observed event can be controlled?

Congruity

Are the data that have been developed congruent with observations generally observed in interrelationships in the biological sciences?

Reproducibility

Are these the first observations of their kind, or have other observations been made to either confirm or challenge them?

Analogies

Are the data that have been developed similar or dissimilar to comparable compounds, processes, or situations?

Bias

Are the conclusions that have been drawn based on the data developed? on those of other scientists with similar skills? (It should be remembered, however, that different data can be interpreted by persons with various biases in a completely intellectually honest manner.)

Extrapolation among species (e.g., from animal data to man) and the sensitivity to stresses from high dose-response measurements to low dose-response estimates complicate the interpretation of experimental results. The best animal study can only provide conclusions about cause-and-effect relationships in that species at that level of exposure. Proceeding up the anthropological scale from the mouse to the ape may or may not result in a similarity of response.

In the evaluation of these data, the scientist participating in occupational medical research must consider that the risk of a population in a given species of experiencing an adverse effect from environmental exposure depends on the dose levels of the substance by the particular pathway, the potency for producing the pathology in question, and the value that represents the spontaneous generation of that disease. A key factor, then, seems to be the spontaneous generation of the disease in the species. If a substance has a high potency, the effect can easily be detected at low doses. At moderate potency, it may be detectable by valid experimentation and, at low potency, it is generally undetectable by experiments. Risk estimates can be made, nevertheless, in spite of confounding factors.

The reviewer of the data is challenged by the theory that, for certain events (cancer and mutagenesis), it may not be possible to prove or disprove the existence of threshold levels of safety for some chemicals.

Government policy is currently based on the concept of "no threshold" for these events, if the substance has been proved active at some dose. It has been implied that there is a dose-effect response relationships, which goes from doses of high levels down to zero, although not necessarily in a linear manner.

A variety of models have been used for this extrapolation but, to date, there is no experimental validation of any such extrapolation to low doses. This should not be interpreted as saying that extrapolations of this type are invalid, but only that their validity has not been proved or disproved with the current limitations of scientific understanding and resources.

Recent data suggest that, as far as the radiation model of one-hit theory is concerned, repair is possible and results of equivalent doses given singly or in divided doses are incongruent. As regards the production of neoplasms, one can accept or reject the maximum tolerated dose procedure recommended by the National Cancer Institute in evaluation of chemicals administered by the oral route, according to their views as to the appropriateness of a technique in which normal detoxication and repair mechanisms are overridden. Therefore, scientists with different theories may interpret similar data in a dissimilar manner.

Dissemination of Information

When occupational medical research indicates that there is a hazard in a process of product, it is incumbent on the person possessing such knowledge to disseminate the information to all those who have need to know. As previously pointed out, such information is generated by one of four groups: academia and nonprofit institutions, government agencies, industry and professional research groups. Under TSCA reporting of substantial risks to the U.S. Environmental Protection Agency (EPA) is the responsibility of only those persons who manufacture, process or distribute in commerce the chemical substance and then only providing there is information to support the conclusion that such substances or mixtures present a substantial risk to health or to the environment.

There are different mechanisms for reporting the results of scientific inquiries for the different generators of research data. Those in academia are most apt to disclose their findings to the scientific community by means of presentations at scientific meetings or publication in appropriate scientific and medical journals. In this regard, a word of caution is required. The premature release of unconfirmed observations can be detrimental to the public. Segments of the press are irresponsible and seldom bother to correct a story that is later shown to be in error.

First reports would undoubtedly be made to the sponsor of the research, since this is the party with the greatest interest. If the findings are sufficiently significant to the public health, they should be made known to the appropriate government agencies as well. Data generated or sponsored by one governmental agency would, through the usual channels, be distributed to other agencies that have responsibility to the particular enterprise that will be most affected, to the industrial associations of which the enterprise is a part, to labor groups employed at the enterprise, and to the public. Data generated by industry would, in the natural course of things, reach the member of management who is responsible for occupational and environmental health and those in management who are responsible for decisions regarding control and process of the product. The workers should be informed at an early date, as should the consumer of the products, the appropriate government agencies and, if relevant, the community at large. Information should be given to other enterprises engaged in similar activities so that the problems recognized regarding the process and product can be controlled.

Research organizations that have been retained to study the problems would report their findings to their sponsors; they should, of course, ensure that transmittal of this information be made in a timely manner to the EPA administration.

Implementation of a Course of Action

While the use of new chemicals and processes has increased spectacularly, frequently leading to major innovations that have been beneficial to society, their use is justifiable only if the perceived benefits are greater than the perceived risks. If the balance is unfavorable, further attempts must be made to reduce risks, while still maintaining the benefits. If reduced risks cannot be achieved, then the product or process must be abandoned. This is a decision made, in most cases, by industry, but sometimes precipitated by government when the degree of risk is viewed by the government or other concerned groups to be at a different level.

Before this stage is reached, the responsible management—represented by those charged with environmental and occupational health and product safety—has implemented a plan whereby all who have the need to know have been informed of the nature of the problem and the basis for the significant course of action. The suspected hazard has been reported to the Administrator of the EPA, new warning statements prepared and decisions made as to the introduction of whatever protective devices will be required for safe handling. The administrative structure of such an organization is described in Figure 1.

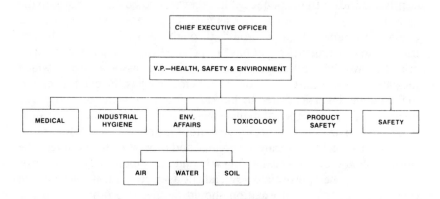

Figure 1.   Industrial hygiene administrative structure.

Regulatory policy for mutagens, carcinogens, and similar substances with long-term effects has evolved as a two-step process: first, to determine if the substance does produce profound effects in animals, if not in man; and, second, if it is active at any level, to maximally restrict its use within the politically acceptable limits to which society agrees.

Measurement of risks from occupational processes and use of chemicals is based on the best scientific information available; however, there are basic limits to knowledge concerning risk estimation, which derive from resource limitations and uncontrollable variables.

The collective judgment of scientists and those considered to be experts may, then, be used to resolve problems of diverse interpretation of measurements and hypotheses explaining them. This accepts the concept that the whole body of knowledge of science, as embodied in the experience of its practitioners, will be able to provide a procedure for resolving issues arising from scientific uncertainties. This, then, results in a "best estimate" and, hopefully, confidence limits. However, the role and judgment of scientists in the societal process have been questioned by some.

While the individual scientist has a personal responsibility for his own work, the conclusions he draws from it may be his own and not necessarily those of others who have examined the data. This distinction is important when the evaluation of scientific facts contributes to the making of large-scale decisions that control the development of technology and the public health. As Harold Lasky pointed out in "Limitations of the Expert," it is one thing to urge the need for expert consultation at every stage of policymaking; it is another to insist that the expert's judgment must be final.

In this regard, Kantrowitz has stated that he does not personally believe it is possible for scientists who have deeply held moral and political views about a question to simultaneously maintain complete objectivity concerning its scientific components.

It is not the purpose of this chapter to review the decisions that are faced by a government involving technology and value judgments. Such decisions involve extrapolation of known scientific facts and currently available technology that are of sufficient moral and political importance that a divergence of opinions is bound to appear. Preliminary observations are sometimes rapidly translated into decisions. When public health is at stake, we should be prepared to act on the side of conservatism. However, immature scientific evidence creates instabilities and uncertainties. If later studies indicate the conclusions or results were invalid, then public disaffection results toward the regulatory machinery and decreasing confidence in the scientific community that supported this conclusion.

SUMMARY

Occupational medical research develops data that must be subject to evaluation by the best scientific methods available. Determination of the meaning of the results obtained requires critical peer review; this should result in a strategy for dealing with the problem of dissemination of information to those who have the need to know.

Separation of issues into scientific and nonscientific decisions enables experts to render scientific decisions while leaving what is decided to be good or not to the democratic process of societal judgment.

REFERENCES

1. Calkins DR, et al: Identification, characterization and control of potential human carcinogens: a framework for federal decision-making. J Nat Cancer Inst 64:169–76, 1980.
2. Ladimer I: Medical experimentation: legal considerations (Part V). Clin Pharmacol Ther 1:674–82, 1960.

CHAPTER 10

OCCUPATIONAL SAFETY AND HEALTH ADMINISTRATION/NATIONAL INSTITUTE FOR OCCUPATIONAL SAFETY AND HEALTH RIGHT OF ENTRY: WHOSE RIGHTS?

Robert D. Moran, JD

Vorys, Sater, Seymour and Pease
Washington, DC

The answer to the question posed in this paper's title can be stated simply: It is each employer's right to remain free of government inspection of his private commercial premises. This right is guaranteed by the Fourth Amendment to the Constitution. All OSHA's inspection authority flows from the statute which, of course, is subordinate to the Constitution.

Since all which follows is based on Fourth Amendment law, it is important to bear in mind the strong language employed therein by our nation's founders:

> The right of the people to be secure in their persons, houses, papers, and effects against unreasonable searches and seizures, shall not be violated....
> ...and no Warrants shall issue, but upon probable cause supported by Oath or affirmation, and particularly describing the place to be searched, and the persons or things to be seized.

The Supreme Court long ago held that this protection extended to businesses as well as homes [*See v. City of Seattle,* 387 U.S. 541, 543 (1967)]:

The businessman, like the occupant of a residence, has a constitutional right to go about his business free from unreasonable official
entries upon his private commercial property.

The word "unreasonable" means any entry which is without proper consent of the owner or his agent unless it has been authorized by a valid
search warrant [*Camara v. Municipal Court,* 387 U.S. 523, 528–529
(1967)].

Eleven years after the two decisions cited above, OSHA argued before
the Supreme Court that its inspections should not be barred by the
Fourth Amendment. The court, however, sustained the employer's
Fourth Amendment rights and held that an OSHA inspection would only
be permissible if the business operator consented to the inspection or a
valid warrant authorized the inspection [*Marshall v. Barlows, Inc.,* 436
U.S. 307 (1978)]. This case pointed out that warrantless OSHA inspections were unconstitutional because they leave OSHA field personnel
with "almost unbridled discretion" as to "when to search and whom to
search" (*Id.* at 323).

Consequently, it is the method of selection of the business to be inspected which must meet Fourth Amendment criteria, i.e., how did
OSHA decide to send an inspector to a particular business establishment.
If OSHA does not employ proper criteria in deciding "whom to search,"
the Fourth Amendment rights of its inspection target will be violated.

Unfortunately, many employers are unaware of their Fourth Amendment rights and fail to require from the OSHA inspector an explanation
as to how he happened to pick this particular business at this particular
time. Caught by surprise and intimidated by the inspector's government
credentials, the employer grants the inspector permission to enter the
workplace. Having allowed the inspector entry, the employer has consented to the OSHA search and has waived his Fourth Amendment
rights.

In recent months, however, an increasing number of employers have
become familiar with their rights to privacy and are routinely declining
OSHA permission to conduct a warrantless inspection of their business
establishment. In such cases, OSHA has two choices: it can give up on
the idea of conducting the intended inspection, or it can make application for an inspection warrant. If it chooses the latter, it must make a
sworn, written application to the local court, and the application must
demonstrate to the satisfaction of the court that the selection of the business establishment for OSHA inspection was based on sufficient "probable cause" under the Fourth Amendment.

Courts have developed two tests to determine whether the probable cause necessary for the issuance of a warrant exists. Both apply in the OSHA context.

The first is the traditional test used by the courts for hundreds of years to restrain police officers seeking to conduct a search for evidence of a crime. It will be designated, for purposes of this chapter, as the "specific evidence" test. Under this test, OSHA must present information to a magistrate that will enable the magistrate to make an independent evaluation of the probable cause to suspect that an existing violation of law is present on the premises targeted for inspection. OSHA is held to this test in employee complaint situations. This test requires that OSHA's warrant application must state the specific conditions about which the employee has complained, so that the magistrate will have sufficient information to make an independent evaluation of the basis for the inspection [*Weyerhaeuser Company v. Marshall,* 592 F.2d. 373 (7th Cir. 1979)].

The manner in which OSHA area offices handle employee complaints is hardly calculated to satisfy this "specific evidence" test. An executed complaint form is delivered by mail or hand to the OSHA office; or an OSHA employee completes a form based on information received over the telephone. Without further inquiry, an inspection of the business establishment named in the complaint will generally be scheduled. If the employer permits the inspection, OSHA has no constitutional problems; but if permission is not given, OSHA will have difficulty obtaining a warrant under the "specific evidence" test. Traditionally, in such cases OSHA's application for a warrant simply recites the receipt of the complaint and claims that the complaint justifies the selection for OSHA inspection of the named business. Some magistrates who are unfamiliar with the applicable case law may issue warrants on receiving such applications, but the law is clear that a warrant cannot be issued on this basis.

Presumably, the only individual with factual information as to the condition which precipitated the complaint is the employee (or other person) who filed with OSHA. The OSHA office, however, does not independently verify the validity of the complaint and OSHA (with no facts) is the applicant for the warrant, *not* the employee with personal knowledge of the situation.

The Supreme Court has consistently held that the application for warrant must be drawn in such a manner that the issuing magistrate has *facts* on which he can make the requisite probable cause determination. A warrant based upon the applicant's "suspicion," "belief" or "mere conclusion" is insufficient to satisfy the warrant clause of the Fourth Amendment [*Giordenello v. United States,* 357 U.S. 480 (1957); *Aguilar v. Texas,* 378 U.S. 108 (1964)]. Since the applicant for warrant is an employee of OSHA, who has not verified the complaint, he has no more than a "suspicion," "belief" or "mere conclusion." In addition, an employee complaint generally does not list "violations." The form that OSHA uses asks the employee to list the "hazards" at the worksite. If OSHA makes no attempt to verify whether the "hazards" were "violations," it cannot supply the magistrate with the required information. In many cases, OSHA does not verify if the person who filed the "employee complaint" was an employee of the company. The magistrate, in such a situation, cannot have facts before him sufficient to justify issuance of a warrant under the "specific evidence" test. The Seventh Circuit made it clear in *Weyerhaeuser (supra)* that a magistrate cannot simply "rubber stamp" an OSHA warrant application.

The second "probable cause" test provides that a warrant may be issued if the specific place of business that OSHA seeks to inspect has been chosen for inspection on the basis of a systematic plan for inspecting all similar businesses. This will be called the *Barlow's* test. The *Barlow's* test has two prongs: first, OSHA must show the magistrate that it has a systematic plan; second, OSHA must show that the specific targeted plant for inspection was selected as a result of the unbiased implementation of that plan.

OSHA has had trouble establishing "probable cause" under the *Barlow's* test. First, OSHA has had difficulty developing a plan to meet the *Barlow's* general administrative plan requirement because statistics which show "hazard rates" for various industries and establishments are often dated and thus inaccurate. In addition OSHA has trouble meeting the second prong of the *Barlow's* test, requiring it to show cause for inspection of a particular workplace. As we have seen, employee complaints are often not considered sufficient probable cause for the issuance of a warrant to authorize such an inspection. Furthermore, OSHA does not maintain lists that indicate injury frequency for specific workplaces. OSHA's difficulty in satisfying both the "specific evidence" and the *Barlow's* tests for the issuance of a warrant reemphasizes the benefit to an employer who wants to remain free of OSHA inspection of refusing to allow entry without a warrant.

Since some magistrates will "rubber stamp" an OSHA warrant application, despite the law to the contrary, many OSHA inspections are conducted on the basis of warrants that are not valid. If an OSHA inspector arrives at a workplace with such a warrant, the employer can block the inspection by declining to honor the warrant and refusing to permit the inspector into the plant. The inspector will not try to force his way into the workplace; his only recourse is to file a motion with the court which issued the warrant for an order holding the employer in "contempt of court" for refusing to honor the warrant. The court must then hold a hearing on this motion, and the employer may defend on the grounds that the warrant is not valid because probable cause was not established at the time it was issued. It is through this process that the law applicable to OSHA warrants is developing.

An inspection pursuant to a warrant does not give the OSHA inspector a free rein once inside the workplace. The warrant limits the scope of the inspection to the areas listed therein. The inspector may not enter areas not specified in the warrant and may not perform acts not authorized by the warrant. For example, if the warrant does not state that the inspector can take pictures or examine company records, the employer is within rights to prevent the inspector from engaging in these activities. The warrant defines only what OSHA *can* do. However, if the inspector engages in activities not listed on the warrant and the employer does not object, the employer may be held to have given permission. While the warrant allows OSHA to gain access to the workplace, the vigilant employer may use that same warrant to restrict the inspection once it has begun.

In sum, the law greatly limits OSHA's right of entry by outlawing warrantless searches in the absence of employer consent. In practical terms, the employer who is familiar with Fourth Amendment rights is frequently able to limit the scope of OSHA inspections, or to avoid them altogether.

NIOSH

The National Institute for Occupational Safety and Health (NIOSH) is separate from OSHA. NIOSH is part of the Department of Health and Human Services, and it is authorized by Congress to conduct research in the field of occupational safety and health. NIOSH is *not* concerned with the enforcement of the OSHAct; it is not a regulatory agency. NIOSH primarily tests for health hazards in the workplace, and OSHA often relies on NIOSH findings to formulate recommendations for health and safety regulations.

The power of NIOSH to enter and investigate the workplace is limited to the parameters of the *Barlow's* decision; therefore, NIOSH, like OSHA, must obtain either a valid warrant or the employer's consent before entering the plant. In a recent case, an employer questioned whether NIOSH had the power to obtain inspection warrants because NIOSH had no specific internal regulations authorizing it to do so. The court, however, held that NIOSH does in fact have this power even in the absence of regulation [*Establishment Inspection of Keokuk Steel Castings, Division of Kast Metals Corporation,* 8 OSHC 1730 (S. Dist. Iowa 1980)]. In addition, NIOSH has the power to seek warrants through an *ex parte* procedure (*Id.* at 1733).

Which tests must NIOSH meet before it may obtain an inspection warrant? The "specific evidence" test for showing probable cause does not apply to NIOSH because it is not an enforcement agency. Rather, NIOSH must meet the *Barlow's* test in order to show probable cause. NIOSH must, therefore, show the magistrate that: (1) it has a reasonable legislative or administrative inspection program; and (2) the desired inspection of a specific plant fits within that program. NIOSH must also show independent evidence, through affidavits or otherwise, that there is reason to believe a health hazard exists. Recent cases have also held that NIOSH may obtain employee medical records, either through the employee's written consent or through Subpoena Duces Tecum [*General Motors Corp. v. Director at NIOSH* 9 OSHC 1139 (6th Cir. 1981)]. Absent these two documents, the employer is not required to hand over employee medical records to NIOSH.

CONCLUSION

Marshall v. Barlow's has both defined and limited the right of entry of both OSHA and NIOSH. Neither agency may enter the workplace without employer consent or a valid warrant issued by an independent magistrate. As we have seen, it is often difficult for these agencies to show the probable cause necessary for the issuance of the warrant, either under the "specific evidence" or the *Barlow's* test. Employers who are aware of the Barlow's case are able to exercise their Fourth Amendment rights in a way that often prevents OSHA and NIOSH from entering particular business operations.

CHAPTER 11

INVESTIGATIVE AUTHORITY OF THE NATIONAL INSTITUTE FOR OCCUPATIONAL SAFETY AND HEALTH

Howard Walderman, LLB
> Office of the General Counsel
> U.S. Dept. of Health and Human Services
> Rockville, Maryland

The National Institute for Occupational Safety and Health (NIOSH) is the lead federal agency in the national effort to plan, direct and conduct occupational safety and health research. While the Occupational Safety and Health Administration (OSHA) is the standard-setting and compliance agency, NIOSH provides support to OSHA through scientific research. NIOSH was created by 29 U.S.C. 671(b), and is authorized to perform all of the functions of the Secretary of Health, Education and Welfare (now Health and Human Services) under 29 U.S.C. 669 and 670 [see 671(c)].

Under the authority of 29 U.S.C. 669(a), NIOSH conducts occupational safety and health research. These activities are initiated by NIOSH, and any site visits to places of employment for NIOSH-initiated research are governed by the NIOSH regulations in 42 CFR Part 85a. The statutory research authority is expressed in extremely broad terms. Thus, NIOSH is directed to conduct (1) occupational health and safety research [669(a)(1)]; (2) research to explore new problems in the field [669(a)(4)], and (3) industrywide studies [669(a)(7)].

Pursuant to 29 U.S.C. 669(a)(6), NIOSH is directed, on proper request, to determine the potential toxicity of any substances used or found in a place of employment. These investigations are known as health hazard evaluations (HHE) and are initiated on a written request by an employer or authorized representative of employees. The NIOSH regulations that govern HHE, including what constitutes an "authorized representative of employees," are contained in 42 CFR Part 85.

To enable NIOSH to perform the research activities enumerated in 29 U.S.C. 669(a), section 669(b) confers on the Secretary of HHS and consequently on the Director of NIOSH, the same rights to enter and inspect workplaces and to question employers and employees as those provided to the Secretary of Labor for enforcement and compliance purposes under 29 U.S.C. 657(a). In the event that there is a lack of cooperation in an investigation, section 657(b), again by virtue of 669(b), confers on the Director of NIOSH the authority to compel testimony and the production of evidence by use of an administrative subpoena.

When NIOSH is refused entry, whether in an HHE or a NIOSH-initiated research study, it seeks an administrative inspection warrant from the appropriate U.S. magistrate under the authority of *Marshall v. Barlow's Inc.,* 436 U.S. 307 (1978). *Barlow's* held that an inspection under the Occupational Safety and Health Act (OSHAct) could be conducted only with the employer's consent or under a valid warrant. The Fourth Amendment provides that "no warrants shall issue, but upon probable cause, supported by oath or affirmation, and particularly describing the place to be searched...." *Barlow's* also said that probable cause in the criminal sense is not required. For purposes of an administrative search such as the OSHA inspection, probable cause justifying the issuance of a warrant may be based not only on specific evidence of an existing violation of health and safety standards but also on a showing that "reasonable legislative or administrative standards for conducting an inspection are satisfied" with respect to particular premises.

The "probable cause test" is clearly met in the case of an HHE request by an authorized employee representative specifically alleging employee exposure to potentially toxic substances. Reasonable legislative standards for conducting an inspection under these circumstances are set forth in 29 U.S.C. § 669(a)(6). Moreover, reasonable administrative standards implementing those legislative standards are set out in 42 CFR Part 85. When NIOSH receives a request that meets the requirements set out in 42 CFR § 85.3-1 to conduct a HHE at a specific place of employment, a further determination is made as required by § 85.4 as to whether or not there are reasonable grounds to justify conducting an investigation. Thus

NIOSH's administrative standard is to conduct an inspection at each place of employment for which a valid HHE request is received and for which there are reasonable grounds to justify conducting an investigation. Based on the legislative standard in section 669(a)(6) of the OSHAct and this administrative plan to respond to requests, there is probable cause justifying the issuance of a warrant based on the test set forth in *Barlow's*.

SPECIFIC CASES

Matter of Establishment Inspection of Keokuk Steel, 493 F.Supp. 842 (S.D. Iowa 1980), aff'd., No. 80–1486 (Eighth Cir. Jan. 13, 1981).

NIOSH received a request from the United Steelworkers of America to conduct a HHE at Keokuk Steel Castings in Keokuk, Iowa. When NIOSH was advised that it would not be admitted to the company to conduct an evaluation, NIOSH sought and obtained, *ex parte,* an administrative inspection warrant. Company officials attempted to condition the NIOSH entry under the warrant on NIOSH agreement to do no personal sampling, conduct no employee interviews on company time or on its premises, and review only records required to be kept by OSHA regulations. NIOSH considered this as a refusal to comply with the terms of the warrant. Keokuk filed a motion to quash the warrant while NIOSH asked that the company be held in civil contempt. The Southern District Court of Iowa denied the company's motion to quash, and because of an ambiguity in the sampling procedure prescribed by the warrant, refused to hold the company or its officials in civil contempt. The court ordered Keokuk to permit authorized employees of NIOSH to enter the workplace to conduct the inspection as set out in the warrant. This was the first written judicial decision involving an administrative inspection warrant obtained by NIOSH to conduct a HHE.

The Eighth Circuit affirmed the lower court ruling on all points. Specifically, the Appellate Court held that NIOSH is not precluded from seeking an inspection warrant merely because regulations do not specifically provide that the agency shall do so when refused entry and that NIOSH may obtain such a warrant on an *ex parte* basis. The court also held that NIOSH met the standards for probable cause necessary to obtain an administrative inspection warrant set forth by the Supreme Court in *Barlow's*. The request by the employee representative identified certain potentially toxic substances used in the workplace and alleged that certain symptoms were experienced by employees where those substances were in use. In light of this request, the Appellate Court found that the lower court's conclusion that an inspection was warranted appeared reasonable. Finally, the Eighth Circuit found that the District Court did not err in ordering Keokuk to permit its employees to wear the personal sampling devices on a voluntary basis or in permitting private employee interviews during regular working hours as authorized by 29 U.S.C. 657(a)(2).

Matter of Inland Steel Company,
492 F.Supp. 1310 (N.D. Ind. 1980)

NIOSH obtained, *ex parte* from a District Court judge, an administrative inspection warrant to conduct a HHE requested by an authorized representative of Inland employees. The written opinion issued with the warrant included the first judicial determination upholding NIOSH authority to conduct medical examinations of affected employees. Section 657(a)(2) of Title 29, U.S.C., expressly authorizes private interviews with employees, but is silent with respect to medical examinations. The Office of General Counsel, DHHS, has advised NIOSH that these examinations are reasonably considered a part of the private interview, and NIOSH rules provide for medical examinations of consenting employees (42 CFR Part 85). The court included NIOSH medical examinations of employees within the scope of the warrant. However, the court refused to include NIOSH access to employee personnel or medical records within the warrant's scope, concluding that, under the OSHAct, an administrative subpoena is the exclusive method for NIOSH to obtain access to records. There is judicial authority to the contrary, since OSHA has obtained access to records under a warrant in the circuit where this case was decided.

Probable cause for NIOSH-initiated research is expressed differently than that for a HHE. In the affidavit supporting the warrant application in NIOSH-initiated research, the background of the study is explained and those "neutral criteria" for selecting the particular establishment for an initial walk-through inspection or for inclusion in an in-depth study are set forth.

Matter of Pfister and Vogel Tanning Co., 493 F.Supp. 351 (E.D. Wisc. 1980).

NIOSH obtained, *ex parte,* an administrative inspection warrant to conduct an initial walk-through inspection of the Pfister and Vogel Tanning Company. The inspection was to determine whether to include the company in the in-depth portion of the NIOSH study of occupational exposure in the leather tanning and finishing industry. The company allowed NIOSH investigators to enter and inspect "under protest" which in the Seventh Circuit, on a company motion to quash, permits the court to review the validity of the warrant. The Eastern District of Wisconsin denied the motion to quash and dismissed the action. This was the first judicial decision to consider the validity of a warrant to conduct NIOSH-initiated research. In challenging the warrant, Pfister and Vogel contended that there was a lack of probable cause in issuing the warrant and the warrant was overly broad. The Court found that (1) the government application and supporting affidavit set forth the necessary neutral criteria required by the *Barlow's* case for selecting the company for inspection and (2) as in *Marshall v. Chromalloy,* 589 F.2d 1335 (7th Cir. 1979), the warrant was not overly broad, given the nature of the NIOSH investigation.

The company also challenged the warrant because the supporting affidavit stated that the NIOSH study concerned nitrosamines, and a previous OSHA investigation at the company, which had lasted over a period of months, had found only trace amounts of these compounds. The company contended that NIOSH should be restrained from conducting duplicate research and further that the warrant should be quashed because the OSHA study was not made known to the magistrate at the time NIOSH applied for the warrant. NIOSH claimed that the OSHA tests were inconclusive and that the equipment NIOSH used was more sophisticated. The court stated that while the agencies should have coordinated their efforts, NIOSH was not bound by the results of the OSHA study. The court expressed concern with the failure of NIOSH to call the OSHA study to the attention of the magistrate when the warrant was obtained. Nevertheless, the court refused to speculate as to whether the magistrate would have decided otherwise had he known of the OSHA study. Needless to say, one of the outgrowths of this case is an increased sensitivity by NIOSH to acquire and analyze the results of any OSHA or state OSHA activities that have occurred at the particular plant.

Aside from its right of entry, NIOSH has the right to gain access to records relevant to its investigation by subpoena. Subpoenas are issued by the Director of NIOSH pursuant to 29 U.S.C. 657(b) and enforced under that section.

U.S. v. McGee Industries, Inc., **439 F.Supp. 296 (1977),** *aff'd mem.* **568 F.2d 771 (Third Cir. 1978).**

This case involved the subpoena of information in a NIOSH-initiated study. Here, the court ordered enforcement of a subpoena issued by the Director of NIOSH for trade secret formulas of a manufacturer relevant to a NIOSH study that did not involve the manufacturer's facility. The court stated that as long as the evidence sought by the subpoena is not plainly incompetent or irrelevant to any lawful purpose, it is the duty of the District Court to order its production. In subpoena cases, NIOSH shows that issuance of the subpoena is authorized by the statute and that the information sought is relevant to an investigation that NIOSH is authorized to conduct.

With respect to health hazard evaluations, the subpoena cases have involved production of medical records in individually identified form. NIOSH wants medical information in individually identified form:

1. to link company medical data on employees with exposure data that the company maintains;
2. to use past medical histories and past diagnoses of physicians on specific employees to assist in analyzing the results of medical examinations on the same individual employees conducted as part of a NIOSH cross-sectional medical study;
3. to verify the company data, it is appropriate from a scientific point of view to have personal identifiers; and
4. to avoid repetition of uncomfortable medical procedures or those which carry a health risk that may already have been performed by the company in its occupational health program.

E. I. du Pont de Nemours v. Finklea, **442 F.Supp. 821 (S.D. W.Va. 1977)**

The court held that although the medical records of employees were protected by a constitutional right of privacy, disclosure of the records in compliance with a NIOSH subpoena would not abridge that right where disclosure was sought in connection with an authorized investigation. The court relied on *Whalen v. Roe,* 429 U.S. 589 (1977), in which the

Supreme Court upheld a state statute requiring physicians and pharmacists to report the names and addresses of persons who receive prescriptions for certain drugs. The Court issued an order which, among other things, restricted NIOSH use of the records to occupational health research and required that the records be returned to du Pont at the conclusion of the study; the company was ordered to maintain the records separately from its record system for 30 years.

NIOSH also relies on a decision that ordered enforcement of an administrative subpoena issued by a medical disciplinary board for individually identified medical records of patients of the accused physician, *Schachter v. Whalen,* 581 F.2d 35 (2nd Cir. 1978). The court rejected the physician's argument that he was protecting the privacy of his patients by refusing to turn the records over to the board.

United States v. Allis-Chalmers Corp., 498 F.Supp. 1027 (E.D. Wisc. 1980)

NIOSH received a request from the United Auto Workers to conduct a health hazard evaluation at an Allis-Chalmers foundry in Wisconsin. After conducting an initial inspection, NIOSH requested access to the company's medical records of affected employees and issued a subpoena for those records. When the company refused to comply with the subpoena, NIOSH sought enforcement. The Eastern District of Wisconsin ordered Allis-Chalmers to comply with the NIOSH subpoena. The court rejected the company's contention that NIOSH had no subpoena power and applied the general rule for determining the validity of an administrative subpoena, i.e., that the investigation be within the authority of the agency, the demand not be too indefinite and the information sought be reasonably relevant to the investigation. The court found no evidence that NIOSH would improperly use or disclose the records. It also noted that the NIOSH investigation of employee health "carries greater weight than the privacy interests of the employees." The court stated that the interest of Congress in protecting the health of the employee outweighs the interest of the employer in protecting whatever privacy rights these very same employees might have in nondisclosure of their medical records. The court did impose certain conditions on NIOSH use of the records to assure confidentiality, including a requirement that NIOSH not disclose records to the public and that only those employees who are working on the project may have access to the company's medical records.

The Court in *Allis-Chalmers* followed the *du Pont* decision and the decision in *U.S. v. Westinghouse* and rejected that in *General Motors Corp. v. Finklea.* The latter two decisions have been reviewed on appeal.

U.S. v. Westinghouse Electric Corporation, 483 F.Supp. 1265 (W.D. Pa. 1980), remanded, 638 F.2d 570 (Third Cir. 1980).

In *Westinghouse*, the District Court applied the general rule for enforcement of the NIOSH administrative subpoena and rejected the reasoning of the lower court in *General Motors* (discussed below). The court recognized the government's interest in protecting the health of the worker and the obligation of the court to assist the government in performing the functions constitutionally vested in it by Congress.

On October 21, 1980, the Third Circuit Court held that under a balancing test, the NIOSH need for the company's medical records sought here for occupational health research prevailed over the general privacy interests of the Westinghouse employees. However, the Court of Appeals remanded the case to the District Court, so that NIOSH may give notice to the employees whose medical records it seeks to examine and to permit the employees to raise a personal claim of privacy as to particular documents in their medical files if they desire. The court declared that most if not all of the information in the files will be the results of routine testing such as X-ray, blood, pulmonary function, hearing and visual tests, and that these testing results are not generally regarded as sensitive.

General Motors Corp. v. Finklea, 459 F.Supp. 235 (S.D. Ohio, 1978), *rev'd and remanded sub. nom. General Motors Corp. v. Director, NIOSH,* 636 F.2d 163 (Sixth Cir. 1980).

The District Court held that NIOSH is authorized to make inspections and to question employers and employees, to carry out its functions and responsibilities. It held further that NIOSH can require an employer to submit medical information regarding its employees, but that the company is not required to submit the information in individually identified form without employee consent, unless there is some compelling need. The District Court also held that an employee is entitled to be heard in a due process hearing to determine whether or not identified medical records may be examined without his consent.

On December 30, 1980, the Sixth Circuit held that NIOSH is entitled to enforcement of its subpoena of medical records as maintained by General Motors (GM), i.e., in individually identified form. The court rejected GM's contention that NIOSH is not authorized to issue subpoenas. The court also decided that the physician-patient privilege (established by state law) was not applicable to this case which involved enforcement of a federal law and thus presented a federal question. Finally, the court held that retention of the names and addresses in the subpoenaed material would not infringe on the employees' constitutional right of privacy, since, with proper security administration, medical information would not be disclosed in individually identified form. The Appeals Court remanded the case to the District Court instructing it to formulate "security provisions that will insure the proper disposition of the subpoenaed records." The District Court previously found that NIOSH maintains its records under reasonably secure conditions (459 F. Supp. 235, 237). The Sixth Circuit declined to follow the employee notification procedures set forth in *U.S. v. Westinghouse* decided by the Third Circuit two months before.

CHAPTER 12

THE CONFLICT OVER WORKPLACE INSPECTIONS

Nicholas A. Ashford, PhD, JD
Center for Policy Alternatives
Massachusetts Institute of Technology
Cambridge, Massachusetts

At the core of the difficulties concerning occupational health and safety inspections is a conflict between two social goals. One goal is that of protecting employers from "unreasonable searches" (that is the language of the Constitution — "unreasonable searches" — not all searches). The other goal is that of protecting employees from health and safety hazards. This may seem like a simple conflict, but I hope to persuade you that the dilemmas that follow from it are complex.

It is important, at the outset, to identify the kind of information that is discoverable on government entry into a private firm. First, there is the kind of information that pertains to violations of health and safety standards or violations of the general duty clause. Second, there is additional information that can be acquired with regard to exposure to health and safety hazards. And, finally, a third kind of information is the discovery of illness, such as occurred in the workers in the Kepone plant. These three kinds of discoverable information have very different consequences for worker health and safety and for the liability of the firm. It is useful to separate them because, from their differences, the firm has different incentives and disincentives to be candid or obscure in divulging discoverable information.

What are the purposes of government entry into the workplace? Does the government inspect firms to discover violations or to reduce injury and disease? I would submit that the primary purpose is to reduce injury and disease by using the inspection mechanism as one of many tools, including education, training and technical advice. If the discovery of violations reduces injury and disease, then the greater purpose is served. The goal of discovering hazards and illness in the workplace is not discovery for its own sake; the success of inspections must be measured by determining how effectively they reduce injury and disease. It is in this broader context that the efficacy of inspection plans must be examined.

Once having decided how useful the inspection activity is, we must ask four basic questions to decide whether workplace inspections that are conducted in an effective manner to reduce injury and disease justify the possible consequences for the private firm.

First, how substantial is the likelihood of an unreasonable search? Should private firms fear harassment from the Occupational Safety and Health Administration (OSHA) or the National Institute for Occupational Safety and Health (NIOSH)? It is difficult for me to understand how there could be major harassment problems with an agency that has been so scrutinized in Congress' authorization process as OSHA. Why would OSHA voluntarily earn itself a reputation of harassment? OSHA may sometimes make unwise citations, and occasionally it may employ less-than-well-educated inspectors, but there is no reason for OSHA intentionally to harass firms. One should admit for the sake of argument, however, that there may be some need to protect firms against unreasonable governmental searches and harassment as a matter of principle.

The second question is: What is to be gained by requiring the government to secure an *ex parte* warrant? What is to be gained by the *Barlow's* decision, which seems to anticipate use of the *ex parte* warrant, provided certain criteria are met? The offered answer is that it provides a neutral magistrate who judges the correctness of a government entry. It is said to be a safeguard of due process. Again, as with the first question, the reason for concern with due process is to protect against alleged unreasonable searches and harassment.

The third question is: What is to be gained by having the government secure a warrant that is not *ex parte* or one that may begin *ex parte* but is resisted and, hence, ends up being litigated in the courts before OSHA's entry? Certainly, advance notice to the employer that he will be inspected tells the employer what he needs to know, and it gives him time to correct the hazards. In all fairness, however, one should ask whether this is bad. If the employer is motivated to clean up the workplace so that it is pristine when the OSHA inspector arrives, is that not what we want? The point is not to collect $42 worth of fines, which is the average OSHA fine for a violation, but to secure a change in firm behavior. If the advance warning secures a change in firm behavior, should not the warning be used? This is a seductively simple question with an implied affirmative answer. But let us look a little deeper and ask whether the correction of hazards in anticipation of an inspection really causes the kind of anticipatory response that the element of surprise causes.

In my view, when the outcomes of an inspection with advance notice are examined and the losses measured against the gains, a great deal is lost by advance notice. With inspections preceded by warrants, there may well be an immediate correction of the violation — in the short run a firm may clean up or remove the hazards — but responses that a firm might exhibit over the longer term will be lost. In addition, if firms believe they can wait to address hazards until they are actually inspected, they will do little until the time comes.

The longer-term, and more comprehensive responses, I have called leveraging. Leveraging is a kind of voluntary response that occurs in anticipation of possible government scrutiny. You are all familiar with the leveraging effect of the Internal Revenue Service (IRS). After we file our income tax returns on April 15, few of us will be audited, but the possibility of auditing will cause a lot of us to do more careful and extensive reporting.

We cannot measure the success of inspections performed by environmental agencies merely by the number of violations they find or by how fast a cleanup occurs after a warning of inspection is delivered by the firm. The overall safety of the workplace must be examined to measure the success of inspections. And the element of surprise is an effective, rational approach to governmental scrutiny.

The issue of advance notice has yet another implication if the warrant serves to limit the scope of permissible inspection to the specific health or safety hazards cited in the reasons for the government desiring an inspection. An OSHA general inspection strategy, like the IRS auditing procedure, places people in an anticipatory response with regard to many regulations. However, if OSHA is asked to state exactly what it will inspect, the scope of its scrutiny is limited and the leveraging anticipatory response is eliminated. Very often, when the private sector asks the government what it is going to regulate, they really want to know what the government is *not* going to regulate. If firms know what the government is not going to regulate, they need not spend resources in areas where they know an agency will not inspect. This is the reason why general inspections and randomizing our less-serious hazards for scrutiny is a legitimate part of an administrative plan.

The fourth question is: What is lost when a particularized warrant-preceded inspection is used? One answer discussed above is the entire leveraging advantage of a regulatory plan, a plan that has worked very well in other agencies, including food inspection, meat inspection and drug regulation. Aside from this opportunity cost, there are other losses. The employer is allowed time to remove sick employees, to send sterile workers home so they are not discovered and to send complaining employees to a different plant. Exposure and medical records may be removed or changed. While these cases may be rare, the invitation provided by the particularized warrant significantly affects the leveraging capacity of a plan.

Advance notice allows a firm to remove toxic substances temporarily but not permanently. An operation may be stopped—exposure to toxic materials may be ceased temporarily while the OSHA industrial hygienist is there. The element of surprise that operates without a warrant minimizes temporary cleanups and provides the possibility for more permanent changes.

In addition, if the scope of an inspection is limited by a warrant, we eliminate the possibility of finding the significant hazards that are present in every workplace. We do not have very safe workplaces, and, until we do, we need flexibility to identify the hazards.

Finally, a warrant, especially one that is contested, disrupts management-labor relations. It is no accident that the safest companies are also the most profitable companies; good management is good management. It is unlikely that the larger, cleaner corporations purposefully resist the OSHA inspector who appears without a warrant. When a firm resists inspection, it gives a signal to its employees that they do not count for much, and that the firm has something to hide. The disruption that results from that kind of resistance may be a high price to pay for privacy. Resistance to inspection may soothe the firm's sense of constitutional righteousness, but it may cause disruptive changes in management-labor relationships.

What does the employer gain by insisting on his right to privacy? He really gains not the protection of his right of privacy, but a reduction of liability that would emerge from discovery of damaging information obtained in nonscheduled, general inspections. With warrants, a reduction in the employer's possible liability for workers' compensation or products liability suits occurs, but a reduced protection for employees also occurs. In my view, this is a serious compromise of the social goals laid out in the Occupational Safety and Health Act (OSHAct). Employers who vigorously insist on their right to privacy are often the same employers who submit employees to physical examinations as a condition of employment and who dismiss workers for being pregnant or for having impaired lung function. Of course, we always need to be careful that we protect ourselves from any power structure—whether a union, a trade association or the government. However, when we bend over backward to adhere to the firm's right to privacy on the one hand and invade the employee's rights on the other hand, then I begin to worry about the moral and ethical justifications.

I would turn to the question that I think addresses the real dilemma of workplace inspections: How do we increase the incentives of employers to improve the workplace and at the same time minimize employer liability? Discovery of toxic substances, illnesses and OSHA violations can end up contributing to the financial detriment of the employer. Employers know that violations of OSHA standards can serve as evidence of negligence in a products liability suit. Discovery of health and safety hazards may move unionized workers to make more demands in their collective-bargaining efforts—and this will cost the employer. The issue, then, is really that of financial interest and not that of the right to privacy.

How do we motivate a good-faith employer to welcome the OSHA inspector, to welcome inspection of employees' health and employers' medical records? Employees may be found to be sterile, and the employer may find himself at the wrong end of a costly lawsuit or a heavy worker's compensation case. How do we encourage honesty when the price of honesty may be corporate liability? The best solution is to remove the necessity of the dilemma: When there are no toxic substances that cause health problems, there will be no need to fear liability. There will be no need to bar the OSHA inspector or the union's industrial hygienist. We have not yet, however, solved the problems of health and safety hazards in the workplace. When we solve those problems, employers will have nothing to hide and there will be no dilemma about inspections.

What is needed is better production technologies, preventive production technologies. Where will we get these new technologies? I believe the ingenuity that built this country will allow us to embark on a new era of production. We will not encourage new technologies by forcing OSHA to use a warrant of particularity with limited scope. We will do it by providing a gentle signal to firms to change their behavior. That is the essence of the OSHAct. It is a signal to firms to begin changing the way they produce products, a signal to change the technology of the workplace.

Today, OSHA is in a double-bind situation. It is told to perform more regulatory impact analyses, and then it is deprived of the funds to perform them. It is told to have more competent inspectors, and then its training facilities are limited. Critics claim that OSHA is causing unnecessary litigation, and yet it is the corporations who file the lawsuits.

And litigation often makes for poor law and poor social policy design. If the outcome of *Barlow's* is more employer resistance to OSHA inspections and more litigation, what is the likely consequence? The result may be more polarization between management and labor and more employee complaints.

In the *Barlow's* decision, there exists an invitation for a change in the law. The court stated it had not yet found evidence of extensive, significant employer resistance. With significant employer resistance to warrants, the court indicated it might reexamine the question of warrants for OSHA inspections. It would seem, therefore, that it is not in the interest of either the worker or the firm to resist regularly the OSHA inspection, unless real government abuse is likely—and that is very unlikely.

DISCUSSION: CHAPTERS 10-12

Mark A. Rothstein, Howard Walderman, Robert D. Moran,
Nicholas A. Ashford, Kenneth S. Cohen, Marie Geraci,
Larry C. Drapkin, John J. Sheehan, Mary Alexander and Carin Clauss

MR. ROTHSTEIN: I have a comment and a question. With respect to the issue of probable cause, I'd like to try to explain to our nonlawyer members of the audience what we are talking about here, and I hope that, when we get to our later presentations on more technical and scientific issues, they will keep us poor lawyers in mind and provide the same kind of explanation.

When the government wants to conduct a search for a criminal suspect, it must show that there is some probable cause for belief that the suspect actually committed a crime. In the administrative area, there is no such requirement. When we talk about administrative probable cause, all that we are referring to is some neutral objective basis for selecting an employer to be inspected.

Now, some questions have been raised about whether OSHA's [the Occupational Safety and Health Administration] inspection program plan is good or bad. That seems to me to be missing the point. If OSHA were to select employers to be inspected in alphabetical order or simply on the size of the employer based on the number of employees, it is certainly arguable that it would meet the administrative probable cause standard. It would be a silly way to inspect, but it would be neutral. That's what the warrant is supposed to do — to ensure that employers are not singled out for harassment purposes or picked on because there is a personal vendetta between the area director and the individual employer.

With regard to scope, once probable cause is established for a general inspection (that is, the employer fits within this neutral scope), most courts have held that the employer's entire plant can be inspected. Yet, oddly enough, if there is further evidence of the need to inspect, such as the filing of an employee complaint or a fatality, many courts will limit the scope of the inspection.

Where there is neutral objective evidence of the need to inspect an employer, the inspection should not be required to be more limited in scope than in a situation where there is no specific information.

Now, with regard to my question, the Reagan Administration has publicly placed a great deal of emphasis on changing the direction of OSHA enforcement from one of confrontation to cooperation. And it seems to me that, from what I've been hearing and reading, most of the effort to increase cooperation has been directed to OSHA; that is, the agency must be more cooperative. And my question is: In the area of warrant resistance and inspection resistance, is there any indication that, should OSHA become more cooperative, employers who have been resisting inspection the most energetically would then permit inspection and, thereby, display the same type of cooperation with OSHA that OSHA has been urged to display to employers?

MR. WALDERMAN: Since no one is here from OSHA to answer, I'll comment from my point of view. The issue of cooperation depends on many factors. For example, NIOSH is conducting now what is called the National Occupational Hazard Survey (NOHS) II. You liked NOHS I; you will love NOHS II. This is a survey to determine what substances are present in the workplace. The earliest survey was started in 1972 and took a number of years to complete. It preceded the *Barlow's* decision. There were very few refusals. In NOHS II—I recall the statistics—they had over 200 entries into places of employment in California. Two outfits requested warrants, and only one resisted the warrant.

I don't know why employers are cooperating in this survey. It could be that the NIOSH [National Institute for Occupational Safety and Health] survey is really a small intrusion. No one gets cited; no reports go to OSHA. So it is difficult to say at this point whether cooperation has increased as far as NIOSH is concerned. There really hasn't been that much resistance over the years, and I doubt if there is going to be now that we have the judicial precedents.

MR. MORAN: I am exclusively involved in representing employers. My experience is that, at some point in the future—I would say that it is at least two or three years away—if the Reagan people actually do follow through and review the regulations, upgrade the inspectors' training and evaluate their inspectors, then employers may well change their views. But at this point I would never advise an employer to consent to an OSHA inspection without a warrant.

Employers are particularly concerned about OSHA instruction CPL2.39, which says, "It would be appropriate use of our resources to attempt to reduce injuries through the criminal provisions of the Act in addition to the civil provisions." Now as long as they are going to publicly announce that they are coming in looking for criminal violations, then I don't think any reasonable employer would be wise to permit an inspection without a warrant.

Of course, if OSHA does get a civil warrant, the employer is insulated from that criminal possibility. I could cite actual chapter and verse where there has been a civil inspection and then OSHA referred the case to the U.S. [District] Attorney for criminal prosecution. All that could have been avoided if the employer had simply declined to permit the warrantless search. As long as things like that are on the books, I would never advise the employer to consent to a warrantless search.

DR. ASHFORD: The innocent have nothing to fear, and I just really don't see an abusive agency in this administration. The agencies are going to be under close scrutiny and the signals have been given to the business community that not much will be done. The sad thing about that is that if we are going to embark on a new period of what is called reindustrialization, which means a lot of new capital investments, are we yet doomed to another generation of badly designed technology where safety does not enter into the design of the plant simply because the signals had been reduced to mushroom level?

Frankly, Bob, the encouragement of obstreperous resistance of legitimate government authority isn't going to get us to a position where industry is benefited in the long run. I just don't see it.

MR. WALDERMAN: I am not here to defend OSHA. Although NIOSH has a great deal of communication with OSHA, probably someone from OSHA should be up here defending OSHA or setting forth OSHA's point of view. I only wanted to comment on the use of the criminal enforcement provisions of the [Occupational Safety and Health] Act.

Recently, I attended another occupational safety and health conference and learned that criminal proceedings constituted an infinitesimal percentage of OSHA matters. The majority of criminal matters were brought by the Justice Department on its own when a particular disaster happened to occur.

DR. COHEN: I would like to address this to Dr. Ashford or anyone else on the panel. With regard to the preplacement examination, should an employer unnecessarily expose (in a workplace safe by a current standard of care) the discoverable hypersensitive or hypersusceptible worker?

DR. ASHFORD: Well, there are two answers, one a personal answer; the other a legal answer. Personally, I believe that the employer owes a duty to the most sensitive employees. They turn out very often to be the canaries of the workplace. They just develop the disease earlier, maybe in some cases to a greater extent than other workers. The concept of hypersensitivity has been abused as has the concept of accident-prone workers. From a legal perspective, the OSHAct requires that *no* employee suffer material impairment. Now, I would say the employer is under a legal obligation to protect all workers.

MS. GERACI: I would like to direct my questions to Robert Moran. First of all, in places where employers have refused entry to OSHA and NIOSH, would you comment on the right of employees and how they can obtain their constitutional right to a safe and hazard-free workplace? Second, you mentioned several ways employers can refuse entry for an investigation and have stated that OSHA and NIOSH have to clean up their acts. Have you volunteered any of your suggestions on how they can improve their system of targeting, data collection and statistics?

MR. MORAN: The last question first. I have, in fact, regularly made a number of suggestions over a period of about ten years. The recent OSHA book I wrote contained innumerable suggestions. I also have made suggestions to the new leadership in recent months, and to the transition committee even before that.

So as far as the rights of employees are concerned, what the employer does about his constitutional rights would not adversely affect the employees' constitutional rights or even statutory rights. If they are concerned about hazards in the workplace, they have a statutory right to obtain an inspection or at least get an explanation as to why OSHA doesn't plan to make an inspection. And, of course, they have tort rights under our public liability statutes. So, the two may well come into conflict, as rights sometimes do, but it is certainly not required in each case.

MR. DRAPKIN: We talk a lot in these meetings about workers' rights to obtain data on their personal exposures, workplace hazards and medical examinations that have been conducted on them. I think some acknowledgment of these rights exist. But, many of the people here say that workers themselves should not have direct access to some information, particularly medical records. But when we argue that we shouldn't give the information directly, we are excluding many people from its access, unless an agency like the government (e.g., NIOSH) can come in and do independent studies. Oftentimes, unions can't afford the resources or the personnel to do the evaluations that the people here are suggesting need to be done in order to understand the information gathered.

So it seems we've come full circle. We say these rights exist, yet we cramp them by not allowing NIOSH or the proposed agency to come in and do the study. I would like to hear comments on that.

MR. WALDERMAN: We take the position that NIOSH does have a right to come in and do studies. It has always been curious to me, personally, that companies can claim to be protecting the privacy rights of employees by not furnishing NIOSH access to medical information. I tried to make that argument a few times in the courts. I think that it was recognized in the *Allis-Chalmers* case, in which the district court stated that the right of NIOSH to have this medical information in order to conduct these studies outweighs whatever privacy rights the employer wants to assert on behalf of the employees.

There is some confusion in the status of the law, because one circuit court has decided that an employee should have a right to object to NIOSH gaining access to particularly sensitive documents. The dilemma is how to reconcile the right of society to have the research conducted with the individual's right of privacy. I do not understand how a company can interpose that it is trying to protect employee privacy rights when the government is trying to protect employee health.

MR. MORAN: Can I add one thing to that? I've had at least one company ask me that, if they permitted or requested a NIOSH survey, could they be assured that the data would be withheld from OSHA? I suggested that they speak to the NIOSH people and get some sort of assurance that they would just do the research and report to the company and whatever public health authorities might be involved and not turn the information over to OSHA for enforcement purposes. They could not get any such assurances. So, as a result, NIOSH was not permitted to come in. Under the system that the Act created, NIOSH was to do research and evaluation to help people eliminate hazards in the workplace, and OSHA was to be the enforcement agency. I think it is a sad commentary on that system that, over the past three or four years, they both became handmaidens to each other and, so, I would suggest that is why a number of employers have resisted NIOSH surveys, and I think that is detrimental to the Act's purposes.

MR. WALDERMAN: The results of the research conducted by NIOSH are public information. Anyone can have access to them. Therefore, I don't think we can restrict OSHA's access to the information. It is sent to OSHA under NIOSH regulations and has been sent since the beginning of NIOSH. Whatever information NIOSH develops is going to be sent to OSHA. That is a fact of life. I just can't imagine NIOSH conducting research and turning it over only to the company, because NIOSH research findings not only affect the employees of that particular company, they affect employees in similar situations in other companies.

DR. ASHFORD: NIOSH is there for research reasons; OSHA is there to inspect. There are two different functions. I insist that the functions are the same and the means are different. If the abuse does not exist, there is no cause for concern. It seems that, as a nation, we're willing to whittle away the ability, the right to privacy and freedom from unreasonable searches (in the criminal context). We want those criminals caught! We want their cars to be searched! We want their persons to be searched! We want them interrogated! We are angry with the Miranda warning. Why does the bulk of society want the real criminal to be interrogated and searched? Because the bulk of society does not consist of criminals.

Now, are the bulk of corporations criminals? No! Why all the resistance? Who is resisting? The bona fide good faith corporation, in my view, hasn't a damned thing to fear. It makes good law practice for those who play on the fear of a few. I think that is all it amounts to.

MR. MORAN: Why do people resist? Some of these inspections, would you believe, last for six months, with five or six OSHA people constantly in the plant. Now they're also enforcing outrageous regulations.

For example, there are violations in this room. The exit signs are the wrong color; they are supposed to be red by OSHA regulation. Now I've heard a public announcement to the effect that OSHA has reformed and is doing things differently.

Within the last six months, I represented a large, very well established, Fortune 500 corporation (not one of your fly-by-night outfits), where OSHA conducted a noise survey. The inspector attached six noise dosimeters to six employees, took their names, addresses and other information. He stood around for eight hours and then took the dosimeters off to record the information to see whether or not there was a noise violation. He had forgotten to put batteries in.

Those are the kinds of people who are out inspecting. The inspector waits around for eight hours with six employees to compile noise results, and he doesn't even know enough to put the batteries in the dosimeter recording instrument. That is the kind of thing that makes employers say, "Do we need this kind of grief?"

MR. WALDERMAN: Well, I think you can beat that citation.

MR. SHEEHAN: I realize that the parameters of the dilemma we are talking about have to do with the issue of cooperation and consultation. The OSHAct itself has both civil and criminal proceedings. I'm surprised at the remark that the criminal proceedings initiated the Act, because we live in a very big society, and consultation is there, and cooperation. They coexist with one another, and I think we ought to be big enough to realize that we can have the two existing at the same time—maybe not at the same place, because sometimes the placement may be a gradual thing. Therefore, it seems to me that if, in the *Barlow's* case, the Supreme Court would induce a real threat on the right of entry either by OSHA or of NIOSH, I would perceive that to be a very substantive attack on the whole OSHA experiment. When Mr. Moran announced this policy of confrontation rather than cooperation, a little bit of a chill ran up my back.

I don't despair of further legislative progress in this area. Under the Mine Safety Act, I'm not aware of the fact that entry has ever been denied. There may be specific reasons why the Mine Safety Act is so structured and crafted that right of entry is not being denied, but I would like to ask the members of the panel (who are all lawyers) if there is something in the OSHAct that can be subsequently changed legislatively to prevent postponement activities from occurring.

In other words, can we get around the problem of the Supreme Court right being fed into the Act legislatively so that we do not have this kind of induced confrontation by constant attempts to prevent entry?

MR. MORAN: In *Barlow's*, the Court cites some examples: animal welfare acts, the Internal Revenue Code, agricultural inspection acts, gasoline inspections. I think that the type of language could be put into the Occupational Safety and Health Act. I see nothing particularly wrong with the wording of the legislation itself. I do see reasons for resisting by virtue of the fact that they are trying to collect evidence of criminal violations, that they are enforcing standards arbitrarily and unreasonably, and that they are sending untrained inspectors. Those are all administrative problems, and I think that, if those administrative problems were cleared up and OSHA did not become such an adversarial organization, employers would have no reason to refuse warrantless searches.

MR. WALDERMAN: MSHA [the Mine Safety and Health Administration] may be having some difficulty entering mines. There have been a number of cases that held that MSHA didn't need a warrant to get into a mine. I think one went before the Supreme Court. Mining has been a pervasively regulated industry and the government has a long history of having access to mine premises. As far as provisions that could be inserted into the OSHA Act, I think the Food, Drug, and Cosmetic Act contains a provision that subjects a company to a fine for refusing to permit the entry of a food and drug inspector.

MR. MORAN: I think the basic premise in this argument is that there is some relevance between the hazard and OSHA inspection. My experience is that OSHA doesn't inspect for true health hazards. They inspect for whatever is obvious or whatever they can add up points with.

I know a company that has had a hearing conservation program since 1958, and they have regular audiometric tests with print-outs of the results. When OSHA came around, they gave the inspector all these print-outs. The company had 650 employees who annually take the audiometric test. Six employees' hearing had decreased from one year to another, and two didn't show up for their annual audiometric test. OSHA issued the company a willful citation. They claimed that the company should have somehow forced employees to show up for the audiometric tests.

The point is, if they had not had a hearing conservation program at all, they could not have been cited for willful violations. OSHA is taking the good employers who are doing the best they can and hitting them over the head, whereas employers who are doing nothing are really getting off the hook or are just getting slapped on the wrist.

DR. ASHFORD: I think it is unwise in the face of the facts to say that is the mainstream of OSHA's performance. Lead, asbestos, benzene and trichloroethylene are real hazards, real dangers. There are deaths from grain elevator explosions. That is where the willful criminal violations come from, where there is disregard for human life and human carnage. I object to little examples being taken to generalize the incompetence that allegedly belies OSHA's efforts. I just don't think it can be done.

MR. MORAN: The case I mentioned is pending in the Circuit Court of Appeals for the Seventh Circuit and, presumably, OSHA's leadership is well aware of it, because it doesn't get to the Circuit Court of Appeals (one level below the Supreme Court) without somebody at headquarters knowing about it. It is a clear-cut example in which the OSHA leadership obviously agrees with its inspector. This is the way it is going to proceed, so it is not one individual inspector. It is something the OSHA leadership is cognizant of.

MS. ALEXANDER: NIOSH has been criticized in the past for conducting studies that are biased because they have not pursued their right to enter facilities of employers who have not wished to cooperate in both in-house and contract research projects. NIOSH has not pursued warrants to gain entry into those facilities. Indeed, some requests for proposals for contract research had, as an evaluated criterion, the ability of contract research organizations to obtain cooperation with the industry to be studied. I would like to hear your comments on that, and I ask this as one involved in NIOSH-sponsored contract research in industrial hygiene and epidemiology for many years. And, could you tell us what you see as the future trend in NIOSH, as to what the policy will be in pursuing warrants in NIOSH contracts and in-house studies?

MR. WALDERMAN: I do understand that NIOSH has been interested in the contractor's capacity in gaining access to the various plants to be studied, i.e., to have the support of a particular industry or trade association in the study, because this cuts down on NIOSH expense in having to enforce the right of entry. We have not, as yet, applied for a warrant to authorize entry by the NIOSH project officer as well as the NIOSH contractor. We would not hesitate to include our contractor in a warrant under appropriate circumstances. As far as NIOSH in-house research, there is no hesitancy on the part of our office if we receive a request from NIOSH that it needs entry into a place of employment and entry has been refused. As to the future, I think we can expect to see the question of entry by a NIOSH contractor litigated.

MS. CLAUSS: I have to express my appreciation to Professor Rothstein for making the responses to Mr. Moran that I would have made if he had not done so.

Let me comment on the BLS data as it relates to this issue. Don't bother asking for BLS data. Mr. Moran is right in one respect, you won't get it because of the pledge of confidentiality. But more importantly, it is not going to do you any good in getting a warrant.

As Professor Rothstein pointed out, the only thing OSHA has to demonstrate is the neutralness of their compliance plan based on BLS data or any other data. The fact that BLS data may be no good is irrelevant to the question of the neutralness of the plan.

There is no fact that OSHA has to prove in getting a warrant. Only when you seek a criminal warrant do you have to show that you have reason to believe a violation is occurring. You don't have that burden in OSHA inspections, and thus it is not going to do you any good to get BLS data.

CHAPTER 13

THE FUTURE OF THE OCCUPATIONAL SAFETY AND HEALTH ADMINISTRATION IN THE COURTROOM: BATTLE OR RETREAT?

Carin A. Clauss, JD
College of Law
University of Wisconsin
Madison, Wisconsin

This chapter will discuss the court's response to a series of attempted retractions. I was going to say retreat, but I did not want immediately to mar my professed objectivity.

The first of these must be the cost-benefit case (cotton dust standard) now pending in the Supreme Court. The simply stated issue before the Court is whether the Secretary of Labor, in issuing a standard under the Occupational Safety and Health Act (OSHAct), must conduct a cost-benefit analysis and determine if the anticipated cost of the standard bears a reasonable relationship to the anticipated benefits. It is unclear exactly what that reasonable relationship should be, since there is no discussion of cost-benefit in the statute and, since no one has perfected a method to cost-out the benefit of saving human lives or freeing human beings from pain and suffering or of determining the savings to employers in medical costs or to society in welfare costs. Most injuries are paid for not by workers' compensation payments (which pay relatively small amounts) but by food stamps, welfare benefits and similar programs. The Court must decide whether to continue to grapple with all these problems in the face of a recent document filed by the new Administration. This document suggests to the Court that it may wish to refrain from further consideration of the cost-benefit issue in light of its announced proposal of notice of proposed rule-making to reevaluate the cotton dust standard and to reconsider the utility of relying on cost-benefit analysis.

It is dangerous to predict what the Supreme Court will do. However, my personal guess is that the Court will go ahead and decide the case. I say this because I am startled that the Administration would use cost-benefit as the basis for asking the Court to stay its hand. This would be effective if the argument before the Court were that cost-benefit is a policy decision. In other words, if the government had argued, "We are not required by the statute to engage in cost-benefit analysis, and we have made a decision not to do so," then the new Administration's reaction would be very clever. Instead, the government said, "We are not permitted to engage in cost-benefit analysis. The statute forbids it." That, of course, is also the position of the union. In essence, the union and the government said, in their argument on January 21, 1981, before the Supreme Court, that Congress has already struck the reasonable relationship between cost and benefit in the OSHAct. The Secretary must adhere to the OSHAct when promulgating standards and cannot further consider cost-benefit analysis.

So the issue is not moot. If the new Administration goes back and conducts a cost-benefit analysis, then the issue with which the Supreme Court will have to deal in the cotton dust case will be the union's contention that conducting cost-benefit analysis and setting standards based on them is prohibited by the OSHAct. I see no real reason for the Court not to go ahead and reach the question.

This case is, thus, different from the *Bakke* case, which involved the affirmative action plan for medical students in which a white student claimed he would have been admitted, but his place was taken by a black. The cost-benefit case is not a claim brought by a single individual. This is a challenge to the conduct used in setting standards and is one that the Court will ultimately have to face. I think the Court might also be interested in expressing a view on the politicization of the legal process in which the Solicitor General makes one argument for the government on one day and is forced to make another argument on another day.

A somewhat more troublesome retraction by the new Administration is the Assistant Secretary of Labor's abandonment of the preference for engineering controls. This position is a major issue in the cotton dust case and was the subject of bitter arguments not only within the old Administration but also in the Court. Employers have argued in court that the standard was defective because there was an equally effective but cheaper solution to controlling the hazard, namely the use of respirators, and that requiring the companies to engage in more expensive engineering controls was arbitrary and capricious. The Court of Appeals discounted that argument and said the preference for engineering controls was a rational legislative judgment on the part of the Secretary, based on the substantial evidence indicating the superiority of engineering controls in controlling hazards. Obviously, however, if we are to believe what we read about the Secretary's statements, that policy decision will be reexamined by the administration, and this review may influence their subsequent standards or lead to reexamination of existing standards.

Unlike the cost-benefit analysis, which presents a purely legal question, for which the Court is better equipped than anyone to answer, the preference for engineering controls may be viewed by the Court as essentially a policy issue. If we consider the Supreme Court's multiple decisions in the benzene case, it is possible to conclude that we may see a rather substantial change in support for engineering controls, and this may be an area that litigants from the union will have difficulty addressing.

Another change in the new Administration is in the area of state occupational safety and health plans. Under the old Administration, I could foresee a great deal of litigation regarding the adequacy of state plans. A complaint had been filed against the Indiana plan, claiming that it was inadequate and requesting that the certification be revoked. Actions were also pending involving two other states. On March 27, 1981, the new Administration withdrew the complaint against Indiana and presumably sanctioned and approved the conduct of that program.

One other issue that I want to mention before I get to what I think may be the most interesting battle of the future is the whole issue of union participation. Cleary touched on this briefly in his chapter.

The Third Circuit Court held in *Sun Petroleum* that the union's right to participate in hearings on issues other than the reasonableness of the abatement date does not give the union the right to object to a settlement entered into by the Secretary and the employer on any issues other than the reasonableness of abatement.

So, for example, in *Sun Petroleum,* the Secretary and the employer agreed on a plan of abatement that the union feels is totally inadequate, and the union does not have the right to a trial presentation of that objection. I thought the dissenting judge in the *Sun Petroleum* case raised a very interesting question. Cleary pointed out that Sun Petroleum's position was that the OSHA Review Commission did not have the right to review the Secretary's settlements or subsequent decisions toward the settlement. The dissenting judge agreed that perhaps the Commission cannot review the merits of the settlement if it is entered into before the Commission has ruled. However, I maintain that in such cases the union should have a voice because a third-party review should be interposed between the Secretary and employer.

The issue was difficult because the Secretary is suing the employer for many reasons in addition to the abatement hazard. There are penalties; there are characterizations of penalties; and the Secretary of Labor obviously has limited resources and did not want to engage in litigation. The Secretary will, on occasion, settle cases that should not be settled. These are usually cases in which the settlement only involves a reduction of penalty, and it is felt that the settlement will save the government money and will get an immediate abatement. Therefore, why go through a trial proceeding to satisfy the union or the Commission, when the reduction in penalty makes sense? However, in those instances in which the decision is made to withdraw citations, thereby making a decision that no violation has occurred and affecting abatement, the union has a much greater interest, an interest that the Third Circuit Court has now said the law does not protect.

There was a subsequent decision in the Fifth Circuit involving the IMC Company. I believe that was an unfortunate case for the union in that it was such a bad decision. It screamed to be reversed. The inspector who filed the citation made a mistake. Moran pointed out not every inspector is absolutely top quality. There were no violations in the view of OSHA and the lawyers in Washington. So, following the filing of the citation and on receipt of the notice of contest, the Secretary made a decision not to file a complaint and the citation was withdrawn.

The Commission claimed that the citation could not be withdrawn without the union's permission and announced that the case should be heard on its merits and that the union could pursue this citation if it so chose. The Court stated that OSHA law gives prosecutorial power only to the Secretary, and where he makes a finding that no violation has occurred, the union cannot independently pursue the citation. This decision is unfortunate because it now means that two circuit courts have found against the more basic question, namely, the union's right to participate at any point in the proceedings.

Although it should be very clear from the law and all other kinds of precedent that the union cannot participate at a point prior to the Secretary's decision to proceed with the case, once the trial is in progress, what rights does the union have? I think this will be a recurring issue that we will see in the course of the next few years, particularly if the current administration moves to retreat in any major area because the union, which will not have a voice in government, will want to be the voice that is heard.

I would now like to turn to my primary concern in this presentation. That is the policy of major employers to exclude women of childbearing capacity from all kinds of production jobs because of their exposure to toxic substances. I want to review this issue not because I am a woman, but because I think it is, first, a basic civil rights issue; and, secondly, I find as a lawyer that it is one of the more interesting issues to face us in the future. These exclusionary policies are very sweeping in scope. They exclude generally all women between the ages of 16 and 50. That is almost 80% of the female workforce in this country. If you are female just entering the workforce, it means you will be excluded from these jobs for the bulk of your working life. These exclusionary policies have been promulgated by chemical companies, lead companies, rubber plants and large numbers of corporations that are using toxic substances in their normal production process.

These employers make exceptions for only those women who can prove they are sterile. It is not relevant that a woman is single, that she is on the Pill, that she is 45 and divorced with three grown children and does not plan to start a second family, or that she is a widow. The reason why these factors are not considered is somewhat derogatory. It is simply because, when a woman tells an employer she does not plan a pregnancy, that does not mean she will not experience a pregnancy.

The plight of these women was graphically brought to the attention of the country in January 1978, when the *Wall Street Journal* ran an article on five women workers at the American Cyanamid Plant in West Virginia. These woman disclosed that they had themselves sterilized in order to keep their jobs. American had announced that women of childbearing capability would be removed from their current jobs when they were exposed to toxic substances and would be given other jobs if that were possible; however, there were not enough jobs for all these women. There were, in fact, only two janitorial jobs available.

This same scenario was played out in another American Cyanamid plant, which manufactured a toxic chemical. The employer decided in July 1978 that this chemical was a teratogen dangerous to the development of the fetus. Because he could not tell which females were pregnant, he terminated all women. However, in December (five months later), he called the women and said, "Gee, I made a mistake." Subsequent scientific studies had shown that this chemical is not hazardous to women. So the former female employees were told that, "when we have a vacancy, we will take you back."

By the way, it is interesting to note that this chemical is also hazardous to males. Nothing was done to remove the men from the toxic substance or to protect them from exposure to the substance; and the men, not surprisingly, filed charges with OSHA.

The third example, which I give to illustrate the magnitude of the problem, involves a company that laid off five women they claimed were exposed to fluorocarbons that were dangerous to women. Two women subsequently were sterilized. One was a 43-year-old sole support of a disabled husband and two children. The second was a widow. They both needed the jobs. Following the sterilization, they returned to work. A few months later, the company discovered that the level of fluorocarbons was not dangerous to women and rehired the others. The two women had undergone needless sterilization operations. What has been and what will be the response of the government to this kind of company policy?

First, the government has made a response under OSHA. The government said that workers are entitled to protection of their reproductive capacity and that includes protection of the fetus, which is part of the reproductive capacity. So, in setting a standard under Section 6(b)(5), OSHA, at least under the prior Administration, considers the adverse effects on reproductive capacity including the fetus; and, where feasible, standards are set at levels low enough to protect both the fetus and the reproductive capacity.

Reproductive effects were an issue in the lead standard. The Secretary noted that the industry's way to deal with this hazard is simply to exclude the women. The Secretary pointed out that, based on the scientific data introduced at the hearings, lead poses a threat to male and to female reproductive capacities; that the danger of defective children is as real if the parent is a man as it is if the parent is a woman; and defines that both sexes need the protection.

In the case of lead, however, it was not *feasible* to set the permissible exposure level at the point necessary to provide protection for reproductive capacity and the fetus, which the Secretary determined was 30 μg/m^3 air. So, as an alternative, the Secretary set the trigger level lower. This means that, under the lead standard, a company must monitor at the 3-μg/m^3 level. In addition, provisions are made for medical removal protection.

In setting the lead standard, the Secretary was careful not to require that the employer remove women because they are pregnant. What is required has nothing to do with pregnancy and reproductive capacity but, rather, that all workers who have excessive lead levels in their blood will be removed. This becomes a specific medical determination that a particular employee is in danger. If this occurs, it is mandatory that the employee be removed temporarily from the job. It is anticipated that, if a pregnant female were working at a station where the exposure was in excess of 30 μg/m^3, she would be in danger and would be removed. It is also anticipated, however, that a medical determination would also be made with respect to a male who was planning to parent a child.

To ensure that this mechanism will work, the regulation has set up a form of medical arbitration board where the determination to remove an employee for medical reasons is initially made by the employer's doctor. If the employee disagrees with that determination, the employer must pay for an examination by a doctor of the employee's choosing. If the two doctors agree, the decision is final. If the two doctors disagree, they elect a third doctor, again to be paid by the employer, and the decision of the third doctor is binding.

Industry challenge to the lead standard included the following: (1) the establishment of a permissible exposure limit that took reproductive capacity and health of the fetus into account; (2) the legality of medical removal protection provisions; and (3) the legality of the tripartite medical review board. The Court of Appeals sustained those provisions of the lead standard. The industry has petitioned for *certiorari,* which at this time is still pending before the Court.

In addition to its position that workers are entitled to protection of their reproductive capacity, the government, in the American Cyanamid case involving the five women who were sterilized, filed a citation under the general duty clause and made two arguments.

First, the employer had failed to make his workplace free of risk because simply removing the women had not removed the toxic substances or the danger not only to female but to male employees. (Now, employers will tell you that they must adopt exclusionary policies out of some sense of morality—to protect the future of the race. But, if that is true, obviously they must remove the hazards, not the women, since concern must include the genetic effects of exposure to the male workers, exposure to the community at large, and to the users of the product.)

Second, the government argued that by conditioning continued employment on sterilization, the company had created a new risk for the female worker and had encouraged her, if you will, to undergo a sterilization operation that had the effect of totally destroying one of her body functions; namely, the ability to reproduce and, in addition, had exposed her to the risks in any medical procedure.

This case was heard by an administrative law judge who did not deal with this difficult issue but said, instead, that the statute of limitations had run out and that the Secretary's action was barred by Section 4(b)(1) of the Act, because the Equal Employment Opportunity Commission (EEOC) was regulating in this area.

Now, obviously, that is an erroneous interpretation of 4(b)(1), since it preempts OSHA only where another agency has safety and health regulating responsibilities; and EEOC, whatever its jurisdiction, does not enter into the area of worker safety and health.

There is currently pending a citation filed against Bunker Hill in Idaho. In this plant, women were terminated and current employment is conditioned on proof of sterility. There is a substantial difference between the American Cyanamid case and Bunker Hill.

Bunker Hill, having witnessed what happened to American, was cautious and did nothing to encourage sterilization. Nonetheless, women of childbearing capability were faced with the choice of either losing their jobs, or failing to get employment, and having themselves sterilized. The Secretary's position in *American Cyanamid* would apply to Bunker Hill as well, because the employer has failed to remove the hazard simply by removing the women which, in the Secretary's view (at least under the last Administration), was sufficient to constitute a violation of the general duty clause of Section 5(a)(1).

In addition to the OSHA response, there has also been a response from the civil rights agencies. I think it is important to at least be aware of that response. Under Title 7 of the Civil Rights Act, which applies to companies that are government contractors, this kind of activity is discriminatory and results in disparate treatment of women. The arguments of these agencies include the following.

First, it discriminates against people because they are pregnant. The clear intent of Congress was to protect women who, because of their role in the reproductive process, had been discriminated against in the past.

Second, it singles out women for this exclusionary policy while leaving in place men who would be equally at risk unless they, too, were sterile or not planning a family. Men would also have a claim that they would not be equally protected, if you want to call being fired protected.

And, third, these rules are too broad under any kind of civil rights analysis because they do not distinguish between women who are pregnant and, therefore, at risk, and women who never will be pregnant.

The company asks how it is possible to make that distinction — how do you determine which women will become pregnant? The Supreme Court dealt with a similar question in another civil rights case involving pension plans. In that case, the *Manhart* case, the company paid women smaller pensions than men for the same contribution on the grounds that 16% of the women would live longer than the men. Because they did not know who would fall in that percentile, they all got less. The Court said they could not do that. If you cannot decide which ones are going to live longer, then you have to treat all women like men. I feel the same rule applies here. If you cannot decide who is pregnant, then you have to treat all women as if they were *not* pregnant and not as if they *were* pregnant.

CHAPTER 14

WORKER CONTROL, PARTICIPATION AND RESPONSIBILITIES IN OCCUPATIONAL SAFETY AND HEALTH ACTIVITIES

Steven H. Wodka

Health and Safety Department
Oil, Chemical and Atomic Workers International Union
Washington, DC

Amid the book burnings, film censorhip, Gestapo-type firings, and the legal flip-flops before the U.S. Supreme Court, the Reagan Occupational Safety and Health Administration (OSHA) claims that it has found the solution to the OSHA dilemma.

The Administration proposes that the threat of an OSHA inspection be removed from those workplaces where labor and management have formed joint committees to deal with health and safety problems. Also, the idea from last year's Schweiker bill has been revived: if there is an inspection, OSHA penalties would be reduced for those employers that have formed such joint committees.

The Oil, Chemical and Atomic Workers (OCAW) Union has had extensive experience with joint committees. Nearly every one of its 450 contracts with the oil industry provide for such joint committees that meet monthly and are made up of equal representatives from the union and management.

However, the union has found that there must be a quantum leap from the joint committee setup to genuine labor-management cooperation before a reduced role for OSHA can be justified.

The experience of the OCAW joint committees has been nearly uniform. If the hazard is minor and inexpensive to correct, there is no problem. It is done. However, any serious or major hazard that is a significant cost item will not be approved by the management members. They will refuse to recommend its correction and they will sit on it by saying that they want to think about it.

Consequently, the union must either resort to the grievance process or file a complaint to OSHA. The grievance process typically takes several weeks to exhaust. Assuming that the case goes to arbitration, several more months of delay are incurred in scheduling an arbitration, holding the hearing, filing briefs and awaiting the decision.

Worst of all, most arbitrators are not skilled in dealing with highly technical health or safety matters. Arbitrators are trained and experienced in contract interpretation, not in industrial hygiene.

Arbitrators are easily frightened by the jargon of safety and health. The United Auto Workers (UAW) recently arbitrated a General Motors' (GM) exclusionary policy that restricted fertile female workers from work areas where there was lead exposure. The UAW argued that the exclusionary policy was discriminatory against women workers and did nothing to protect the male workers who were also at risk. The UAW's position was that the workplace should be cleaned up. GM simply argued that if the women were not excluded, then the arbitrator would be responsible for the spontaneous abortions and defective offspring that GM claimed would otherwise result. The arbitrator was bowled over by GM's argument and he upheld the exclusionary policy.

The criteria that determine whether a health and safety hazard is a violation of the contract or a violation of OSHA can be vastly different. Recently an OCAW local at a Texaco refinery in Illinois arbitrated the company's failure to test adequately the refinery for explosive quantities of gases prior to the issuance of hot work permits. The criteria for the arbitration was whether the plant manager, in originally denying the grievance, had acted arbitrarily. This was a very hard test to meet, and the local lost the arbitration. The local then filed a complaint with OSHA. OSHA investigated and issued a general duty clause violation on the company's inadequate gas testing and insufficient training of the gas testers.

Therefore, OCAW's experience has shown that joint committees are fairly ineffective in dealing with major problems. However, worker control over the occupational health and safety program in a plant can bring significant improvement *without* the need for intervention by OSHA.

The best example of this can be found in the case of the Kawecki-Berylco Industries (KBI) plant in Hazelton, Pennsylvania. This plant, which manufactured pure beryllium metal, was organized by OCAW. In

1971 the union determined by a preliminary survey that the permissible beryllium dust limit in the plant was being exceeded by 1000 times. It also determined that 10% of the workers either had symptoms of beryllium disease or were actually suffering from this debilitating lung disease caused by beryllium. The union spent the next two years educating the rank-and-file workers as to the hazards of beryllium.

In 1973 the local union successfully negotiated a three-part health and safety program into their collective bargaining agreement. First, a union-designated independent industrial hygiene engineer was given access to the plant to measure the dust levels and make recommendations for engineering and work practice controls. Second, a union-designated expert team of physicians performed regular medical surveillance of the workers to determine the effects of their beryllium exposure. Third, two local union members were trained in dust sampling and given their own sampling equipment. These employees had the right to leave their jobs and sample operations when unusual conditions occurred. Their samples were analyzed by an independent laboratory. The company bore all the costs and expenses for this entire program.

The program worked. With the workers aware of the levels they were exposed to and the adverse medical effects on their fellow members, the workers were in a position to move the company along to install the controls. This was done with the assistance of outside experts whom they could trust. Keep in mind that the local did not have control over the company's financial resources that were needed to abate the hazard. By 1979 the plant was meeting the current OSHA limit for beryllium of 2 $\mu g/m^3$ and was very close to meeting a proposed OSHA limit of 1 $\mu g/m^3$.

But in September 1979 the KBI plant shut down. KBI could no longer competitively compete for the federal government's contracts for beryllium metal. KBI's costs were higher than its competition, Brush-Wellman of Cleveland, Ohio. Brush's beryllium operations had been in violation of OSHA standards. OSHA had cited Brush, but the cases were under contest and Brush was not required to abate. Today, all of the federal government's contracts for pure beryllium metal go to Brush-Wellman, the apparent low-bidder.

The workers at the KBI plant never filed a complaint with OSHA. That was the *quid pro quo*. As long as the company was responding to the union and its experts' recommendations to clean up the plant, OSHA would not be brought in.

But the real question is "how did the union get the company to the point of responding in good faith?" The key was equal access to information. For years, the company had its own industrial hygienists who had spewed forth all sorts of claims that the plant was safe. When the union was able to credibly counter the company's claims, the stage was set for change to occur.

Again the key elements of this program of joint labor-management cooperation were:

1. access to the plant by a union-designated independent industrial hygiene engineer with all fees and expenses paid by the company;
2. the right of designated union members to leave their jobs and perform air sampling, particularly during upset conditions (these members were trained at company expense by the independent industrial hygiene engineer, and did not lose pay when sampling);
3. periodic medical surveillance by a union-designated team of medical experts, all of which was paid for by the company (provisions were made in the contract to provide care for those workers for whom the physicians recommended no further exposure to beryllium).

OCAW stands ready to drop the adversarial approach with any company willing to adopt this type of program.

However, the actions of the Reagan Administration do not create the type of climate that promotes cooperative approaches. As stated, the key is equal access to information. So what has this administration done?

*   aborted the proposed labeling rule,
*   destroyed thousands of booklets on the hazards of cotton dust, and
*   made thinly veiled threats to vacate the access to record standard.

Lastly, the general weakening of OSHA as an effective enforcement agency fails to provide an incentive for an employer to voluntarily do anything. In the view of OCAW, this is the true objective of the Reagan Administration.

I believe that this question of incentives is of vital importance. Our system of health and safety regulation in this country simply does not provide an incentive to employers to correct hazards before they maim or kill workers. This observation is true whether the regulating is done by the Carter Administration or the Reagan Administration.

This is not trade union rhetoric. I draw attention to the "Final Report of the Interagency Task Force on Workplace Safety and Health" (1978):

> Even if all injury cost components are properly identified, the expense per worker of preventing injuries may often be greater to the employer than the cost of the injury itself [p. II-12].

OCAW submits that this basic economic fact of life is the reason why we have not seen a dramatic drop-off in the rate of occupational deaths and injuries in the last nine years.

The basic problem is that the regulatory structure of the Occupational Safety and Health Act is so weak that it makes little difference to employers, especially in the oil and chemical industry, whether the OSHA inspector comes knocking at the door or not.

The reason why employers assume a combative stance with OSHA has nothing to do with a lack of confidence or trust in OSHA. It simply pays to fight OSHA. In the United States, businesses exist to make a profit. We have established a regulatory system that makes it profitable for a business to fight OSHA rather than to comply. Why, then, should we be surprised at the polarization between industry and OSHA, or at least between industry and labor? The Interagency Task Force made the same finding in their report:

> employers may *profit* by contesting (rather than promptly complying) whenever the present use of abatement money is worth more than the proposed penalty plus the legal costs of delay [p. III-15, emphasis added].

This point is even more true in an economic climate when money is losing its value at a rate of 16–18% per year. It pays for a business to utilize its money to turn a profit rather than fulfilling a health or safety objective that will show a much lower rate of return.

Only when we have an economic and regulatory system that can reverse this equation will the occupational health dilemma come under control.

DISCUSSION: CHAPTERS 13 AND 14

Kenneth S. Cohen, Steven H. Wodka, Theodor D. Sterling,
Carin Clauss, David R. Denton and Janet Miller

DR. COHEN: Mr. Wodka, you addressed the concept of economic incentive. Aside from the minor fines that are proposed in OSHA [Occupational Safety and Health Administration] citations or the minor costs of delaying actions, do you feel that employers will find incentives in the negligence actions which are arising against employers that are far more amplified in their dollar figures, or the actions brought by third-party litigants for the now-neonate who is no longer a fetus and who is immune from the immunity between the employer and the employee?

MR. WODKA: Obviously, the third-party litigation that's going on in the asbestos field right now is definitely going to have an effect. The problem is that it is so limited in its scope. I think the figures show that since World War II, about two million workers have been exposed to asbestos, yet I would say that no more than 10,000 cases have been filed as third-party actions against the suppliers of asbestos. It is too limited to have the very broad effects that we need.

The interesting thing is that all those third-party cases are based on the failure of the asbestos suppliers to warn the employees who were exposed. The Chemical Manufacturer's Association in the last few weeks urged the Reagan Administration to pull back the labeling standard. You would think that these third-party suits would resolve any question as to the necessity of informing employees of hazards. Obviously, however, the CMA feels that the litigation that's going on now is not a significant incentive to justify a nationwide labeling standard.

DR. STERLING: I have been puzzled for some time by the attention paid by the courts and the administration to risk-benefit analysis. To some extent, this is my field, and I should like to comment. My field experience with risk-benefit came about when the Secretary of Agriculture asked that a committee of the National Academy of Sciences be guided by risk-benefit considerations in making judgments as to whether or not a particular herbicide registration should be cancelled. I was asked to be on that committee because I was familiar with operation research statistics.

Now, the committee did not fulfill that request because there was no standardized literature available dealing with risk-benefit that could be applied in any way to this problem at all. There is a field of operations research and there is a field of statistical decision theory wherein certain consequences of actions can be evaluated in terms of gains or losses incurred. But these meet, in some way, criteria that occupational health problems do not meet by any stretch of the imagination. The first requirement is that gains or losses can be reduced to a particular scale. Another criterion is that gains and losses must be incurred by the same party. If these criteria are met to some extent and one is strongly met, then benefit analysis is possible. But I fail to see how in environmental areas or occupational health areas, the criteria are met. So I'm puzzled that the courts would even consider seriously such arguments.

MS. CLAUSS: Hopefully, the courts will not seriously consider such arguments. There are no criteria in the statute for an application of a cost-benefit analysis. So it is an absolute nightmare. I am sure Rehnquist would have to find it an unconstitutional delegation since there are no criteria spelled out for the courts to apply. My personal guess is that the Supreme Court will agree with the government.

MR. WODKA: I think cost-benefit analysis will be the "voodoo economics" of the 1980s. The guidelines are set up in such a way as to make the Office of Management and Budget (OMB) an integral part of the rule-making process, and there is no way a protective health standard will receive OMB approval. There was a short interview published recently with James C. Miller, III, who is the chief regulation hatchetman at OMB. He basically ruled out any significant new regulations in the next four years.

Under regulatory analysis, proposed rules will be repeatedly sent back to the agency until the agency, e.g. OSHA, does them the way OMB wishes. OMB is an integral part of the process. No new rules will be released by OMB until regulatory analysis meets its approval. I believe that it's going to put the "kabosh" on the possibility of any new kinds of standards coming out in the next four years. Ms. Clauss' last statement was of interest to me because, if this new administration felt that the Supreme Court was on its side, it would have allowed the cotton dust case to go ahead with the presumption that the textile manufacturers would win. But obviously, they have real doubts that the Supreme Court ruling on that issue will be favorable to the industry.

DR. DENTON: Mr. Wodka, you seem to expect that because of current regulatory environments there may be increased attention paid to occupational issues by labor and industry. OCAW [Oil, Chemical and Atomic Workers International Union] has been admirably impressive in protecting the health and safety of the workers, but as was pointed out yesterday by one of the speakers, too many unions have been willing to trade their members' health and safety for increased wages and benefits. I'm wondering if, given the current economic environment, with the current trends in the new Administration, with concern for jobs and inflation, whether you have any reason to believe that the bulk of the American labor movement is going to pay any increased attention to the occupational health and safety issue as opposed to concentrating efforts continuously on the economic and benefits issues?

MR. WODKA: The economic question has always been there, even before the current down-slide in the economy.

I have been in countless negotiations in which our people have a health and safety demand on the table and management comes up with the economic offer. If you are running out of time and are up against a deadline, are you going to strike over the health and safety issue? Invariably, our people chose not to. They hope that the government, namely OSHA, will take care of the problem. I only expect this to worsen in the next few years. For example, some of the plants in the Midwest and Northeast are hanging on by a thread in terms of economic conditions. If a really effective inspection is made with fines and violations requiring hundreds of thousands of dollars for fixup, it might be the thing that tips them over. That is the constant dilemma that workers face. Workers are the ones constantly making the choice between working or living for a while after they finish their working life. That is a burden we have that the employer doesn't.

MS. MILLER: Mr. Wodka, do you feel workers have some responsibility to comply with regulations or employer policies? What leverage does the employer have when the employees do not comply? It is not "macho" to wear hearing protection or to use other personal protection devices. What leverage does the employer have to get employees to comply?

MR. WODKA: I don't know of any worker that wants to see his fingers cut off or have a beam dropped on his head when he is not wearing a hard hat. People don't go into work to commit suicide. I think your question really deals with discipline. The employer has rules, how does he make sure they are followed?

Under every one of our collective bargaining agreements, and I'm sure in your industry as well, the employer has the right to enforce discipline. The whole grievance-arbitration set-up is geared toward helping the employer in that regard. We have plants right now where there are hundreds of disciplinary grievances backed up for arbitration, and these have nothing to do with health and safety. There is no way that the local can arbitrate them all. It has to decide which are the worst cases. In the disciplinary case in which the worker loses a couple of weeks pay for an unjust suspension, the worker has to absorb that. So the system is geared towards the employer.

I think the real problem is the attitude that management takes in that particular plant. When the employer requires respirators, for example, is he trying to engineer out that exposure or is he simply placing the burden on the workers to take care of the problem? It makes a big difference in the compliance of the workers. They will wear respirators when they know that the employer is trying to correct the problem.

MS. CLAUSS: Education is another way to ensure employee compliance. The employers should educate their employees and alert them to the hazards. You get more voluntary compliance if workers know what the problem is.

MS. MILLER: I was surprised to find what I consider a high number of employees who do not show good compliance. I would like to know if others have had similar experiences.

MS. CLAUSS: The law books are filled with hard-hat cases. The employees allegedly refused to wear hard hats. But I think that the employer first has an obligation to provide a strong education program and second to enforce the safety program through discipline. In all of the cases that I've seen, the employer was not willing to discipline the employees. They were not reprimanded or suspended for failing to observe the safety rules.

DR. COHEN: Leaks into the lay literature regarding the new administration's suggestion that engineering controls be replaced in favor of personal protection should be regarded as reasonable only by those people who have never themselves had the opportunity to wear personal protection for a reasonable period of time. Although the need for personal protection is very real in many cases, those who have not worn a respirator for more than an hour should try it sometime. When wearing ear muffs, one very rapidly feels as though a bolt has been inserted through both ears and the nut on one end is being cinched down tight. After a series of these experiences, I then recommend to Administration staff that they begin to evaluate the need for personal protection.

It is not easy to address these concepts to top management when engineering controls are an ultimate decision. A technique which I use to illustrate the point is to present a management seminar wherein the Chairman of the Board wears a respirator for a limited period of time, and with subsequent pain and discomfort comes a rapid installation of ventilation requirements rather than the institution of personal protective devices.

CHAPTER 15

THE FUTURE APPROACH OF THE OCCUPATIONAL SAFETY AND HEALTH ADMINISTRATION TO STANDARD-SETTING

K. W. Nelson, MS
ASARCO Incorporated
Salt Lake City, Utah

My views as to how OSHA should set standards in the future arise from contemplating, in particular, the current arsenic and lead standards, both of which have caused ASARCO and the smelting industry in general a great deal of difficulty. I should like to review my criticisms of each of these standards and offer my suggestions to OSHA toward revisions of those standards should they be revised, and toward a better approach to setting future standards.

The lengthy preamble to the Occupational Safety and Health Administration (OSHA) lead standard includes the following statement:

> The signs and symptoms of severe lead intoxication which occur at blood lead levels of 80 μg/100 g and above are well documented. The symptoms of severe lead intoxication are known from studies carried out many years ago and include loss of appetite, metallic taste in the mouth, constipation, nausea, pallor, excessive tiredness, weakness, insomnia, headache, nervous irritability, muscle and joint pains, fine tremors, numbness, dizziness, hyperactivity, and colic.

I would disagree strongly that severe lead intoxication cases at blood lead levels of 80 μg/100 g and even 100 μg/100 g are "well documented."

In all the years that I have reviewed blood lead results among our lead workers and in discussions with our medical director and our plant physicians, I have never heard of any case with symptoms of "severe lead intoxication" such as those described at blood lead levels of 80 μg/100 g, nor have I read of such a case in the literature. In fact, I have knowledge of dozens of instances of individual blood leads well over 100 μg/100 g without any clinical signs of illness. This is not to say that ASARCO thought such blood leads were good. Our goal has always been to keep levels below 80 μg/100 g which I believe to be a safe level provided hemoglobin is in the normal range.

To quote again from the lead standard preamble:

> OSHA concludes that workers exposed to lead leading to blood levels in excess of 40 μg/100 g will develop physiological and pathological changes which will grow progressively worse and increase the risk of more severe disease.

The author of that statement ignored the epidemiological and other evidence that workers with blood lead levels over 40 μg/100 g do not develop significant changes that worsen.

The preamble is replete with slanted statements that paint a terrible picture of the dangers of lead. It is clear that the authors of the two statements I have quoted were novices with respect to the toxic effects or material impairment to health that might be caused by lead.

In view of OSHA's opinion that only blood lead levels below 40 μg/100 g are safe, there is a curious aspect to the lead standard. It is found in the section that covers the requirement for continued removal of the worker from exposure to lead in those instances in which blood lead values do not drop to a specified level within 18 months of reassignment. Following 18 months, a final medical determination can be made that will permit the employee to return to his former job status despite what would otherwise be an unacceptable blood lead level. The employee is no longer subject to automatic removal pursuant to the blood lead levels specified in the standard. Presumably, his blood lead level could rise to 100 μg/100 g without removal, as long as he is, in the opinion of physicians, well.

So here we have a standard requiring removal at lower and lower blood lead levels over a five-year period, only to have the scheme negated *if* blood leads do not fall to a specified lower level within 18 months. Clearly, this is inconsistency. Our own company lead hygiene program was aimed at keeping blood leads below 80 μg/100 g, but the physician helped make the decisions on actual removal from the job. This is also the case in the lead standard proposed for the European Economic Community countries. The decision to remove a man from exposure is made by the medically responsible person and is not based solely on a blood lead value.

So, my first admonition to OSHA is that the professionals involved in standard-setting need to be honest, objective and, most of all, expert in the appraisal of the toxic effects of exposure to a hazardous substance so as not to stack the deck of evidence in support of standards that are unnecessarily stringent.

As to the arsenic standard, I maintain that it also is unnecessarily stringent. The airborne inorganic arsenic limit, regardless of particle size or inorganic chemical form of the arsenic in the particulate, is 10 $\mu g/m^3$ air. Visualize a piece of arsenic smaller than a typewritten period, divided into smaller particles, and dispersed in a cubic meter of air. In the course of a work day, one might inhale about 10 m^3 of air. It is difficult to imagine that a carcinogenic effect results from inhaling such minute quantities of a nonradioactive, natural substance.

Compare the 10-$\mu g/m^3$ arsenic limit with the OSHA 1-ppm limit for vinyl chloride. One ppm amounts to about 2600 μg vinyl chloride/m^3 of air. Compare the arsenic standard to the OSHA-proposed limit (rejected by the Supreme Court) for benzene of about 3200 $\mu g/m^3$ air, or 1 ppm. Surely, arsenic is not 250 or 300 times as potent a carcinogen as vinyl chloride or benzene.

One of the most difficult and litigious problems of OSHA standards is the matter of feasibility. Whenever ASARCO argued that, with the known and affordable technology, levels of 10 $\mu g/m^3$ of arsenic or 50 $\mu g/m^3$ of lead were unattainable in our smelters, we were told by OSHA that the same claims had been made by vinyl chloride producers during the hearings on the vinyl chloride standard and, yet, the standard is being met. That may be, but surely OSHA could see the difference in control problems between vinyl chloride production, which is accomplished in closed vessels in relatively small tonnages, and copper and lead production, which is accomplished in vast works in large tonnages. No engineering consultant, OSHA's or industry's, testified that the current limits for arsenic and lead could be achieved in smelters by feasible engineering controls. Furthermore, evidence was presented during the arsenic standard hearings that ASARCO's study of arsenic exposure and its relation to lung cancer incidence indicated that a threshold of response *did* exist, for in the lowest exposure grouping there was *no* statistically significant excess of lung cancer. However, OSHA could not accept the results of that study in view of its dogmatic "no threshold" policy for carcinogens.

I might point out another inconsistency: OSHA feared the carcinogenicity of arsenic, yet for more than a year after the standard's effective date, it permitted employees with exposures up to 50 $\mu g/m^3$ to dispense with wearing respirators. If, indeed, 10 μg was the safe limit for arsenic, why did OSHA deliberately permit exposures up to 50 $\mu g/m^3$?

In enforcing both the lead and the arsenic standards, OSHA has insisted that permissible exposure limits be attained by engineering controls, even when it was clear from all the evidence that neither standard could be met at all times and at all jobs by affordable controls.

I also advise OSHA to abandon its "no threshold" and "engineering only" policies and, in addition, to permit the employer to be the judge of feasibility of engineering controls. In both the lead and arsenic standards, biological monitoring is the best way to assess effective exposure levels. Why not use these procedures as the bases for performance standards and set reasonable engineering control levels as goals, not as mandatory requirements. It is much simpler to institute engineering controls if suitable technology is available and costs are affordable. However, lacking these, employers need the option of initiating exposure control via respirators combined with biological monitoring.

Finally, I would recommend to OSHA that, in addition to not slanting the evidence and abandoning the no-threshold, engineering-control-only attitudes and allowing the employer to make feasibility judgments, OSHA should heed the advice and wisdom of industries with long experience in controlling occupational health hazards, should avoid assuming an adversary role with employers, and should settle as many disputes as possible with employers at the local or regional level to prevent a jam of cases before administrative law judges and the Review Commission.

An ideal way to reduce the burden on OSHA and employers alike would be to implement an OSHA certification of hygiene programs in plants. This would eliminate the need for periodic plant surveys by OSHA inspectors. OSHA could monitor these programs through periodic review of the air sampling, biomonitoring and medical records of the plant. This could be supplemented by plant tours and visual inspections, but extensive sampling could, for the most part, be eliminated.

CHAPTER 16

CARCINOGENS AND STANDARD-SETTING

Paul Kotin

Health, Safety, and Environment Department
Johns-Manville Corporation
Denver, Colorado

The science and the politics of government regulation reach their highest level of complexity in the setting of standards for carcinogens. This room is full of hazards, including carcinogens. Man never has been and never will be free from exposure to carcinogens. Clearly, however, not everyone in this room is going to die from cancer, either because competing causes of morbidity and mortality will preempt death from cancer or, more commonly, because the carcinogens in this room and in other environmental situations are either biologically unavailable or exposure to them is at a level that will produce no effect.

Despite this fact, science and public policy are somewhat incompatible in the area of regulating carcinogens, and the best example of the clash is the scientific controversy over no-effect levels or thresholds for the action of carcinogens. If a scientific position such as zero threshold is translated into the public policy that regulation and standard setting should have zero risk as their goal and that is to be the exclusive basis for carcinogen regulation, a conference of this sort may be an unwise and perhaps even an unethical expenditure of resources. I say that because the obvious corollary of a zero-risk regulatory posture is that all resources must be committed to research and development of engineering controls and to product design to eliminate all exposure to carcinogens. Eliminating all exposures to carcinogens would, of course, eliminate many of the legal and ethical dilemmas in occupational health.

At the same time, redirecting and refocusing resources would, first of all, obviate the need for expensive, laborious and highly controversial risk analysis and risk-benefit calculations. Second, such vexing and, for the present, unanswerable ethical and moral questions as the dollar value of a human life would be avoided. Third, the process for allocating responsibility for the control of environmental carcinogenic hazards would be clarified. Fourth, and very important, many millions of dollars would be liberated for financing other cancer problems that are now being addressed inadequately; these problems include determining a universally agreed-on method of identifying carcinogens and determining how naturally occurring carcinogens should be handled.

Yet, as we are all aware, existing science and technology, to say nothing of current social values, preclude the setting of standards for carcinogens on a zero-risk basis. The rubric for the regulation of carcinogens is not the simple "presence" of a carcinogenic agent in the environment, for there are numerous carcinogens that are not subject to regulation, tobacco and aflatoxin (a peanut contaminant) being two examples. This reality makes standard-setting for carcinogens a complex process, requiring the synthesis of scientific data, cultural and ethical values, and social policies. With the added dimension of the judicial system in which to resolve disputes, standard-setting for carcinogens has, in fact, become the most time-consuming, costly, and controversial of all regulatory efforts in the past several decades. For numerous reasons, spokespersons for all the constituencies concerned with carcinogen standards have contributed to current controversies.

The dilemmas relating to occupational health, specifically occupational carcinogenesis, are due to the following. First, the end point of effect of an adequate exposure to a carcinogen is cancer, the most dreaded of all human maladies, since it is frequently lingering and often fatal. Cancer is also dreaded because public ignorance of the disease, its causation, symptoms, treatment and so on, is greater than for most other maladies.

Second, relatively recent public awareness of the scope of environmental carcinogens has been combined with public disappointment and even anger that preventive and therapeutic cancer controls continue to elude science and medicine, despite repeated euphoric claims from the medical community of major breakthroughs and the imminent control of the disease.

Third, some constituencies have promulgated the belief that the world consists of two mutually exclusive segments: one, the purveyors and disseminators of environmental carcinogens; and the other, the exposed populations who are targets of the carcinogens. The first segment is categorized as devoid of ethical principles and legality if you will. The second segment is told repeatedly that exposure to any amount of a carcinogenic agent assures the future induction of cancer and that every exposure to a carcinogen is preventable. Failure to prevent public exposure is assumed to be the result of the dereliction of responsibility by somebody—industry or government usually. Little mention is made of the inescapable fact that substances, processes and products meeting important societal needs or intrinsic to voluntary behavior (some of which may be highly destructive) are among the most common sources of environmental carcinogens. Examples include hydrocarbons from motor vehicle exhausts, diagnostic X-rays, tobacco smoke and certain chemical and physical agents societally deemed necessary, such as estrogens used in the treatment of menopause. Cancer, therefore, is not considered within the context of comparative risks, yet it is incompatible with reality to take any element of life, including cancer, and sequester it from the totality of risk to determine regulatory policy.

The fourth aspect of the controversy surrounding standard setting for carcinogens has been the disregard by all parties of the need to rely on judgment evaluations, since the current state of knowledge about cancer in all its aspects, from carcinogen identification and carcinogenic mechanisms through cancer epidemiology and natural history, ultimately up to cancer prevention and treatment, is superficial and riddled with gaps. Clearly, then, scientific data comprise only one of the elements in setting standards for carcinogens. Even if, by some miracle, areas of scientific ignorance were obliterated, society, through its spokesmen, the regulators, would still be faced with the necessity of blending the facts of carcinogenesis with the values of society. Ignorance only compounds the difficulty in addressing this need.

Emphatically, then, the societal value that must guide the regulator is risk. The entire rationale for regulation, including standard-setting for carcinogens, is the existence of a risk. Rowe has defined risk in very flowery terms as "the potential for realization of unwanted consequences of an event," pointing out that assessment of risk entails identification of the cause of the risk, measurement of its effects, determination of risk exposure and definition of the consequences of exposure. I would define risk more simply as "the potential for harm."

To be of use to a regulator and to protect society in a responsible manner, risk must be quantified. Quantitative risk assessment is a statistical process that attempts to use data from epidemiological studies and/or laboratory tests to predict the number of cases or deaths that would result from the exposure of a population to a specific carcinogen or array of carcinogens in different areas.

The most recent assay into quantitative risk assessment was, of course, the HEW document entitled "Estimates of the Fraction of Cancer in the United States Related to Occupational Factors." Despite the great differences I have with the contributors to the document from the National Cancer Institute (NCI), the National Institute for Environmental Health Studies (NIEHS), and the National Institute for Occupational Safety and Health (NIOSH), one cannot overstate the importance of the task they undertook. Even though the estimates were less than completely computed, it would be folly to say that the *idea* of estimating future risk is not important.

Quantitative risk assessment is an essential component of cost effectiveness of cost-benefit calculations, two current buzzwords that mean something different to every individual. What quantitative risk assessment does — and I can assure you I am not as callous as what I am going to say sounds — is that it allows cancer deaths to be translated into value terms, including monetary ones. I believe that once a risk to a carcinogen is quantified, some weighing of risk against cost or benefit is inevitable.

In looking at the essential components of risk assessment, the regulator is faced with a series of data. In the biological universe, he knows that the several classes and categories of carcinogens vary importantly in their chemical and physical properties, their potency, their interspecies effects, their biochemical pathways, and in their initial biological effects and mode of action on the long road to cancer. All of this is relevant to the regulator.

Also relevant is (1) pathogenesis, which is the sequential series of changes from the initial carcinogen/cell interaction to overt production of cancer; (2) morphogenesis, the structural progression of tissue changes ending in cancer; and (3) the natural history of the cancer, which is the clinical course of the cancer. They vary among agents, species and dose levels. In addition, carcinogens vary markedly in their ability to evoke repair mechanisms in cells and tissues.

The importance of these elements to the regulator is that an inevitable progression along any one of these individual paths does not exist. Equally important is that the sequence is dose dependent. Regrettably, dose-relationship studies can be done in the laboratory only under very costly conditions, but where they have been done, such as at the University of Toronto, they have demonstrated dose dependency as a critical factor in each of the sequential steps of evolution of a cancer. Furthermore, they have demonstrated that, up to a point, each of the steps is capable of being a terminus short of cancer.

Other factors relevant to the regulator are that the scope and weight of the biological evidence for carcinogenicity differs; the chemical, physical and product states of carcinogenic agents differ; and exposure experiences in various environmental compartments differ. Still unresolved, though, is the issue of exposure to mixtures of carcinogens of varying percentages or to residual carcinogenic contaminants.

The multidimensional heterogeneity of carcinogens suggests that risk will be different for different agents under different circumstances of exposure. It almost forces individual assessment at times. But let me immediately put to rest the implication that I or anyone can sanely believe that every potential carcinogen can be individually tested in a manner fully addressing societal pressures and needs. The most comprehensive effort to address this complexity has been that of the Occupational Safety and Health Administration (OSHA) in developing a generic standard for cancer. Despite some highly questionable science, which in some parts is even humorous, the document published in the *Federal Register* is of great value as a starting point for discussion.

The problems facing the regulator are further compounded by the realization that, for each of the preceding variables, the level of knowledge varies significantly. Therefore, there exists a hierarchy of levels of proof, each level having an index of certainty that can be of utility to the regulator.

For some carcinogens, there is a full array of human epidemiological data, animal data and biochemical data on mechanisms of action, and the agents can be quantified as well as identified. The data are concordant in these cases, and the evidence of carcinogenicity is conclusive. Polynuclear aromatic hydrocarbons, asbestos and aromatic amines would fit into this category. In other cases, even though high-risk human environments have been identified, there is ignorance of or great uncertainty as to the agents responsible for the high risk. Because of the uncertainty, the observations cannot be verified by laboratory work. I think the woodworking industry and perhaps the rubber industry would be examples.

Still other situations exist in which animal bioassay studies reveal carcinogenicity, but human epidemiological studies fail to confirm an analogous human effect. The use of polymeric or metallic prostheses would qualify for this category, as would beryllium, but more controversially. Another area consists of those instances in which there are negative animal data but adequate and valid human epidemiological studies have identified a risk. I really could not care whether or not an animal ever gets cancer from arsenic or a mouse ever gets leukemia from benzene, if the human data demonstrate a risk. However, to regulatory agencies, the lack of confirming animal data constitutes a point of vulnerability.

Finally, there is a large area of positive animal data showing carcinogenicity, but corresponding human epidemiological studies are inadequate or, more frequently, have not been done at all. Examples of this area would be carbon tetrachloride, DDT, heptachlor, chlordane and Mirex. So the animal data alone are all that the regulator has to go on, even with all the limitations of that type of data.

I have described, obviously, a less-than-ideal situation in the scientific or biological component of risk analysis. As someone who has devoted his whole professional life to carcinogenesis, I am terribly disappointed that in 1981 there is not much more to tell a regulator than a competent person in the field could have told that same regulator in 1951. So how can the regulator use this limited information in reviewing the socioeconomic and ethical dimensions of risk analysis? Scientific, social, political and economic practitioners and theorists have developed several techniques for approaching risk analysis, one of which, zero-risk advocacy, I have already discussed. Since the complete elimination of risk is neither achievable nor practical, a more common approach has been the development of a calculus for risk assessment. Wilson at Harvard and Rowe at American University, among others, have described techniques for determining relative risk and have actually calculated risk comparisons and have published them in long tables with numerical values for the relative risk of most human endeavors and experiences, including exposure to carcinogens. The importance of such calculations is in the use of "comparative" as the operative word, in recognition of other environmental hazards no less threatening to life than cancer.

Risk assessment analysis has been challenged as being a tool for those intent on retarding progress in the establishment of any health standards, not just standards for carcinogens. In the absence, though, of efforts to evaluate the probability of human exposure, the size of the population potentially exposed, the circumstances in which exposure occurs, the intensity of exposure, the potency of the agent, the effect of cofactors, etc., regulating carcinogens really becomes a vehicle for the expression of a particular social philosophy. Values rather than facts become the primary determinant, and this is as true for attitudes toward speeding in a car as it is for attitudes toward carcinogens. Several examples of contrasting societal attitudes to carcinogens emphasize the political, social and economic forces at work so that the "facts" of carcinogenesis are only one component of decision-making and regulation.

Cigarette smoke is not only the most important cancer risk so far identified for man; it is, moreover, the most ubiquitous and thoroughly documented carcinogenic agent. The role of tobacco, both quantitatively and qualitatively, in causing cancer can be seen in the fact that it is believed to be a prime factor in cancer of the buccal cavity and pharynx, of the esophagus and larynx, and of the bladder, in addition to its role in the causation of lung cancer. With respect to lung cancer, it has been noted that more than 20% of *all* cancer deaths in the United States are caused by smoking and, if no one smoked, 81% of lung cancers could be prevented.

Despite the persistent assertion on the part of the tobacco industry that "the case against cigarettes" has yet to be proven, epidemiological data and experimental laboratory data are concordant and conclusive, and cigarettes unquestionably belong in the category of potent carcinogenic agents. Yet, regulation of the hazard has been limited to a pallid warning posted on the product and in advertisements. The ultimate absurdity is that tobacco is not only excluded from regulatory health legislation, but it is simultaneously subsidized as a crop to ensure its continuation in commerce. One can legitimately conclude that politicoeconomic forces, rather than scientific data, represent the major elements in government policy and action in this case.

Another case is that of aflatoxin B1, a product of the mold *Aspergillus flavus,* a carcinogenic contaminant of an array of crops used for human and animal consumption, most notably peanuts and corn. Aflatoxin has been verified as a human carcinogen in studies in Africa where liver cancer is highly endemic. Animal studies have also verified its carcinogenic properties. Human exposure in the United States occurs directly through the ingestion of peanuts and indirectly through consumption of milk from cows fed contaminated corn. Not only is aflatoxin carcinogenic, it is recognized as the most potent carcinogenic agent known. Despite this, the agent is approved for human consumption and there are no warnings as to carcinogenic risk on the products in which it occurs.

Saccharin is an example of an agent identified as a carcinogen through feeding studies in laboratory animals, in this case rats. Human studies are either nonexistent or inadequate. The data and conclusions from the animal studies have raised all the traditional concerns associated with the experimental demonstration of carcinogenicity, such as the significance of lifetime feeding studies, the ability to extrapolate from high dose to low dose and from experimental animal species to man, the criteria for potency determination, the significance of increased incidence of background tumor occurrence in experimental animals, and so on. Despite the inconclusiveness of the research, a ban on the use of this agent was clearly mandated by the Delaney Amendment. Because of public outcry, however, since some constituencies claim health benefits from the sugar substitute, Congress intervened in the proposed regulatory action. The current status of this agent is that a warning label is required on saccharin-containing products. It is of particular interest that saccharin has given rise to a new dimension in attitudes by the public toward environmental carcinogens. In this case, the public's attitude was "tell us the risk and let us decide," a definite conflict with the Delaney Amendment as it currently stands.

I have emphasized the complexities of carcinogens because the "simplicities" generate little or no confrontation. To sum up:

1. Carcinogens must be regulated, and regulated at the lowest possible, achievable or desirable level of risk. All three adjectives have one thing in common: they are surrogates for risk analysis and cost-benefit calculations. They also mean that consideration needs to be given to the feasibility of proposed regulations. All too often delays in effective dates of legislation or resorting to litigation have been, in large measure, the result of technical inability to comply with regulatory demands, not just the result of political pressures by industry. In fact, regulators themselves have frequently bemoaned the impossibility of achieving technical targets or date commitments contained within legislation.

2. All constituencies should be involved in the decision-making efforts leading to carcinogen standards. This is in no way meant to dilute government's role in regulating carcinogens in the environment. Rather, in carrying out their responsibilities, government regulatory agencies must avoid simplistic generalizations about carcinogens and their control and strive, instead, to find a delicate balance between scientific knowledge and public will. In the area of science, the major element should be the determination of the impact on morbidity and mortality of an environmental carcinogen. Prevention can be successful only if it has a foundation of accurate knowledge. Government, then, must be a generator of data, either through research undertaken by such government facilities as NCI, NIEHS and NIOSH, or by contractual arrangements with academic and independent research communities. Next, government must be an evaluator of the data accumulated, taking into account all the caveats and limitations both of epidemiological research and experimental laboratory data, and establishing the quality and quantity of data necessary to determine the carcinogenicity of an agent. Thus, government agencies must establish standardized methods for determining what, if any, research is needed, for designing and implementing the necessary studies, and for analyzing the data and the conclusions drawn. Once the body of scientific knowledge is reviewed, evaluated and categorized according to criteria uniformly applied to each suspect agent, it is possible for the regulator to address the other component in formulating regulations, that of public policy. The use of advisory committees composed of representatives from all concerned parties (government, labor, industry, consumer and environmental groups) is an appropriate approach in this respect.

3. Regulatory agencies have a special responsibility for educating Congress and the public, both as to what is achievable under the current state-of-the-art and as to the necessity for sufficient support in order to recruit and retain competent career staff, thus ensuring the highest level of quality in agency programs.

4. Regulations must have both credibility and substance, so that public acceptance is possible, even if there is no public agreement on all aspects of the regulations. In addition, failure to comply with a standard should have a clear set of substantive penalties, making it difficult not to comply.

5. Government agencies must be prepared to reexamine their positions on the basis of newly generated data and must also evolve mechanisms for the dissemination of the less exciting, though equally important, data on those agents in the environment that represent no hazard to the public.

CHAPTER 17

COST-BENEFIT, SOCIAL VALUES AND THE SETTING OF OCCUPATIONAL HEALTH STANDARDS

Marvin A. Schneiderman, PhD

Clement Associates, Inc.
Washington DC

No matter wither the constitution follows th' flag or not, th' Supreme Court follows the iliction returns.

Finley Peter Dunne [1]

Standard-setting, like the Supreme Court, in this country does follow the election returns. I think this is a fact of life. The fact that Finley Peter Dunne saw this at the turn of the century says to me that there is no new wisdom. I find the complexities and problems referred to in the last chapter are complexities that have perplexed me too. I see the new administration placing emphasis on different complexities. I see it more concerned with economic costs and less concerned with health costs than am I. I will consider the various kinds of costs, risks and benefits later.

Not too long ago, I heard a physician who is the safety and health director for a very large company—a company, by the way, that has claimed to have a very good safety record—talk about what he would like to see as a basis for health and safety standards. He said he would like to see performance standards. Performance standards are common in public health administration. They are the last stage of the kinds of things that public health administrators talk about. We talk about structure, process and outcome. Performance standards equal outcome. We want to see standards that lead to a healthful outcome.

This particular health director was very critical of most of the regulatory activities of the federal government. In particular, he was critical of the U.S. Environmental Protection Agency (EPA) for establishing mileage requirements for automobiles that had to be reached by a certain time (e.g., 27 mpg by 1982). I asked him, "Doctor, you are an advocate of performance standards, but you are critical of the EPA for requiring that automobiles reach 27 miles per gallon by 1972, or something of that sort. Isn't 27 miles per gallon by 1982 a performance standard?" He could not see that as a performance standard because of the time constraint. At first I couldn't see that he couldn't see it. What I finally realized was that he and I gave different meanings to the same words. He saw X miles per gallon as a performance standard but, to him, the requirement that this standard had to be achieved by year Y was not appropriate. To me, it was not a standard at all unless there was some time by which it had to be achieved.

Now, I must give you my definition of a performance standard. It is a measurable goal to be achieved in a finite time. I must have the finite time because I want to know when I should measure whether I have achieved my goal. The standard needs to be measurable in order for me to be able to measure it at that time. Sometimes a performance standard has so distant a horizon that I must have intermediate standards so that I might test whether I am on my way toward reaching the goal. For example, say that my performance standard for asbestos workers is: a cigarette-smoking asbestos worker shall have, by the year 2010, the same death rate from lung cancer as, say, a cigarette-smoking university professor or asbestos company vice president. I have to set my goal for the year 2010 (or later) because the effects of anything done now will take at least until then to be fully felt or seen. So, I must have intermediate standards.

Intermediate standards can be performance standards, too, but of a different kind. My intermediate standards may be the achievement of 12 asbestos fibers per unit volume of air in the workplace by year Y, going down to 1 fiber by year Y + 10. Now in setting these standards, I am making some assumptions. I am assuming that illness is dose-related. I am assuming that one particle at year Y + 10 will achieve the health (or illness) levels I wish to achieve by year 2010. My assumptions can be wrong.

Now, back to my company health director. That he did not have a time scale does not make him a villain. That he did not see the need for intermediate standards does not make him a villain. I am going to assume no villains. That may not make for good drama, but it may make for improved health. And while not looking for villains, I also want to stay out of peoples' minds and motives (and I want the same consideration). The reason I should not want to look into anyone's motives is a simple one. If I look into your motives and I am right in my assessment and I uncover your hidden motives, you are going to be angry with me because I have uncovered something you preferred to hide. However, if I look into your motives and I am wrong and I assert that you have a given motive that you do not have, you are going to be angry with me because I implied something about you that was not true. So, if I make you angry with me — no matter whether I am right or wrong — it seems to me there must be something wrong with the process. Making you angry is wrong because it only leads to a cutoff of communication. If you are angry with me, you will not talk to me. So I think the search for motives and the ascribing of motives is a way to cut off communication. If we want to develop appropriate health standards, we must talk with each other.

Dr. Kotin discussed the problem of identifying carcinogens; what one does with them once identified, the difficulty with the animal data, and the difficulty with epidemiological data. People generally assume that the best data are the epidemiologic data because they refer to humans. That is not a bad assumption, but it sometimes misleads. Karstadt [2] of Mt. Sinai in New York, put together data for a recent meeting at the Banbury Center in Cold Spring Harbor, New York.

Dr. Karstadt looked at the work that the International Agency for Research on Cancer (IARC) had done in identifying carcinogens. IARC reviewed 442 materials in the first 20-odd volumes that were published in its reviews of materials tested for carcinogenesis. These materials are not a random selection of materials that exist in the universe. They were tested because someone suspected they might be carcinogenic, or they looked like carcinogens, or they were in extensive use, or something of that sort. Of the 442 materials that were reviewed, there was sufficient information on 142 for the IARC to conclude that they were animal carcinogens. IARC has a formal, published definition as to what constitutes "sufficient" information to declare a material a carcinogen. Of these 142 materials, information on human exposure was available for 60. In some cases, "some information" meant case reports or a little exposure information but not very extensive information. Of those 60, a scientific review group for IARC found that there was sufficient information on 18 to indicate that these materials (or processes) were definitely human carcinogens. There were 18 others that they indicated were "likely to be" human carcinogens. These were mostly materials for which there was somewhat less information. For an additional 18, the human information was so inadequate as to preclude any conclusions. I do not know why the remaining six were not reviewed.

Now, that meant that on 82 materials shown to be carcinogens in animal testing, as reviewed by an international expert committee, there were no human data whatsoever. Dr. Karstadt reviewed the whole list of 442 materials and found that about 70 of them were in active commerce in the United States. (There was some overlap of these 70 and the 60 reviewed by the IARC.) She wrote to all the companies who were making or using these materials and asked them what they had done or were attempting to do with respect to determining whether these materials might be human carcinogens. She received a very good response from the companies. Epidemiological studies were underway on about six of these materials. For eight others, the companies reported completed epidemiologic research. So, for 20% of known animal carcinogens in active commerce in the United States, there were additional data on the possible effects on humans. There was some information for 13 additional materials, but it was not of high quality. In almost every other case (60% of the materials), there were legitimate reasons why these materials were not being studied in humans: small group of employees were at work with the material; the material was being used in a wholly contained system; the material had been used last year and was not being used now; the material was being phased out. This is not a condemnation of the companies who were using these materials. If there is good reason for not obtaining human epidemiological data, then we will have to depend on the animal data for setting standards to reduce risks to exposed humans.

Now, to return to the performance standard suggested earlier, Figure 1 [3] shows different occupations and the probability of death before age 65 for persons in these different occupations. These are British data. We do not have equivalent data for the United States. The vertical scale gives the probability of death before age 65 for English and Welsh males alive at the age of 15. Quite obviously, the best profession to be in is teaching. Teachers have the lowest expected mortality (except for teachers of music). Only 16% would be expected to die before age 65. If our courts ruled that pensions should be proportional to life expectation (which they have not), then retired teachers should be getting lower pensions than they are now getting because they are going to live longer than anybody else. In fact current economic thinking would argue that teachers are going to bankrupt the Teacher's Insurance Annuity Association (TIAA) and the Social Security system. To avoid this, we should have fewer teachers, lower pensions for teachers, or else we should make their jobs more hazardous. I don't like any of those "solutions."

The next group of people who do very well are managers and public officials. Among public health inspectors, only about 20% are expected to die before the age of 65. It was expected that 28.8% of all workers

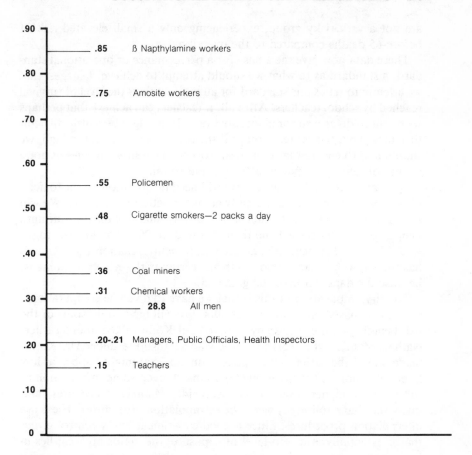

Figure 1.   Probability of death by age 65: England and Wales, 1971. Men age 15 by occupation and exposure [3].

would die before age 65. This is an important number because approximately half of all cancers in men occur after age 65, and 70% of the men who live long enough to ever go to work can be expected to live past 65.

The highest risk group that Roderick looked at were the beta-napthylamine workers, of whom 85% would be expected to die before the age of 65. The next highest risk group was amosite asbestos workers, then policemen. As you might expect, a policeman's lot is not a happy one. Policemen (in the United States) do get shot at and, in Great Britain they have other problems. Of the two-pack-a-day cigarette smokers, it is expected that 48% would be dead by age 65. Coal miners have excessive risks, too, as do sailors and members of the military. Chemical workers are not a very risky group, experiencing only a small elevated risk of before-65 deaths compared to the overall average.

These data now give me a base for a performance or operational standard, a standard as to what we should attempt to achieve. I suggest that we attempt to set as our standard for all male workers the level of survival reached by school teachers. After all, if teachers can achieve that perhaps we ought to do as well for all occupations. That might be a little difficult to achieve and, therefore, perhaps if the teacher standard is too high, we should aim to ensure that industrial workers have survival rates as good as that of managers, public officials, and so on.

If we are going to look ahead, we will need information on the toxicity of materials before we can possibly get information on humans. We will need this for two reasons. First, it is ethically unacceptable to wait until people have been exposed and then followed for 20, 30 or 40 years to see what illnesses they develop; and second, in many situations, as Karstadt has shown, we will never know with any certainty the effects on humans, because the data can never be gathered.

The first implication of this is that we have to do some computations, most likely based on laboratory results. We will have to do some of the risk-benefit or pure-risk analysis that Paul Kotin talked about earlier. Nathan Mantel, who worked at the Cancer Institute under Dr. Kotin, made one of the earliest attempts to estimate what might happen at low levels of exposure (of experimental animals) given some information at substantially higher levels of exposure [4]. Mantel's procedures, and many that have followed, were not extrapolation procedures. These are interpolation procedures. Since a good experiment has a control group that is essentially at a zero level of exposure, the major problem lies in knowing the shape of the response curve in the region between the zero dose and the higher doses. Since 1961, when Mantel and Bryan published their paper [4], they have been challenged from two sides. The first challenges came from people who argued that the standards for safety resulting from the Mantel-Bryan procedure require doses that are much too low. That is, it is argued that it is not possible to achieve those levels. They argued that Mantel-Bryan was asking for essentially zero exposure. One can ask for zero exposure, but one is not going to get it. The argument then was "If you are not going to get it, why ask for it?"

So at first Mantel-Bryan was challenged as being too rigid. Soon, however, Mantel-Bryan was also attacked as not reflecting known biology and, hence, as not being rigid enough. Richard Peto of Oxford was among the people who objected on these grounds. Peto has written a thoughtful overview on the multistage nature of cancer that leads to the conclusion that Mantel-Bryan underestimates risks at low doses [5].

Nicholson [6] summarized some of the issues in asbestos exposure (Figure 2). His dose measure is duration of exposure. The vertical scale gives relative risk. The graph shows this risk in relation to the duration of

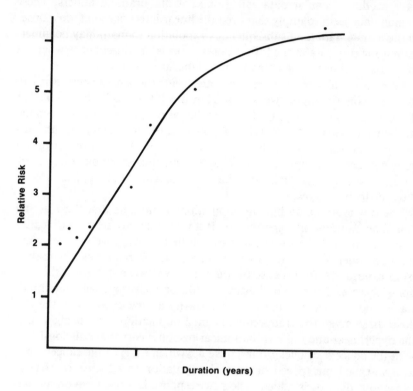

Duration (years)

Figure 2.   Asbestos workers' (amosite-exposed) relative risk to lung cancer vs duration of exposure [6].

exposure. The figure shows that for some relatively short durations of exposure we get what appears to be a linear response (duration perhaps being a measure of exposure). However, the risks from the lowest exposures of the shortest durations sit well above the linear curve, implying greatest risks per unit dose at the very lowest doses. The Mantel-Bryan model, however, implies just the opposite. It implies risks at low doses below what the linear curve would suggest. The Nicholson data would imply that Mantel-Bryan were not rigid enough.

Nicholson's data are not the only data that show this kind of unexpectedly high response at low doses. Dr. Ashford referred to this phenomenon in his chapter of this book when he described the "canaries in the workplace"—people who respond to very low levels of exposure and who often leave the job after only a short period of employment. These "early leavers" may be the ones that are at highest risk. Archer [7] appears to have produced similar data with respect to the uranium miners. These human data seem to imply that even the linear interpolation to low doses in the animal data (and subsequent extrapolation to man) may be underestimating response at the lowest doses. This is of consequence when we make risk estimates for humans from laboratory data.

Mancuso et al. appear to say they are seeing the same hypersensitivity phenomenon among the workers at Hanford [8]. Land [9] has looked at those data very carefully. He finds that he cannot distinguish between the straight-line curve, the curvilinear downward or curvilinear upward response curves. That is, the data, even after we have information on large numbers of people, are inadequate to distinguish among these alternatives. Nonetheless, it is important to recognize the implications of these different curves.

The observations on human populations remind us that these are all data from nonrandom experiments. If it is true that we see hyperlinearity at low duration of exposure, it may well be that what we are seeing are the consequences of the most sensitive persons removing themselves from exposure early because they are most sensitive and are still developing disease. Thus, hyperlinearity could be consistent with the multistage model of cancer that implies linearity at low doses—had the low doses been given to a randomly assigned population. The straight line, the hyperlinear and the less-than-linear response curves are all consistent with the concept of background dose d_b, which leads to the cancers seen in a so-called unexposed or control population to which the effective exogenous dose d_e is added. The new response is a response on a dose-response curve for which the dose is $d_b + d_e$, i.e., the response is "additive in dose." What is not consistent with additivity in dose is the concept of threshold. Threshold must assume that every new material operates in a unique manner and produces cancers of a kind that have all been seen before by a mechanism that has never occurred before. The issue of threshold in risk assessment is important, for if there are thresholds, then there must be some (low) dose below which there is absolute safety—for all persons.

Figure 3 [10] gives the distribution of work experience, time on the job of Dow Company workers. Dow has a U-shaped distribution of the duration of employment of workers. For short durations, there is a "burn-in" phenomenon. Some people who come to work quickly find the job distasteful or not what they thought it would be and, so they leave quickly. These are the ones represented on the left side of the figure.

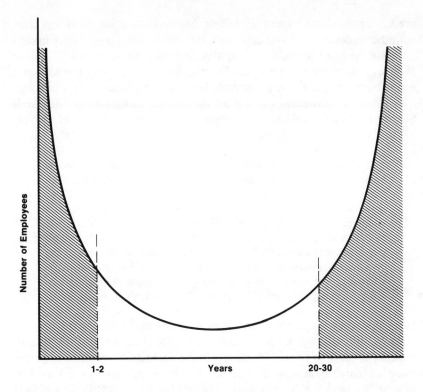

Figure 3.   Duration of employment, example [10].

They may also be the "canaries," the early-warning people. Perhaps they are saying, "Oh, this job is for the birds. It makes me sick. I am getting out of here." These could be people of higher sensitivity, people who really cannot stand the job because of health reasons and who leave it early.

The right side of the curve is composed of people who made a career of working at Dow. These are the ones who are usually found in the retirement group. They are the ones you can follow for 25, 30 or 35 years. Most of the epidemiologic studies that we do are based on these people, and they exclude the canaries. If epidemiologic studies look only at people who have been exposed two years or more, something of that sort, they disregard the early leavers. Employed populations are not random samples; these are not random fallings out. These worker populations are self-selected for duration of employment. If the job makes you sick, you are not likely to stay. Some of the statistics that we use that assume randomness, as in dose-response computations in experiments, are not applicable to these situations. So we have some problems then with the epidemiology, and extrapolating from mouse to man may underestimate the risks to highly susceptible humans, the "canaries."

Now, I must turn to cost-benefit considerations, having just talked a little about how hard it is to estimate benefits, which here is risk averted.

A Brookings economist at a meeting that the Congressional Research Service held at the Library of Congress on risk assessment told the following story on himself. He said,

> You know, I've been looking at risk and the risk assessments people have done and at their cost-benefit analyses. I know, through birth and early training, that this was the best of all possible worlds. However, after having looked at about a dozen cost-benefit analyses where people had proposed something be done (mostly with respect to health effects), I discovered that not only was this the best of all possible worlds, but that it was also a perfect world. Not one of the cost-benefit assessments indicated that it was cost-beneficial to do what was proposed. The estimated cost of doing something was always too high compared to the possible benefits gained.

Does this story imply some problems in cost-benefit analysis?

I was involved, when I was still at the National Institutes of Health, in looking at some of the health effects of the accident at the nuclear power plant at Three Mile Island. I think, by the way, there were very few deleterious health effects, except perhaps for the psychological effects and worries and concerns. However, in terms of immediate measurable physical effects, I think there were very few. Having read the *New Yorker* series on Three Mile Island [11], I began to think of the contrast between what went on at Three Mile Island and what went on following the release of the safety valve at that chemical plant in Seveso, Italy. In Seveso, the engineering design was such that when the pressure and temperature reached a certain level, the safety valve blew. The stuff in the reactors spewed over the valley. Much of it was contaminated with dioxin, a very toxic material. It may very well be that it will not be possible to live in the portion of the valley just north of Milan for some time. Birds and small animals have died. Women have had abortions out of fear of producing deformed children. Chloracne has developed. Cancer is feared.

So in Milan we had an engineering design that saved the plant but may have led to some substantial damage to the community. Now, look at Three Mile Island (TMI). We had a partial meltdown, ruin of the core of the reactor, and several million gallons of highly radioactive water poured into a containment building. There was some minor release of radioactivity to the outside, but total damage to the outside community was very minor. The engineering standards to which TMI was built included community concerns, while the ones at Seveso did not. It costs a lot of money to build a nuclear power plant to those good engineering standards, to include a containment building and backup systems so that even three or four human errors and some poor plumbing didn't lead to a meltdown. It costs a lot of money to do that.

What did we buy at that cost? I ask this question because I must know the answer if I want to know how to do a cost-benefit analysis with respect to nuclear power. I think we bought the health and lives of a very large number of people in that area, plus some of us downwind in Washington, Baltimore and Philadelphia. We bought the fact that had we really not had all that engineering safety design built in we may very well have killed a substantial portion of the population in the area. A TMI with many fatalities certainly would have also killed the nuclear industry. The nuclear industry is pretty sick at the moment, and a seriously fatal accident would no doubt have killed it. One could not be a Teller and say "The only serious injury at Three Mile Island was to me. I had a heart attack."

What else did we buy? We may have bought 20 years of less dependence on foreign oil. For the remainder of this century, we are gong to have a nuclear industry in this country, not a terribly large one, but one that is going to produce some energy. So we are not going to have quite the same dependency that we might have had on imported oil. If we don't have that same dependence on imported oil, we may not get into a war in the Persian Gulf to protect our oil supplies. On July 5, 1981, the nuclear power industry published a two-page advertisement in the Sunday Magazine section of the *New York Times* making these same arguments. They look to a permanent nuclear power industry, while I look toward a 20-year "window" during which time I hope we will be able to develop alternative renewable sources of energy. Nonetheless, they are now advertising (At whose cost? To whose benefit?) that nuclear power will keep our sons out of a war. If literally not completely demolishing the nuclear industry leads to saving of the nuclear industry, which then leads to our not being too dependent on the Persian Gulf oil, which leads, in turn, to our not having to have a strong military posture in the Persian Gulf, which may lead to not having a war, we must put all that into our benefit computations. Those are benefits derived from the extra costs of making that a safer plant than had it been built according to Seveso-like standards. I do not know how to do that benefit computation, by the way. How much is an averted war worth?

Recent talk that I have heard about how to set standards leads me to one further concern. The suggestions are now being made that standards be set "for the good of society." I think that says if, in total, most people benefit, it is okay for some people to be hurt. I do not accept that. My advice is to be wary of people who tell you they are going to do something for the good of society. It is rarely for the good of all society. I come from the religious and ethnic group who were part of the "final solution" that was going to do something for the good of society. Hitler was going to do something for the good of society and for the world by wiping us out. He was going to create the final solution. (Solutions are assumed to convey benefits—especially final solutions.) That solution was not of much benefit to me, my coreligionists or my family.

The United States is not a uniform, homogeneous society. If we do something for the good of somebody, it is a morally acceptable good if it does not lead to a bad for somebody else. Keep an eye on the guy who is going to do something for the good of all society. I have more trust and am much happier with someone who says he is going to do good for himself and his friends and his associates. At least there is an honest man. And I want to deal with him. I want to do good for myself and my friends, and I have a very large ego that encompasses a lot of people. I have a hard time convincing myself that it is necessary that some workers give up their health in order that businesses and/or other workers may prosper.

Figure 4 is a risk-cost-benefit matrix. It is a messy matrix, or at least it deals with messy ideas. The figure shows three rows and three columns labeled R (risk), C (cost) and B (benefit).

There are nine combinations of these shown in this figure. For every combination of two items, there are constraints imposed by the other. For example, one can have a cost-cost situation. That is where you say I want to achieve a given benefit (or reduce risks to a given level) and now you want to find which of the costs is lowest to achieve that benefit. Or, one can have a benefit-benefit situation. That is where you will specify a maximum cost, and the problem is then to see which course of action will buy the greatest benefit. That's like saying, "I'm not going to spend more than $1.98 on this. How much benefit can I buy for $1.98?" For every combination, one item (cost, risk, benefit) can be fixed, and one can work on the other two. To do this, of course, requires that one know a great deal about the consequences of all actions. By and large, we do not know these. In most of the market situations in which the economists operate, we often assume full knowledge, complete mobility and the opportunity to make certain kinds of changes that really do not exist. Hence, the matric implies some messy muddling about.

	Costs	Risks	Benefits
Costs	C	C	C
Risks	R	R	R
Benefits	B	B	B

Problem: Which one to fix and which two to manipulate?

Figure 4. Costs, risks and benefits – a messy matrix.

There is a final remark that I wish to make concerning industrial safety, health, and the choice of a job. Recall that in Roderick's data [3], teachers had the best survival rate. I was taught in my economics classes that education bought higher incomes. My teachers told me that more highly educated people got more money. But that did not conform to my own observations. My next-door neighbor is a liquor dealer. I have a PhD. He is a graduate from high school and his income is substantially higher than mine. And, of course, the income-education correlation is not unity. It does not always hold. There are other forces at work. At last year's meeting of the American Statistical Association's Social Statistics Section, a series of papers looked into what more education did buy. Even on the average it does not buy much higher income. What it often does buy is a less hazardous job.

In my county, school teachers and policemen get just about the same pay. School teachers are all college graduates. I doubt if very many policemen are. By and large, what we are getting in the social classes differentiation that comes out of education are safer jobs. Recall Figure 1 (Roderick's data). When one looks at these data and asks what occupations are associated with more education, we find these are also the occupations with lowest risk. There is, in the United States, an inverse correlation of education – and risky jobs – with cigarette smoking. Cigarette smokers are disproportionately represented in the higher risk occupations and have less education on the average. This is not the time or place to discuss research that seems to be social-class–oriented — with more concern for the ills of the social class that the researcher belongs to — but I would hope that some sociologists (and others) might look into this. Are any of the "way-of-life" theoreticians in cancer causation at high risk as a consequence of industrial (or other) job exposure?

SUMMARY

The data presented here lead me to three conclusions:

1. We cannot rely on epidemiology alone to provide us with the data necessary to protect people's health. We need to rely heavily on laboratory data. To make these data more meaningful, we need to do more basic research, which will teach us more about mechanisms which, in turn, will lead us to better mathematical models of the diseases we are interested in. The epidemiology contains strong hints that the linear dose-response models may understate risks at low exposures. This is a disturbing concept.

2. Cost-benefit is still an infant art. Current proposals to use cost-risk-benefit analysis is, in my mind, pushing the technology. There have always been external forces pushing the technology with respect to a knowledge field. The field is not new. It has been around since Jeremy Bentham. If, in 200 years, it has grown only to infancy, there must be some serious growth-inhibiting factors operating. Perhaps people like Mantel, Lave, Zeckhauser and others can find the hormones to make this field grow to maturity more quickly. In the interim, it strikes me as less than fully rational to use infant ideas to make adult decisions. In the interim, we must use best available technology to reduce risk. I don't think we achieve zero risk, but we can certainly keep it in mind as a goal. Perhaps we should look for "virtually safe" levels in the manner of Mantel and Bryan in which they define virtual safety as a risk of 1×10^{-8}.

Wilson [12] has suggested a lifetime risk of 1×10^{-6} as an appropriate "safe" level. Adding carcinogens to our environment at a dose that gives lifetime risk for each new carcinogen of 1×10^{-6} would increase cancer incidence in this country each year a bit more than the total annual number of acute lymphocytic leukemia cases now seen in children. To me, that is too much to add. I have suggested here that the risks for all workers should be equivalent to the ones in the white collar professions. Those are obviously achievable (for some segments of the population) performance standards. I think it will be hard to achieve for all segments of the population as long as the white collar people think it is okay for blue collar people (the lower classes) to be at higher risk.

A labor leader once asked me, "Why is it or why should it be that lawyers, doctors and statisticians, newspaper reporters, and so on who are cigarette smokers have so much lower lung cancer rates than the guys who work in the industry that I represent?" And the answer is I don't think it should be, and it doesn't mean I think doctors, lawyers, etc. ought to be at higher risks. I think other people ought to be at lower risks than they are now.

3. To achieve the risk levels that are desirable, there need to be intermediate, short-term, measurable standards that can help tell us if we are really moving toward our long-term goals of a society safe for all of us. Man must not be wolf on man. (The etching by Georges Rouault that depicts this state is in the Fogg Museum of Harvard University.)

DISCLAIMER

The opinions expressed here are those of the author, and are not necessarily those of Clement Associates, Inc.

REFERENCES

1. Dunne FP: The Supreme Court's decisions. In: Mr. Dooley in peace and war. 1898.
2. Karstadt M, Bobal R, Selikoff IJ: Availability of epidemiologic data on humans exposed to animal carcinogens. In: Peto R, Schneiderman M, eds. Danbury Report 9: Quantifications of occupational cancer. Huntington: Cold Spring Harbor Lab, 1981.
3. Roderick H, Sussex University, England. Personal communication.
4. Mantel N, Bryan WR: Safety testing of carcinogenic agents. J Ntl Cancer Inst 27:455–470, 1961.
5. Peto R, Roe FJC, Lee PN, Levy L, Clack J: Cancer and aging in mice and men. Brit J Cancer 32:411–426, (1975).
6. Nicholson WJ, Perkel G, Selikoff IJ, Seidman, H: Cancer from occupational asbestos exposure: Projections 1980–2000. In: Peto R, Schneiderman M, eds. Danbury Report 9: Quantifications of occupational cancer. Huntington: Cold Spring Harbor Lab, 1981.
7. Archer VE: Health concerns in uranium mining and milling. JOM 23:7: 502–505, (1981).

8. Mancuso TF, Stewart A, Kneale G: Radiation exposures of Hanford workers dying from cancer and other causes. Hlth Physics 33:5:369–384.
9. Land CE: Estimating cancer risks for low doses of ionizing radiation. Science 209:1197–1203, (1980).
10. Hoerger F: Indicators of exposure trends. In: Peto R, Schneiderman M, eds. Danbury Report 9: Quantifications of occupational cancer. Huntington: Cold Spring Harbor Lab, 1981.
11. Ford D: A reporter at large (Three Mile Island Parts I and II). New Yorker April 6 1981:49–120; April 13 1981:46–109.
12. Wilson R: Analyzing the risks of life. Tech Rev 81:4:40–46, (1981).

CHAPTER 18

A LABOR PERSPECTIVE
ON STANDARD-SETTING FOR CARCINOGENS

Stanley W. Eller
Workers' Institute for Safety and Health
Dayton, Ohio

Cancer has reached epidemic proportions in the United States. More than 55 million Americans now living will eventually develop cancer — one in four Americans, according to current rates. A report issued by the American Cancer Society estimated that 765,000 people would be diagnosed as having cancer in 1979 alone [1]. The report noted that 395,000 persons were expected to die that year from cancer — more than 1000 individuals every day.

Cancer is the second leading cause of death in the United States, exceeded only by heart disease. It has become a part of the national conscience. Regular press reports and television coverage and the fact that cancer will strike two out of every three families have brought cancer into everyone's living room. This attention has frequently added as much confusion to the public's understanding of cancer and its causes as it has been useful in clarifying the likelihood of a cure. The publicity that surrounded the recent decision to withdraw saccharin from the market is probably one of the best examples of confusing the public about the risks of cancer. The confusion began when the Food and Drug Administration (FDA) first announced that saccharin was being banned, based on the results of animal studies, which showed that high doses of saccharin caused tumors in rats. The public, however, was not informed that the FDA regulatory action was required under the Delaney Amendment. This amendment states that any substance shown to have carcinogenic potential must be banned from human consumption. The scientific debate that followed the ban focused instead on the likelihood of an individual ever consuming enough soda pop to equal the dosage of saccharin that caused the tumors in the rats. The beneficial uses of saccharin to the diabetic and in controlling obesity were also used as arguments to discredit the FDA's regulatory action. In addition, severe doubts were cast about the ability to extrapolate from animal studies to humans, which also served to create public cynicism. The scientific community did little to assist the nonscientist to understand the basic issues. With scientific opinion sharply divided over the safety of saccharin and the need to regulate its consumption and with only limited awareness of the issues, the public was asked to choose which scientists they were going to believe.

The saccharin case, however, is not the only example of publicity campaigns that have reduced the public's anxieties and fears by minimizing the product's risks and exhorting the virtues of chemicals. The ubiquitous occurrence of carcinogens in the environment has been used to challenge the need for workplace controls. The fact that people take risks every day of their lives is used as an argument against those who would strive to create a workplace and environment as free of manmade hazards as possible. "Better living through chemistry" is one corporation's slogan, designed to convince us that without chemicals we would be reduced to a standard of living reminiscent of the Stone Age. These efforts, however, have only served to confuse the public and to thwart regulatory activity.

Industry has claimed that the widespread usage and profitability of chemicals that are suspected carcinogens justify opposition to the regulation of these potential carcinogens in the workplace and the environment. The proliferation of new products and processes throughout the petrochemical industry has created a nightmare for regulators and for industry. Until recently, there have been no testing requirements for the introduction of new chemicals into the workplace, which has exacerbated this problem.

CAUSES OF CANCER

Why does one person develop cancer while another person does not? The fact that this is what often occurs confounds researchers and epidemiologists in their efforts to predict accurately who will develop cancer and who will not. Certain cancers, such as breast and cervical cancer, appear to have a familial or hereditary connection. Certain lung, stomach and liver cancers may be caused by lifestyle factors, such as smoking, eating and drinking. Environmentally related cancers may be the result of radiation, naturally occurring substances and manmade substances that have been introduced into the environment. Other cancer-causing substances have migrated into the water supply, food chain and air from the workplace. Certain cancers, such as bladder cancer in dye workers, mesotheliomas in asbestos workers and angiosarcomas in vinyl chloride workers have a direct occupational relationship. There is, however, a significant indication that each cancer has many causes.

Most persons are exposed daily to a wide range of carcinogens that may interact with one another in a synergistic fashion, resulting in an unexpectedly high incidence of cancer. One of the best examples of this effect is the interaction between asbestos and smoking and the resultant development of lung cancer. An asbestos worker who smokes is 50 times more likely to develop lung cancer than a nonsmoking asbestos worker. This does not diminish the fact that a nonsmoking asbestos worker is at a greater risk of developing lung cancer than a worker not exposed to asbestos.

Until recently, the majority of the National Cancer Institute's (NCI) resources for research went into finding a cure for cancer. Surgical procedures, radiation treatments and chemical drugs were all developed as the weapons of modern medicine to aid the cancer victim. The nation has focused on the elusive pursuit of the magic bullet to cure cancer. Unfortunately, however, despite the enormous expenditures over the past 20 years on research into methods of treatment, the survival rates of patients afflicted with many of the major types of cancer have improved only slightly during the past 20 years. The reality is that today most cancers are fatal within two to three years of onset, despite limited success in slowing the progression of the disease [2].

The emphasis on treatment and cure arose from the belief that only 1–5% of all cancers were work-related and that the majority of cancers were caused by genetic or lifestyle factors. In 1978, however, NCI, the National Institute of Environmental Health Sciences (NIEHS) and the National Institute for Occupational Safety and Health (NIOSH) released a report [3]. The report estimated that at least 20% of all cancers are occupationally related, and it could be as high as 40%. Although this report has not had as significant an impact on the focus of cancer research as some would like, it has brought scientific and public attention on the contribution to cancer risk of environmental and occupational exposures.

Although research for a cure continues, there is increasing emphasis on finding means to prevent cancer.

THE SCIENTIFIC BASIS FOR
DETERMINING CARCINOGENICITY

Traditionally, a cluster of cancer cases has been observed and then an investigation conducted to determine the probable cause. If a causative agent could be determined, then a basis for controlling the substance was established. Latency periods of 20–40 years and the likelihood of multiple exposure have complicated investigations, often making them imprecise. To supplement available data, researchers study the carcinogenicity of substances in animals. These studies routinely take several years to perform and often do not take into account the synergistic effects of multiple substances. In addition to animal studies, researchers have developed screening procedures using bacteria. The validity of these animal and bacterial tests in determining a substance's potential for causing cancer in humans has been a subject of major controversy.

One of the basic purposes of human and animal studies is trying to determine the exposure threshold above which a toxic response occurs. For noncarcinogenic substances, this kind of information is useful in determining a permissible exposure level for workers. For carcinogenic substances, however, a threshold may not exist, and there may not be any safe level of exposure [2, p. 5218].

The issues were addressed in some detail during the 1978 hearings held by the Occupational Safety and Health Administration (OSHA) on its proposal to regulate toxic materials that pose a potential carcinogenic hazard to humans.

CONTROLLING EXPOSURE
TO WORKPLACE CARCINOGENS

The resolution of the scientific issues discussed during these hearings provided the basis for OSHA's Cancer Policy for the Identification, Classification and Regulation of Potential Occupational Carcinogens, issued January 22, 1980 [2]. This policy called for the classification of toxic substances into one of two specific categories for regulation or onto a candidate list for testing.

In general, Category I included those toxic substances found to be carcinogenic in humans or in two or more mammalian species of test animals or in one animal species, if the results of the study in one species had been replicated. Category II included substances reported to be carcinogenic but for which the evidence was only suggestive or only positive in one species [2, p. 5206]

In addition to adopting a classification system, OSHA also proposed model standards to serve as bases for rule-making decisions. The model standards included three provisions:

> (1) worker exposure to Category I Potential Carcinogens will be reduced to the lowest feasible level; (2) permissible exposure levels will be achieved primarily through engineering and work practice controls; and (3) medical examinations and personal protective equipment required by individual standards will be provided at no cost to the employee [2].

The cancer policy also included a requirement providing for no occupational exposure to Category I toxic substances where suitable substitutes exist that are less hazardous to humans [2]. The basis for these regulations is "the widely accepted scientific conclusion that at the present time there is no satisfactory scientific way for demonstrating a safe level of exposure to a carcinogen" [2, p. 5256].

RISKS AND BENEFITS

OSHA has always been concerned with the importance of demonstrating the existence of risks and the likelihood of benefits in setting standards for toxic substances. Permissible exposure levels and control methodology were important issues in the 1978 hearings. OSHA has consistently worked to set standards that control exposure to carcinogens to the lowest feasible level. The U.S. Supreme Court's decision to overturn OSHA's benzene standard, *Industrial Union Department, AFL-CIO v. American Petroleum Institute, et al.* [65 L. Ed. 2d 1010, 100 S.Ct. 2844 (July 2, 1980)], will have a significant impact on the ability of the agency to regulate carcinogens in the workplace in the future. The Court concluded that:

> Section 3, Paragraph (8) of the Occupational Safety and Health Act of 1970 requires the Secretary of Labor to find, as a threshold matter, that the toxic substance in question poses a significant health risk in the workplace and that a new, lower standard is therefore "reasonably necessary or appropriate to provide safe and healthful employment and places of employment" [4].

The Court further concluded that OSHA had not demonstrated the benefits that would be derived by lowering the existing standard from 10 to 1 ppm [4, pp. 1587,1595,1597].

As a result of the benzene decision, OSHA proposed revisions in its cancer policy.

> *Section 1990.111(h)*
> This paragraph states that for Category I Potential Carcinogens, exposures will be reduced to the lowest feasible level primarily through engineering and work practice controls. The words "to the lowest feasible level" are deleted. In consequence, no binding requirement for exposure level is included in the Cancer Policy and the level will be set on a substance by substance basis taking into account all relevant evidence and statutory provisions.
> Obviously included within this would be consideration of the significance of the risk present by each substance [5].

Despite claims by the high court that "OSHA is not required to support its finding that a significant risk exists with anything approaching scientific certainty" [4], the rulings suggest such is not the case. In the benzene case, OSHA presented evidence detailing the poor quality of the epidemiological data with which it had to work. The dissenting opinion of the Court found that OSHA had fulfilled its burden and had substantiated the risk using the best available data [4].

Benzene is not an isolated case. There is a significant dearth of reliable data with which to perform scientifically conclusive epidemiological studies [3,6]. By requiring OSHA to set standards based on benefits, the high court is essentially demanding that OSHA count the bodies before it initiates a rule-making procedure. By challenging OSHA's legislative authority to establish standards for carcinogens at the lowest feasible level, the Court has thrust itself in the midst of the debate as to whether or not there is a safe level of exposure to carcinogens.

COST-BENEFIT

One issue the U.S. Supreme Court did not resolve in the benzene case is whether OSHA is required to perform cost-benefit analyses in setting workplace standards. The cost of regulations is not a scientific issue but one of politics and economics. From the political standpoint, the question is how much job-related injury, illness, cancer or death the public is prepared to accept. This assumes, of course, that someone can accurately predict just what the statistics will be. The economic question is whether society is willing to pay the increased cost of goods and services that will be passed along to the consumer as a result of engineering controls and other less well defined costs.

Industry's attempts to determine the economic impact of regulatory controls have not always produced the most glowing results. During the hearings on proposed regulations to reduce the exposure of vinyl chloride from 500 ppm to "no detectable level," industry predicted implementation would result in costs as high as $90 billion and losses of up to 2.2 million jobs [7]. Claims were made that the standard was not necessary and that compliance would be too expensive and beyond industry's capability [7, p. 107]. Within one year the vinyl chloride/polyvinyl chloride manufacturing industry had successfully met the new standard, and this was accomplished without major economic dislocation or plant shutdowns [7, p. 109].

Efforts have been made to restrict the cost debate to the cost controls and the impact on inflation, but preliminary studies indicate that society as a whole absorbs a significant amount of the cost of illnesses. In a recent report to Congress on occupational diseases, estimates of the total gross earnings loss from disabling occupational diseases increased from $6.4 billion in 1972 to $11.4 billion in 1978 [6, p. 53].

A comparison of workers' compensation claims and benefits for injuries and illnesses is revealing.

A 1972 Social Security Administration survey of the disabled indicated that about 3% of those reporting work-related disease received workers' compensation benefits compared to 43% of those severely disabled with occupational injuries [6, p. 67]. There are two primary reasons: (1) few workers perceive their disease as being work related; and (2) under the workers' compensation system, a worker must prove causality, which is extremely difficult in occupational disease cases.

The 1975 survey of closed workers' compensation claims indicated that slightly more than 60% of the compensation claims for occupational disease were contested, compared with about 10% of the accident claims [6]. Of those occupational disease cases that were contested, causality was the issue in 73% while the question of work relatedness comprised only 21% of the accident contested cases, with 56% of the cases being challenged on the extent of disability [6, p. 70]. These data demonstrate that much occupationally related disease is not compensated by the workers' compensation system and, therefore, must be compensated by the worker's family and by the health care system. The improvement of databases for occupational diseases and cancers is crucial for establishing the need for regulations. Any discussion of cost-benefit must include the costs to society as a whole.

Much ado has been made by industry, which complains of the over-regulation and high cost of meeting OSHA's stringent standards. This is, indeed, nothing more than a smokescreen or tempest in a teapot. Many of OSHA's standards originated as national consensus standards under the auspices of industry-dominated organizations such as the American National Standards Institute. The same is particularly true for all the nit-picking safety standards about which industry so vehemently complains. The compliants, in reality, substantiate that OSHA has begun to enforce industry's own presumably acceptable standards. OSHA's compliance record unequivocally indicts some industries for not having voluntarily complied with their own standards.

CANCER AND STANDARD-SETTING

With the notable exceptions of the standards for lead and cotton dust, and the Access to Medical Records and Exposure Data, the standards that OSHA has promulgated have focused almost exclusively on the control of workplace carcinogens. In each of these cases, there was significant evidence to support the human carcinogenicity of the substances. The case-by-case approach, however, has resulted in the promulgation of standards for only seven potentially carcinogenic substances in OSHA's first nine years. This approach, which included the constant reexamination of already resolved scientific and policy issues, prevented timely and efficient regulation [2]. In an effort to establish a generic approach to the regulation of toxic substances, OSHA developed its system to classify and regulate potential carcinogens.

Despite the efforts to develop effective standards, the prognosis for controlling workplace carcinogens is not good. Recent litigation has overturned the benzene standard on the grounds that OSHA had not substantiated the existence of a "significant risk." OSHA's standard for arsenic has been stayed and is likely to be overturned on the basis of the benzene decision. OSHA's cancer policy will undergo revisions as necessitated by the benzene decision.

The fates of other standards, such as the lead standard, the noise standard, and toxic substance labeling standard, are in doubt.

A January 1981 Executive Order by President Reagan requires that all these standards be subjected to cost-benefit analysis to ensure that the method of regulation is the most cost effective. Since industry has strenuously argued that personal protective equipment is significantly cheaper, it is possible that the entire effort to regulate exposure to potential carcinogens by substitution and engineering controls—the control methods currently advocated by OSHA's cancer policy—will be rendered ineffectual.

REFERENCES

1. American Cancer Society: 1979 cancer facts and figures. New York, NY: American Cancer Society, 1978.
2. U.S. Department of Labor, Occupational Safety and Health Administration: Identification, classification and regulation of potential occupational carcinogens. Federal Register 45(15) (Book 2):5014, 1980.
3. National Cancer Institute, et al: Estimates of the fraction of cancer in the United States related to occupational factors. Rockville, MD: National Cancer Institute, September 15, 1978.
4. U.S. Supreme Court: Industrial Union Department, AFL-CIO v. American Petroleum Institute, et al. Washington, D.C.: The Bureau of National Affairs, number 17, July 17, 1980, p. 1587.
5. U.S. Department of Labor, Occupational Safety and Health Administration: Preamble to notice conforming provisions of workplace cancer policy to benzene ruling (46 FR 4889, January 19, 1981). Washington, D.C.: The Bureau of National Affairs. Occupational Safety and Health Reporter, January 29, 1981, p. 1109.
6. U.S. Department of Labor, Assistant Secretary for Policy, Evaluation and Research: An interim report to Congress on occupational diseases. Washington, D.C.: U.S. Department of Labor. Submitted to Congress in June, 1980, chapter 1, pp. 11–50.
7. Epstein S: The politics of cancer. Garden City, NY: Anchor Press/Doubleday, 1979, p. 108.
8. U.S. Department of Labor, Occupational Safety and Health Administration: Occupational safety and health standards; national consensus standards; and established federal standards. Federal Register 36 (105) (Part II): May 29, 1971.

DISCUSSION: CHAPTERS 15–18

Robert L. Jennings, Jr., K. W. Nelson, Craig Swick,
Marvin A. Schneiderman, Stanley W. Eller, Paul Kotin and Kenneth S. Cohen

MR. JENNINGS: I am a former OSHA [Occupational Safety and Health Administration] employee who spent a little over a year working on the development of the lead standard. I think Mr. Nelson's comments about OSHA's decision-making on health effects would certainly not be borne out by even the most cursory review of the preamble to that standard. To suggest that there is some blatant slanting of evidence and that consideration of the issues is amateurish and that there is no literature or research to support what the agency did is to brand a very long list of well-respected researchers as amateurs. Those researchers would include Lilis, Fischbein, Rom, Repko, Seppäläinen, Cooper and a host of others in the United States and throughout the world.

To suggest that the agency was just pulling a rabbit out of a hat in terms of its judgment of health effects is not only to condemn OSHA, but it is to condemn the EPA [Environmental Protection Agency], which made similar judgments, and also to condemn the European Economic Communities' decision-making on lead, which was virtually identical to OSHA's on all body systems with the exception of the reproductive system. I, furthermore, would like to point out that the Court of Appeals, which upheld the lead standard against industry attacks, had no difficulty whatsoever finding that the agency's decision-making on health effects was rational.

Now, I have no dispute with Mr. Nelson that the agency tends to look at health effects evidence somewhat differently than perhaps ASARCO and some other lead industries. We had arguments by ASARCO in the lead hearings that a worker could lose two-thirds of his kidney function and that wasn't much of a matter for concern. That is not the way we looked at the evidence when OSHA developed the lead standard.

Finally, Mr. Nelson stated that there was some contradiction in the standard, that there was no special procedure for workers whose blood lead levels wouldn't decline to acceptable levels in 18 months of removal. We felt that there was a substantial number of people in the lead industry with many years of past exposure. They have accumulated so much lead in their bodies that, no matter how long they were removed, the blood lead was never going to get down to acceptable levels. What should be done with these workers? What the agency prescribed was removal of such individuals for 18 months, followed by competent medical evluation before workers could be returned to their jobs.

As a closing comment, I would like to point out that there is a major effort underfoot to gut the lead standard in the same fashion as the cotton dust standard is about to be gutted.

I would urge anyone who wants to know about OSHA's decision-making in the lead standard to sit down and read the record, weigh the evidence, and form your own opinion as to whether the agency took its responsibilities seriously. There were months of rule-making hearings and comment periods before we came up with a responsible social decision on this issue.

MR. NELSON: I still think my criticisms of OSHA's decision on the lead study are valid. First, they were not able to show any documented cases of lead poisoning occurring at blood levels of 80 μg/100 g, at least I didn't see any.

Second, why does the regulation allow a worker to go back to work after being off 18 months with blood lead above the acceptable level under the care of a physician? Why were all of these elaborate schemes of blood lead values developed (which the EEC-proposed regulation does not have) if you are going to throw out the whole business and put workers under the care of their physicians? We can debate this for weeks and we would never agree. Your mind is made up. Mine is too.

MR. SWICK: I would like to ask all the members of the panel what their opinions are on using the Ames or other nonanimal models in setting carcinogen standards, considering both the cost involved and the ability to obtain information more rapidly.

DR. SCHNEIDERMAN: If I were a manufacturer and I had a new material that my research department had produced, and if it looked as if it might have some good market potential, I would run it through the Ames test. If it came out negative, I would continue my development. If it showed oncogenesis, I would stop development under certain sets of circumstances. I would drop it if there were other products on the market that looked as good. If there were no others on the market, I might continue development, looking for a less toxic formulation. If results were negative but it had a big market potential, I would then carry on animal studies. If the animal studies were positive, again depending on the economic potential for the material and the other materials available in the market, I would decide whether it was worth the promotional activity that was necessary to bring it to production.

What I'm saying is that my decision would be modified by both the scientific data and the economic issues that are associated with product development. If I could produce this material in an enclosed manner so nobody was exposed to it, I would produce it—even if it appeared to be mutagenic or carcinogenic. I would pay attention to my marketing people who tell me whether it was worth the efforts to go on with it.

I think any simplistic look in which my decision to develop was based simply on an Ames yea or Ames nay would just be misleading. I have great respect for Bruce Ames. I think he may one day get a Nobel prize, but that doesn't mean I should use his test without my thinking and being intelligent about it and without bringing other data into my decision-making.

MR. ELLER: In a situation where, suddenly, an Ames test has been done on a previously unstudied chemical, with a positive result, it would be of concern for labor and may be a basis for pushing to have some kind of regulation of that substance. We believe that a positive Ames test would be justification for additional animal tests to be undertaken. If something shows up positive on the Ames test, it should be looked at as a first indicator of carcinogenicity, just as you begin to look at effects in susceptible individuals as a first warning that a chemical should be more effectively controlled.

DR. KOTIN: I would emphasize that, by definition, there cannot be a cancer of a unicellular system, so that the Ames test cannot be a test for cancer. What it is is a test for the ability of an agent to interact with genetic material in the cell and produce what we call mutations. It is important to know an agent's ability to do that.

DR. COHEN: Would you comment on how the cost-benefit risk assessments of today will find acceptance with tomorrow's juries?

DR. KOTIN: It is a very profound question and one that can only be answered in great length. But certainly you've asked a critical question because much of the problem that we face today in litigation is the inability to judge the past in terms of the present, if I understand what you are saying. But as to the issue of cost, I think cost benefit is an exercise in futility if you don't have reliable risk-analysis data. Risk analysis can be done with some degree of precision, and when it is done with full recognition of all the caveats the other speakers have pointed out, the problems inherent in risk analysis are significantly less than those of cost-benefit. I think what I would do is take the position that prudence must be the operating mode—in the sense that a product has to be evaluated in terms of its life cycle, from the origins of the raw materials to the ultimate disposition of the product. And I would not limit myself to the manufacturer of the product or the basic materials themselves but would give consideration to all the environmental aspects of the life cycle of the material. Then, if you have any real weak points along the environmental train, you would look for alternatives.

DR. SCHNEIDERMAN: Risk assessment for chemicals is difficult. I heard Bill Nicholson (of Mt. Sinai) talk last week about some attempts at risk assessments following asbestos exposure. He mentioned some of the problems about the forms of asbestos: amosite, chrysotile, crocidolite, etc. Are they equally toxic? There are concerns about the nature of the exposures. Was the textile worker in Manville, New Jersey, exposed to the same kinds of material as the miner up in Quebec? All these issues are of consequence when Nicholson sets down the table in his paper. He also needs to look at the dose-response relationship and the slope of the dose-response curve—what some people want to call potency. What he found was a 150-fold range from the shallowest slope to the steepest slope for ostensibly the same material. I think you've got to recognize that we have these enormous uncertainties to deal with in doing risk assessments. By and large, most people have great difficulty with uncertainties, including juries. How do you deal with this 150-fold range in slope when it is compounded by such things as cigarette smoking? I think it is going to take us some time to resolve these issues. As I remarked earlier, all these ideas really started as utility theory with Jeremy Bentham several hundred years ago. And we haven't solved them in the intervening 200 years.

SECTION 2

WORKERS' RIGHTS
AND RESPONSIBILITIES

CHAPTER 19

OCCUPATIONAL MEDICAL RECORDS: ISSUES SURROUNDING THE DEVELOPMENT OF THE OCCUPATIONAL SAFETY AND HEALTH ADMINISTRATION STANDARD

Robert L. Jennings, Jr., JD*

Occupational Safety and Health Administration
Washington, DC

Everyone agrees with the following three propositions:

1. It is valuable to create and preserve occupational medical records to monitor the effects of worker exposure to toxic substances and to serve as a database for research.
2. Workers should be informed of the contents of their medical records and should have the right to learn what their employers know about their medical conditions.
3. Occupational medical records should be entitled to the same stringent requirements for protection of confidentiality as is afforded to other medical records.

The Access to Employee Exposure and Medical Records standard [1] of the Occupational Safety and Health Administration (OSHA) was developed to assure that occupational medical records are preserved, and that workers, their designated representatives and OSHA have rights of access to them, where appropriate. The overall situation involving occupational medical records, however, is far more complex than these initial propositions suggest, and OSHA's access standard only touched on a small number of issues. This chapter summarizes major provisions of the standard, discusses some of the broader issues that OSHA ducked and closes by advocating some perhaps controversial proposals.

---

*Present address: Baskin and Sears, Pittsburgh, Pennsylvania.

CREATION AND PRESERVATION OF OCCUPATIONAL MEDICAL RECORDS

OSHA Access Standard and Current Employer Practices

Any rational discussion of occupational medical records must first examine the circumstances under which records are created and preserved. OSHA's access standard is founded on the current widespread medical monitoring of employees through medical programs structured and controlled by employers. For example, National Institute for Occupational Safety and Health (NIOSH) data collected in the early 1970s indicated that at least 60% of large chemical and allied products facilities provided some kind of periodic medical examination to company employees [2]. The access standard requires that, to the extent an employer causes occupational medical records to be created, these records must be preserved for the duration of employment plus 30 years. The standard ducks three crucial issues, however:

1. the need to guarantee that appropriate medical monitoring is, in fact, conducted;
2. the need to determine exactly what kind of organization should create and preserve the resulting medical records; and
3. the need to limit the information collected to the minimum needed.

The Importance of Creating Appropriate Records in View of Current Approaches to Occupational Health

Employers have a fundamental ethical duty to assure that employees exposed to potentially toxic substances receive appropriate medical monitoring to ascertain if any ill effects are occurring and to enable long-term research. This moral and social responsibility is no different from the duty of a manufacturer to adequately test its products, either on animals or by other means, for potential hazards to consumers [3]. In the past, occupational medical monitoring has typically focused on whether or not gross impairment has occurred, e.g., is the battery worker clinically lead poisoned? Today, the focus must be placed on subtle biochemical and behavioral changes preceding the frank symptomatology of chronic disease. Much of what we know about occupational disease has been gained over the years through a process of counting the bodies of diseased and dead workers. This unfortunate reality will continue, and medical monitoring can be a prime vehicle to form an accurate count.

The new OSHA administration no longer speaks of the benefits of effectively controlling hazards even when there is no absolute certainty as to their magnitude; but, rather, the administration is proudly championing "formal cost-benefit analysis" and the "least costly alternative" approaches to regulation. Stripped of the flowery economic arguments, the concepts of formal cost-benefit analysis and least costly alternative approach are quite simple. When decision-making is predicated on "formal cost-benefit analysis," industry is completely free to handle a chemical in whatever way it chooses and to continue to do so until so many dead or diseased workers have been counted that the hazard can no longer be politically ignored. Using a "least costly alternative" approach to regulation means that effectiveness is no longer viewed as a relevant consideration; thus, respiratory protection, no matter what its faults, will now be substituted for engineering controls. Both "formal cost-benefit analysis" and "least costly alternative" approach sacrifice worker protection for some supposedly greater commercial or national good.

In any case, the lack of focus on prevention and the requirement for body counts that can be plugged into cost-benefit analysis will mean a growing role for occupational medical monitoring. For many workers, these records may be their only hope for ensuring that changes are made in the workplace.

Options for Creating and
Maintaining Appropriate Records

Almost everyone would agree that an employer who derives a profit from the use of a toxic chemical has the moral or ethical duty to pay for medical monitoring of workers exposed to the chemical. The evolution of company-financed industrial medicine in this country has generally followed a model of the employer creating and managing an in-house corporate medical department. This in-house department either performs all medical monitoring and associated record-keeping itself or manages and reviews the work of quasiindependent local physicians and clinics. This historical evolution is largely a function of the size and diversity of the country as well as the lack of strong central unions and the lack of a centralized national health care system.

At first glance, it may seem perfectly straightforward for the company to completely control what the company pays for, but other options exist. Occupational medical monitoring, record preservation and research could easily be conducted by an expert federal agency like the Public Health Service, funded by a user tax on employers. Local or regional independent medical centers could be established as adjuncts to major universities or hospitals and could be funded by a special insurance policy issued to individual workers. Occupational medicine could be practiced exclusively by physicians and clinics not in the direct employ of any one employer or industry. Labor-management based independent institutes could be established on the model of the Swedish Bygghälsan, or national occupational health service of the construction trades [4]. Bygghälsan is a nonprofit, nongovernmental institution managed by a Board of Directors formed of labor, management and academic representatives. It conducts regular periodic occupational medical monitoring as well as major epidemiological research.

Problems Associated with the Current Emphasis on Employer Medical Departments

The value of alternatives to the typical in-house medical department becomes apparent when one considers the many problems associated with medical record-keeping controlled exclusively by the company. A substantial degree of medical monitoring is compelled as a condition of employment and is conducted in an adversarial context in which the company doctor is serving the company first and the worker only secondarily. Examples of exams required by employment are: preemployment exams, fitness-to-work evaluations after sick leave, disability pension and workers' compensation eligibility disputes. The company doctor who conducts the exam may well end up testifying against the employee in arbitration or workers' compensation proceedings. In addition, whether or not management chooses to admit it, the company has a strong financial incentive to hire and retain not just fit workers, but the most healthy and most physically fit. There is an obvious incentive to discard marginal employees who could adversely affect health insurance or workers' compensation premiums. Workers and their representatives uniformly testify before OSHA that the foregoing dynamics are frequently in play in many corporate medical programs [1].

Another problem is the obvious conflict of interest involved in company-controlled research using occupational medical records. Few companies are completely indifferent to occupational disease, but it is equally true that few companies have shown a great eagerness to find occupational disease in instances where it is not already fairly obvious. OSHA's access standard was promulgated precisely because so many large companies persistently refused to give individual workers and their representatives access to the contents of company employee medical records [1]. The same evidence that documents an occupational health problem may: (1) document a workers' compensation claim; (2) result in union pressures for major expenditures to clean up the plant; (3) prompt adverse publicity; and (4) lead to intervention by NIOSH or OSHA. In short, there is a fairly substantial conflict of interest involved for these medical programs exclusively controlled by the company that performs occupational health research or that monitors workers for subtle changes preceding chronic disease.

A final complication in determining who should control medical records is the need to ensure worker cooperation. Effective medical monitoring for research and other purposes is highly dependent on the full cooperation of the employee. This is particularly true if the physician is focusing on subtle physical or behavioral changes. If the worker perceives that his participation in the monitoring could in any way harm his job security, the worker will often consciously minimize the occurrence of any subjective adverse symptoms. OSHA's 1978 lead standard explored this widespread problem in depth [5]. When a company doctor controlled by the company does the medical monitoring and the company holds the medical records, there are apt to be serious questions raised as to the validity of the medical information collected.

The combination of the foregoing factors is particularly troublesome with respect to a worker's expectations of confidentiality regarding his medical records. The right of privacy is fundamentally the legitimate expectation of each worker, and he or she should feel confident that intimate details of his life will not be disclosed to others or used to his detriment. If a company medical program suffers from or is perceived to have any of the problems listed, it is difficult to imagine how workers can be expected to feel confident that their medical records are confidential.

Tension Between Collecting
Adequate Information and Invasion of Privacy

A final problem, which is not dependent on who creates and preserves occupational medical records, is the tension between the need to collect adequate medical information and the need to meet the worker's expectations of privacy. The more intrusive the medical information demanded from the employee, the greater the consequences of abuse and the greater the perception that the intruder is overreaching. For example, if emphasis is placed in a history-taking on reproductive or personal habits or on family health problems, many workers will inevitably feel that their privacy is being invaded. The dilemma arises from the fact that this sort of information, on occasion, may be extremely important. Consider the risks posed by major occupational exposure to lead to male and female reproduction and to the fetus [6]. OSHA's lead standard, in fact, mandates that medical surveillance inquire into reproductive health matters [7].

ACCESS TO OCCUPATIONAL MEDICAL RECORDS

OSHA Access Standard

Access to occupational medical records by various parties raises numerous issues equally as complicated as those relating to the creation and preservation of these records. Again, OSHA's access standard addresses only a few of these issues.

The access standard provides for unqualified direct worker access to his or her personal medical records, with special provisions concerning terminal illnesses, psychiatric conditions and the identity of confidential informants [1]. Designated representatives, such as employee doctors, lawyers and family members, as well as union officials, are given access only with the detailed specific written consent of the employee [1]. OSHA access is assured without employee consent but only in accordance with elaborate administrative regulations governing (1) when and how OSHA will seek unconsented access; (2) who within the agency can use this information and for what purposes; and (3) how the agency will safeguard these records once in the agency's possession [1,8]. Finally, the access standard is essentially silent on the issue of regulating corporate access to employee medical records [1].

Worker Access and Designated Representative
Access with Worker Consent

Direct worker access to medical records and designated representative access with consent are explained in detail in the standard's preamble [1] and merit little additional comment. The principle is simple: a worker and those of his choosing should have a right to learn what the company already knows about the worker's health status. A limited number of industry groups have the courage to openly oppose this "radical" concept and cloak their arguments in pious concern for employee rights to privacy. These arguments are frivolous and simply represent the company's desire to perpetuate a system by which it maintains absolute and final control over what, if anything, workers are told about occupational disease.

Corporate Access: Legitimate Use versus Abuse

Setting parameters for corporate access to occupational medical records gives rise to extremely complicated problems. On the one hand, the company has numerous legitimate uses for occupational medical records. Some of the adversarial examinations, such as preemployment and fitness-to-work exams, benefit employees indirectly. Periodic screening of employees for subtle changes and research projects is extremely beneficial to workers if its purpose and design are to protect workers rather than to protect the company. The problem is that abuses can easily accompany the legitimate uses. In-house corporate preservation of medical records offers an ongoing opportunity for abuse. The hearings on the access standard revealed that abuse frequently occurs in the form of unrestricted access to records by local plant management [1]. Often, an occupational medical program is part of the corporate personnel or loss control department or is subject to the direction of local plant management. The company doctor may be or may fervently want to be ethical, but if local plant management demands access to an employee's medical record, it will have access. A printed directive from the corporate medical director sitting in international headquarters is not likely to end such abuses on a local level.

The occupational medical community's response to the issue of corporate access has been the following provision in the 1976 American Medical Association's (AOMA) Code of Ethical Conduct for Physicians Providing Occupational Medical Services: "[E]mployers are entitled to counsel about the medical fitness of an individual in relation to work but are not entitled to diagnoses or details of a specific nature." This sounds fine in theory but does not reflect reality. For example, in effective occupational monitoring, certain employees may periodically appear to be at particular risk of developing permanent impairment. This could be due to such factors as poor industrial hygiene or engineering malfunctions at specific job stations, poor respirator fit or maintenance, or inadequate employee training or supervision. Presumably, one would hope that the examining physician would involve management in investigating and correcting the possible causes of these problems. Under the normal course of events, it is simplistic to suggest that the doctor has not informed management of the diagnoses or the specific nature of a worker's medical condition.

Union Access Without Worker Consent

The access standard does not allow union access without consent to individual employee medical records, but neither does the standard forbid this access [1]. Over the past decade, many unions have become increasingly sophisticated and involved in occupational health matters. Some unions have pressed for the opportunity for unconsented access to certain employer-held medical records in order to conduct thorough investigations of potential health problems. As in many cases of legitimate broad-scale research, it may not be possible to obtain consent, due to such factors as the size of the population involved or the absence of mailing addresses. If however, personally identifiable records are to be used, serious questions must be answered as to who will handle the records, for what purposes, and how, so as to guarantee no breach of confidentiality. Determining who will answer these questions is vexing since the company holds the records and may have no incentive to be cooperative.

The issue of union access to identifiable medical records is currently before the National Labor Relations Board (NLRB). The following three administrative law judge decisions were consolidated before the NLRB: *Colgate-Palmolive Co.,* No. 17-CA-8331 (March 27, 1979), *Minnesota Mining & Manufacturing Co.,* Case No. 18-CA-5710-11 (March 13, 1979), *Borden Chemical, A Division of Borden, Inc.,* Case No. 32-CA-551 (April 25, 1979). They were briefed and argued in mid-1980 before the NLRB. See also, *Detroit Edison Co. v. NLRB,* 440 U.S. 301 (1979) (unconsented union access to psychological testing records).

I am not opposed to union access, but I would like to see the use of an Institutional Review Board (IRB) system such as that proposed by the Food and Drug Administration (FDA) in 1979 [9] and commonly used in most basic health research involving humans. Furthermore, as I see little difference between employer and union access to personally identifiable medical records for research purposes without consent, employer-sponsored research using medical records should be governed by an identical IRB system. Burdens not worthy of being placed on the employer should not be placed on the union.

OSHA Access Without Consent

As mentioned earlier, access by OSHA to employee medical records without consent is governed by elaborate administrative regulations that were promulgated simultaneously with the access standard. Unconsented access by OSHA requires that need must be established in detail and that written approval of a former order by the Assistant Secretary and a Medical Records Officer must be obtained [8]. Stringent security procedures and other provisions are established [8]. In the first six to nine months following the promulgation of the standard, there were only a few occasions in which OSHA sought unconsented access to detailed medical records. The efficacy of the administrative regulations remains to be seen, but a firm protective structure has been established on paper. This protective structure is well grounded in several court decisions, including a Supreme Court case that fully supported unconsented governmental access to personally identifiable medical records for public health purposes under limited circumstances [*Whalen v. Roe,* 429 U.S. 589 (1977); *Dupont v. Finklea,* 442 F.Supp. 821 (S.D.W.Va. 1977); *United States v. Westinghouse Electric Corp.,* 483 F.Supp. 1265 (W.D.Pa. 1980); *General Motors Corp. v. Finklea,* 459 F.Supp. 235 (S.D.Ohio 1978), *rev'd.* ____ F.2d ____ (No. 79-3168, Dec. 30, 1980); *Marshall v. American Olean Tile Co.,* 489 F.Supp. 32 (E.D.Pa. 1980)].

REMEDIES FOR ABUSE OF OCCUPATIONAL MEDICAL RECORDS

A final issue involves remedies for abuse of occupational medical records. It is often said that there is no right without a remedy, and this is especially true in right-to-privacy issues. There is no confidentiality of medical records if a worker has no recourse when access to medical records is abused by his employer. To date, there is no governmental regulation of this matter nor is there likely to be any in the near future. (Numerous limited bills concerning the privacy of medical records were introduced in both houses of Congress in 1980, but no bill was enacted. None of these bills applied to medical records held in an occupational setting.) Private lawsuits are generally of little help for two reasons. Damages are extremely speculative but, more importantly, very few state courts would hold a company doctor or employer liable, since most state courts hold that there is no physician-patient relationship in the occupational setting [1]. Policing of the occupational medical community by the AOMA or state or local medical societies is meaningless as evidenced by the failure of any of these organizations, to the best of my knowledge, to act on any of the many specific instances of abuse identified in the access standard's rulemaking.

PERSONAL PROPOSALS

In conclusion, I would like to advance several personal proposals. The list of potential problems I have touched on gives rise to numerous suggestions but, in the final analysis, two problems dominate the issue of confidentiality of medical records. The first problem involves exclusive corporate control over the creation and preservation of medical records by in-house medical departments. The second problem involves the lack of a meaningful remedy for abuse of medical records.

My comments up to this point have presented corporate occupational medical programs in fairly bleak terms. Obviously, there are some excellent corporate programs that cannot be accused of abusing worker medical records. I do believe that these exemplary programs are more the exception than the rule. However, my personal belief is not really important since the issue before us is the right to privacy, and this right is largely one of perception. It is safe to say that many, if not most, workers hold company doctors in even less esteem than they do attorneys (of which I am one). This results from past abuses, from the inherent conflicts of interest involved in in-house medical programs, and from the fact that workers and their unions have no input whatsoever in the structure and management of the medical program. Furthermore, these three factors mean that no matter how well a company's medical program is set up and operated, the output of that program will more likely than not be viewed with suspicion by all of labor, by government, and by the public alike. This tension and distrust is particularly unfortunate because it is preventable.

As to my proposal, everyone pays lip service to the value of labor, management and government working together; this is one area in which this could be done. There is no reason why a corporate medical program must either be conducted in-house or controlled exclusively by management. The Swedish Bygghälsan, or another model, could be used to build a nonprofit, nongovernmental organization that would do precisely the medical monitoring and research the company requests, but would do it in a way that satisfied worker concerns and interests and would be sufficiently independent of direct management control to ensure that everyone would have confidence in its output and in its handling of occupational medical records. If a prestigious company or group of companies wished to experiment and establish such a program in cooperation with a union, I am confident that the Reagan OSHA administration could be squeezed for the dollars necessary to do the initial planning and implementation. In my judgment, companies are paying a high price for their insistence on absolute control over all aspects of their occupational medical programs, and that price will continue to be exacted until an alternative cooperative system is developed. A lack of true confidentiality of medical records is only one adverse consequence of the current system.

One last proposal concerns remedies for abuse of medical records. I do not expect all of the Fortune 500 companies to rush to implement my prior proposal, but a simple step can be taken by all companies to dramatically increase worker confidence in that, whatever else, the company is serious about not abusing their medical records. Whenever the company wishes to collect medical information from a worker, the company should enter into a written contract that states the uses to be made of the medical information and promises absolute confidentiality against other uses in exchange for the worker's willingness to provide the information or to subject himself to the medical examination. This simple contract should outline security procedures and should make any breach of contract arbitrable or subject to legal action. Most importantly, the contract should include stipulated damages for a proven breach of contract, which would represent how seriously the company takes its promise to protect the medical records. How much is a person's right to privacy worth in monetary terms? Pick any dollar figure you feel is appropriate, but perhaps a minimum recovery of $5000–10,000 plus legal fees would be an adequate commitment by the company. Such a contract could easily be drawn up and made a part of every collective bargaining agreement or employment relationship in this country. I feel that each worker has the right to expect such protection because, without it, most workers have no way to deter or respond to abuses of their medical records. Without a remedy, there is no true right, and the simple remedy I have outlined would go a long way toward giving workers confidence that their medical records will be treated with the respect they deserve.

REFERENCES

1. 29 C.F.R. 1910.20, 45 Fed. Reg. 35212-84, May 23, 1980.
2. 45 Fed. Reg. 35255.

3.   Section 402A of the Second Restatement of the Law of Torts (1966), Special Liability of Seller of Product for Physical Harm to User or Consumer.
4.   Occupational Safety and Health Administration, Swedish-American Conference on Chemical Hazards in the Work Environment, March 4–7, 1980.
5.   29 C.F.R. 1910.1025(k), 43 Fed. Reg. 52952, 54440–73, Nov. 14, 21, 1978.
6.   43 Fed. Reg. 54388–98, Nov. 21, 1978.
7.   29 C.F.R. 1910–1025(j)(3)(i)(C) & (ii)(A).
8.   29 C.F.R. 1913.10, 45 Fed. Reg. 35284–97, May 23, 1980.
9.   44 Fed. Reg. 47699, Aug. 14, 1979.

CHAPTER 20

CONFIDENTIALITY OF MEDICAL RECORDS: LEGAL PRECEDENTS AND ISSUES

Marvin Lieberman, JD, PhD
Committee on Medicine in Society
The New York Academy of Medicine
New York, New York

Privacy and accessibility, concealment and openness are pervasive and conflicting considerations in the discussion of the confidentiality of medical records whether in the workplace or in other settings. Our concern about these issues forces us to engage in a delicate balancing act. How do we balance the interests enhanced by safeguarding privacy with the valid interests we have in permitting access to medical records for medical research, epidemiological investigation and furthering occupational health?

Legal scholars, philosophers and economists disagree as to whether privacy itself is a fundamental interest or is a derivative of other rights. Prosser, for example, suggests that violation of privacy represents four fundamental kinds of invasion of different interests of the plaintiff: (1) an intrusion on someone's seclusion or solitude; (2) a public disclosure of private facts that are embarrassing to someone; (3) a presentation of someone in a false light in the public eye; and (4) the commercial appropriation of someone's picture for commercial purposes [1].

Others argue that the right to privacy represents a unique set of interests. The Supreme Court, in *Whalen v. Roe,* has suggested that privacy as a constitutional right deals with "two topics of interest. One is the individual interest in avoiding the disclosure of personal matters, and another is the interest in independence in making certain kinds of important decisions" [2]. The former interest might be broad enough to include access to birth control methods and materials or to an abortion. The latter interest deals more appropriately with questions of liberty rather than privacy. The cases presented in this paper will deal with the first category, the individual interest in avoiding disclosure of personal matters.

Despite the differing views of many scholars, the emerging concern for protection of the right of privacy is a worthwhile effort. I would contend that examples of this effort include: The Privacy Act of 1974, the Report of the Privacy Protection Commission in 1977 and the Federal Privacy of Medical Information Bill, H.R. 5935 which passed the House after surviving scrutiny by three house committees, but died in the Senate in 1980.

Privacy deals with "our accessibility to others: the extent to which we are known to others, the extent to which others have physical access to us, and the extent to which we are the subject of others' attention [3, p. 423]. It is obvious, as Gavison notes, "that both perfect privacy and loss of privacy are undesirable. Individuals must be in some intermediate state, a balance between privacy and interaction in order to maintain human relations, develop their capacities and sensibilities, create growth and even survive" [3, p. 440]. The right to privacy may only have been enforced or should be enforced when unwarranted or unreasonable intrusions occur. She concedes that the law will intervene and should intervene when the intrusion is deemed to be reasonable, but often such intervention will occur only when a related interest is harmed. The notion should sensitize us to the fact that privacy contributes to individual autonomy, mental health and to a pluralistic society [3, p. 424].

Do persons in the United States today enjoy more privacy than in nineteenth-century America? A good case can be made for the affirmative. For example, in that period in a small town, it took a lot of doing to prevent one's neighbor from knowing everything about one. In twentieth-century America, the growth of urbanization and of an impersonal organized bureaucracy enhances our privacy in some areas, yet brings dangers to privacy about which we have never dreamt. These dangers are largely due to the capacity of technology to handle, retrieve and store information. At the same time, as a result of greater tolerance for differing lifestyles and changes in standards of acceptable behavior, certain individuals will reveal voluntarily information they would have considered shocking 100 years ago.

Scholars differ on the source of the danger to privacy. Stigler [4] suggests that the extent to which the issue of privacy is genuine arises out of the growth of government and the expansion of its functions. Coleman [5] has also suggested that recent privacy legislation often is not privacy legislation at all, but is legislation to restrict the rights of the corporate actors and to increase the rights of persons. By corporate actors, Coleman refers to other organized institutions as well as to government. Indeed, Coleman has found in his research that there is increased hostility and alienation from and loss of confidence in corporate actors, including government and corporations.

Stone [6] agrees that abuses spring from plant bureaucracies in the profit and nonprofit sectors. As to the latter, Professor Stone says "such mischief as these nonbusiness corporate institutions threaten is neither more confined nor more controllable because it is not motivated by profit" [6]. The comments by Stigler, Coleman and Stone, while illuminating as to general trends in our society, are not meant to nor can they apply completely to the complex environments of the workplace in which claims of access to and privacy of medical records are asserted by a variety of parties.

The medical record in the workplace is created by the company physician and the company finds itself regulated by government and may be in an adversary position with the worker vis-a-vis the union. Company physicians have asserted claims to privacy when government has asked for access to the medical record without the express consent of the employees. Access claims have been asserted by government for purposes of epidemiological investigation and to protect the employee, by employees who wish to see their medical records to ascertain their exposure to toxic substances, and by unions who not only wish unconsented access to exposure records but are now claiming the same right of access to medical records as that of company physicians and union members [7].

Accusations that a self-serving concern prompts a claim are often heard in the debate among all parties concerning privacy of medical records and access to these records. However, these accusations contribute very little to the merits of the case. If the purpose underlying a claim of right were to be examined from the viewpoint of motivation of the parties of interest only, many of our most cherished procedural rights and freedoms which underlie our system of justice would never have been achieved.

The Privacy Commission has suggested that the patient-physician relationship itself "is an inherently intrusive one in that a patient who wants or needs medical care must grant his doctor virtually unconstrained discretion to delve into the details of his life and person" [8, p. 282]. The Commission, in its recommendation, suggested that three goals be considered: "1) to minimize intrusiveness; 2) to maximize fairness (out of concern to maximize fairness the requirement for patient access to the records emerges); 3) to create an enforcible expectation of confidentiality" [8, p. 291].

House bill H.R. 5935, which passed in December, 1980, but died in the Senate, included provisions to ensure individuals access to their records with certain exceptions, such as in the case of terminal or mental illness.

If the person is terminally ill or has a mental illness, he or she has the right to designate a representative who will have access to the records even though the representative may turn this information over to the patient. The medical facility must describe, in writing, its policy regarding access to records. The bill provided for civil actions for actual damages against federal, state, local and other authorities as well as against medical facilities where the plaintiff's rights are "knowingly or negligently violated."

On May 23, 1980, the Occupational Safety and Health Administration (OSHA) promulgated a standard [9] dealing with access by the employee or his designated representative and by OSHA to employer-maintained exposure and medical records covering exposure to toxic substances or harmful physical agents. The standard includes the following:

- The employer can designate a representative to exercise access rights which right might include a certified collective bargaining agent. The representative has unconsented access to employee exposure records, but the representative must obtain specific written consent to medical records.
- The standard applies to any chemical or biological agent, or physical stress listed in the National Institute for Occupational Safety and Health (NIOSH) "Register of Toxic Effects of Chemical Substances."
- Exposure records must be preserved and maintained for at least 30 years. Medical records must be retained for the duration of employment plus 30 years. OSHA access to exposure records, medical records and analyses of these records is not conditioned on the employee's consent.
- OSHA has established internal procedures including medical records officers to oversee privacy protection within OSHA.
- Where employee assistance records are kept with employers' medical records files, OSHA is to have access to these records.

NIOSH had joined with OSHA in the original proposed rule-making for this standard. NIOSH withdrew, however, choosing to rely on its subpoena powers under 29 U.S.C.E. 657(b) to obtain health records for its evaluation of health standards.

A brief review of the issue of patient/physician privilege in rules of evidence applicable to judicial proceedings provides helpful background and should sensitize us to a major privacy issue raised by the OSHA Access Standard. To what extent has the patient a privilege to prevent the disclosure of information arising out of a patient-doctor, physician-nurse relationship in a judicial proceeding? In the proposed Code of Evidence for New York state [10] for medical care involving physical illness, the patient-physician relation is privileged only if the disclosure would tend to subject the patient to embarrassment, humiliation or disgrace. However, there is no such qualification for psychiatric information. The point is that successful psychotherapeutic treatment depends on the active rather than just the passive cooperation of the patient. The patient's cooperation is unlikely to be secured without strict assurances that confidentiality will be well protected. A dilemma results when a psychiatrist hears that his patient is considering or planning a violent act. What is the responsibility of the psychiatrist? If the patient knew in advance that such information would be revealed to a third party, he might avoid seeking psychiatric consultation.

The proposed New York state Code of Evidence suggests that legal protection of confidential communications is not essential for treatment of physical disease since the patient's desire to secure a successful treatment in the case of a fracture, for example, ordinarily would not be diminished by full disclosure. The access standard promulgated by OSHA has a particular provision dealing with the information about patient assistance programs. These are voluntary programs in industry dealing with alcoholism, drug abuse and other related problems. OSHA will have unconsented access to patient assistance records if the patient information dealing with an assistance program is part of the employee medical record, but will have no access if this information is kept separately.

Under the statutes, government agencies such as OSHA and NIOSH may obtain access to medical records for epidemiological and other purposes by promulgating a standard under regulatory powers or by subpoenaing a record in court. OSHA has chosen the route of a standard whereas NIOSH has sought to use the subpoena route.

Several cases have been brought to the federal court in which NIOSH has subpoenaed medical records. Subpoenas are frequently issued following a request for a health hazard evaluation from a collective bargaining representative of the workers in a particular plant. NIOSH seeks from the company, individually, identifiable health information, including the names and addresses of workers no longer working to enable followup of these employees to determine whether there has been any physical harm or illness caused by exposure to a particular substance. In the cases that have come before the court, the employer has asserted several arguments. The companies indicate that they will be pleased to comply with the subpoena, provided the employee executes a release granting permission to forward this information to NIOSH. In *General Motors Corporation v. Director of NIOSH,* 9 OSHC 1139 (1980), General Motors claimed that enforcement of the subpoena would be contrary to the statute of the state of Ohio which makes privilege patient communications. The U.S. Court of Appeals for the Sixth Circuit has held that since the case involved the federal question, the Ohio privilege statute does not control and federal statutes and the common law governs this matter. The court declined to expand the scope of the law to protect physician-patient relations in the workplace.

The second argument companies have presented is that full enforcement of the subpoena would infringe upon employees' right to privacy and would violate the employees' expectation that information given to the plant physician by them would remain strictly confidential. In examining the conflict between NIOSH's ability to evaluate health hazards and the interest of the patient to have the information kept confidential, the courts have, in most cases, held with the Institute, relying on the decision in *Whalen v. Roe* case.

In *Whalen v. Roe,* the state of New York had attempted to resolve the problem of abuse of a narcotics drug program by requiring that a copy of any prescription involving one of the controlled, dangerous addictive substances be forwarded with the name of the patient to the state Health Department. While the court, on the one hand, asserted the right to privacy, it argued that for public health purposes the confidentiality of medical records could be breached. The Supreme Court stated in *Whalen v. Roe,* 429, U.S. at 602:

> Nevertheless, disclosure of private medical information to doctors, to hospital personnel, to insurance companies, and to public health agencies are often an essential part of modern medical practice even when the disclosure may reflect, unfavorably, the character of the patient.

In *Westinghouse v. U.S.A.,* 8 OSHC 2131 (1980), the U.S. Court of Appeals for the Third Circuit presented a lengthy discussion of the need to protect the individual from unwarranted access to his records. In this case, the court recognized that NIOSH had shown a reasonable need for access to the entire medical file, but the courts did express a concern that it is possible that the medical records might contain "information of such a high degree of sensitivity that the intrusion could be considered severe, or that the employees are likely to suffer any adverse effect from disclosure." It is said that Westinghouse was not justified in its blanket refusal to give NIOSH access to the records [11, p. 2137]. It did not recognize that in a particular file there might be information that an employee may consider highly sensitive which would include records of employees' personal consultations with physicians on a broad spectrum of health matters. The court argued "that we cannot assume that an employee's claim of privacy as to particular sensitive data in that employee's file will be outweighed by NIOSH's need for such material" [11, p. 2139]. In this connection, the court recommended that NIOSH give prior notice to the employees whose medical records it seeks to examine and that it permit employees to raise a personal claim of privacy if they desire. While the court did not specify the form of notice, it suggested that the employee's failure to object in writing by a specified date would permit the courts to assume that the employee had no objection to the release of the data. The mechanics should be worked out by the district court and the procedure should depend on the sensitivity of the information, the number of persons involved, and the need for NIOSH to conduct this evaluation expeditiously. The court argued that a short delay would not unduly compromise NIOSH's investigation.

Connected to the question of unconsented access of a governmental agency to medical records in the workplace is the issue raised by allegations from employees and unions that, at times, employee medical records are not secure from inappropriate scrutiny by company management. It has been suggested that the remedy for this type of abuse would be to change the settings under which occupational health services are provided so that these are jointly controlled by labor and management. Regardless of the merits [12] of these proposals, it is not a realistic alternative. Implementation of such major changes in occupational health auspices would require that its advocates persuade the labor movement to commit a considerable amount of its influence to achieve this goal in the face of other more serious priorities. Violations of the privacy of medical records through unconsented access by management is a breach of the Code of Ethics of the American Occupational Medical Association [13], and it might also be grounds for complaints against physicians who are guilty of such behavior under the state rules regarding medical discipline. Enforcing these rules is dependent on obtaining adequate resources to support skillful investigation and effective hearings on complaints. Too often the critic of the profession lacks the sustained interest, knowledge or resources to do anything more than to bring the problem to the temporary attention of the public. Professional organizations have the responsibility to monitor adequately and to report fully to the public on the effectiveness of the enforcement machinery surrounding professional codes of ethics or state regulation of physician discipline. Without such efforts, the privacy of medical records, indeed, the integrity of the physician-patient relationship cannot be assured.

REFERENCES

1. Prosser W: The handbook of the law of torts. St. Paul: West Publishers, 1971, p. 804.
2. *Whalen v. Roe,* 429 U.S. 589 (1977) at 599–600.
3. Gavison R: Privacy and the limits of law. Yale Law Rev 89:421, 1980.
4. Stigler GJ: An introduction to privacy in economics and politics. J. Legal Studies 9:623, 1980.
5. Coleman JS: An introduction to privacy in economics and politics: a comment. J Legal Studies 9:645, 1980.
6. Stone C: The place of enterprise liability in the control of corporations. Yale Law J 90:1 at 2.
7. Brief of Petitioner, Industrial Union Department, AFL-CIO, in IUD, *AFL-CIO v. Donovan,* 80-1550 (U.S.C.A.D.C. Cir. 1980).
8. Privacy Protection Study Commission: Personal privacy in an information society. Washington, D.C.: U.S. Government Printing Office, 1977.
9. 45 Federal Register 35212 (May 23, 1980); as corrected at 45 Fed Reg 54333 to be promulgated at 29 C.F.R. Sec. 1910.20.
10. New York State Law Revision Commission: Proposed code of evidence for the State of New York, 1980, Section 504.
11. 11 *Westinghouse v. U.S.A.,* 8 OSHC, 1980.
12. R. W. Ackerman: The organizational environment and ethical conduct in occupational medicine: a perspective. Bull NY Acad Med 54:707, 1978.
13. Code of Ethical Conduct for Physicians Providing Occupational Medical Services. J Occup Med 18(8): 1976.

DISCUSSION: CHAPTERS 19 AND 20

Robert L. Jennings, Jr., Larry C. Drapkin, Glendel J. Provost,
James H. Spraul, J. David Watts, Marvin Lieberman,
Mark A. Rothstein, Ana Kimball and Howard Walderman

MR. JENNINGS: I have several comments. First of all, I think possibly some of my earlier comments may have been misunderstood.

I feel that all of the industry arguments against individual worker access that I have heard are frivolous. Those include arguments like, "My God, you can't let workers have access to medical records because that will spark all kinds of medical malpractice lawsuits," and "No company in the country will be able to get occupational medical services." That last argument is made even though there are 20 states or so that impose patient access requirements on physicians in their regular practice of medicine.

Now, the issue of OSHA [Occupational Safety and Health Administration] access to medical records, without consent, in my view is far from frivolous. I can say for myself and for Dr. Bingham and all of the OSHA staff people that worked on the access standard that we found the issues of governmental access without consent to be complicated, very difficult and very troublesome. There is so much that is dependent on the good will and good faith of the particular administrators that are running the governmental program at any point in time.

We tried to set up a system that would protect workers but, obviously, if this administration or any administration in the future wants to abuse medical records, then the requirements that Dr. Bingham established are not going to prevent that abuse. That is particularly troublesome because a lot of companies have told their workers that OSHA abuses these records, and the workers are scared. I do not believe abuses have occurred but there is some potential for abuse.

MR. DRAPKIN: The assumption is made that we are operating in essentially a totally free marketplace, that people can choose to work where and when they want to, and that they are aware of the health consequences of the choices and consider all of this in deciding which job to take.

But the problem is that we really don't have that kind of free marketplace. We have a high unemployment rate. In specific parts of the country, the unemployment rate for groups of people is almost 25%, depending on the community and the group of people. And we are talking about many instances in which people have assumed that there must be a risk of injury, illness or death on the job and that such risks are unavoidable.

I would also like to suggest that in many states we have seen other ways of promoting disclosure of medical records. Mr. Jennings indicated that employers can protect themselves from liability by giving out information. But in states such as California, we have had successful cases brought in which medical records were not disclosed to workers by employers. People can get around the workers' compensation exclusive remedy provisions, which often makes it difficult to get adequate compensation for an occupationally induced illness.

Also, in states like California we have seen other cases that seem to indicate that employers will be required to disclose records. This will apply first to physicians, but cases indicate that these duties are being imposed on many manufactureres as well as on physicians and will be imposed in a few years on employers. This will occur regardless of the exclusive remedy provision. I think that disclosure is a far wiser way to go at this point because, although assumption of risk sounds attractive, I think there are so many constraints operating against it that it will prove to be a deceptive solution.

MR. PROVOST: It occurs to me that even though Mr. Drapkin is, in fact, correct—we don't have a perfect free market—we have the system, at least as far as employees are concerned. Workers cannot jump from job to job simply because their job is dangerous. Workers are not always able to say, "I'm not going to assume the risk of that job." But right now they don't have access to the information. Why not give them access to the information while they are still on their jobs? At least they will have more knowledge about what they are facing in the workplace and will be able to deal better with it. If the decision is, "Well, I've got a car payment and a house payment and I can't get another job," at least they will know to what they are exposed. If the worker later contracts an illness or injury as a result of an exposure, whatever it is, he or she will have a remedy that at least is effective. Workers won't have to settle simply for workers' compensation. This proposal would also provide the worker with information on his exposure. If the employer chooses not to, he accepts the consequences, which will be considerably more substantial and severe than increased workers' compensation premiums.

DR. SPRAUL: I speak mainly as a member of AOMA [American Occupational Medical Association] and particularly as a member of the Board and of the Committee on Ethical Practice for the past four years. I wish to correct an erroneous statement made by Mr. Jennings. There have been actions taken against members and indirectly involving nonmembers with regard to issues of ethical practice.

MR. JENNINGS: I'm glad to hear that. But if they were made publically or if any serious censure was made before the whole occupational medical and public health community over the last several years, I missed them.

MR. WATTS: I would like to take this opportunity to make some comments concerning the professional practice of industrial hygiene. Initially, as you may be aware, both the American Occupational Medical Association and the American Academy of Industrial Hygiene adopted a code of ethics. Nowhere in either code does it state that a professional should work for the employer's interests. In fact, both professions recognize that the primary responsibility is to protect the health of the employees, and I sincerely hope that view represents the majority opinion of those attending this conference. Our responsibility is to practice the profession in an objective manner, being cognizant of the fact that the health and welfare of employees may hinge on our judgment. I would seek employment elsewhere if the opposite were true.

My second comment: As you are aware, approximately 80–90% of U.S. employers have 100 or fewer employees. Therefore, the employer is too small to employ a full-time, in-house occupational health individual and, consequently, most of the services for industrial hygiene and/or occupational medicine are contracted. Based on this finding, I do not understand how Mr. Jennings can make the statement of an abuse of the programs when these services are oftentimes contracted.

MR. JENNINGS: First of all, I think every professional acknowledges his responsibility to conduct himself or herself in the most professional fashion possible. Industrial hygienists and corporate medical directors and medical officers are admittedly charged with protecting workers. That is their paper job description; but I would venture to suggest that there probably aren't too many of your plants or any of the large chemical plants in which the workers view the company doctor or the company hygienist as their principal spokesman or their principal defender on occupational safety and health matters. You are in a conflict-of-interest situation. You report to the company. You report on matters that the company tells you to report on. You don't get up in a public meeting or in OSHA hearings and take positions contrary to the written comments of your own employer. You can't do that. When I was with OSHA, I couldn't get up in a public meeting and expound at length on my personal views if they were contrary to agency policy. In the context in which you operate, there is a fundamental problem in the discrepancy between worker perception and the actuality of the situation.

DR. LIEBERMAN: The failure of a corporate medical director to testify at a hearing contrary to his company's policy does not raise the same type of ethical questions as improperly handing over medical records to management. In the first case, one needs to examine the circumstances of the failure to testify before a judgment can be made. The matter of achieving higher ethical standards in occupational medical programs can be clarified by some comments about the experiences the New York Academy of Medicine has had in attempting to improve professional medical discipline in New York state, which is under the jurisdiction of the state Department of Health and the state Board of Regents in the Education Department. The professional organization (for example the New York Academy of Medicine, for which I work) has a responsibility to do all it can to strengthen the process of physician discipline within the state by assuring that there is adequate funding, by encouraging physicians to volunteer as participants on hearing panels, by communicating to the governor, the legislature and other state officials suggestions on how to improve the procedures. I am happy to say that we are doing this. The decreasing role of the medical organization in directly regulating physician discipline on a voluntary basis stems from the lack of resources to do an adequate job. Legal costs to pursue these cases are enormous. The professional organizations have a responsibility to come forward with proposals for improvement, which could range from voluntary accreditation programs all the way to public regulation.

MR. ROTHSTEIN: Corporate medical directors and industry officials in general have voiced a number of criticisms concerning OSHA access standards. One of the main criticisms runs something like this: "The access standard does not require that any particular record be maintained, only that those maintained must be available for OSHA." The argument is that, based on OSHA's access standard, corporate medical officials will be encouraged to minimize the amount of medical record information that is maintained, thereby decreasing the quality of medical care for employees and the amount of medical research that is done at the corporate level.

My question is in two parts. Number one, do you think that this objection is realistic, and do you expect that this is a common reaction against OSHA's access standard by corporate medical directors?

And, number two, if it is a serious problem, what are the ethical implications of the position being taken by management?

MR. JENNINGS: When OSHA initiated the development of an access standard, it was conceptually a part of an overall package, including labeling of products. It became clear that the labeling standard was going to be extremely difficult, controversial and complicated and would take a long time to develop. And so, the section dealing with access to medical and exposure records was carved out and executed immediately so as to at least give workers access to whatever existed in the workplace.

Now, there were several comments in our rule-making proceeding, particularly from small companies, that went along the following lines. "You're absolutely insane if you think I am going to create medical records if I have to give them to the workers." But not when you got to the big companies, to the prestigious medical programs, and to the big trade associations. They were asked "Are you going to cut back your occupational medical monitoring as a result of this standard?" And they would say, "Certainly not, because our concern is for the worker." These employers did make arguments about supposed adverse effects of the standard, but they said they would absolutely not cut back on medical programs. I have not heard of any cutbacks. I wouldn't be surprised if they have occurred in some situations, but I feel it is a necessary price the agency had to pay in order to open up whatever records existed.

DR. LIEBERMAN: If the result of the OSHA access standard is to discourage companies from maintaining adequate medical records, then one would have to review the regulations to make the necessary changes to ensure that such records are maintained.

I would like to direct a question to Mr. Jennings. Certain physicians have expressed concern about the OSHA access standard. They may or may not be in error. It would be helpful to the discussion if you told us why you call the arguments against the OSHA standard frivolous.

MR. JENNINGS: Perhaps in the space of 20 minutes I may have been a bit too glib in trying to pull all my comments together. When I say frivolous, I am referring to what is covered in about 15 pages of preamble to the access standard in which every single argument in opposition to individual worker access was listed and all of the various considerations were laid out. Quite frankly, I don't think any of those arguments were of tremendous substance or were fundamental problems in terms of decision-making on the standard. OSHA access, however, is a dramatically different situation.

DR. LIEBERMAN: I think that while some physicians are concerned about making their medical records available to patients, several leading medical directors in industry have indicated to me that they really had no major objections to this. OSHA access to patient assistance program information does create some problems for me since the employee expectation of privacy is very conducive to the success of these programs in curbing alcoholism and other problems among employees.

MR. JENNINGS: Well, the fundamental point is that the access standard is not directed toward medical records in totality, but toward medical records that we felt had a substantial relationship to occupational safety and health. And there are numerous companies that have set up employee assistance programs that are distinct and separate from their medical programs. There is no benefit to disrupting those programs by imposing the access standard because the occupational safety and health benefits would probably be very marginal. We didn't feel there was a need to apply the access program to those kinds of distinct and separate specialized programs.

MS. KIMBALL: Mr. Provost, how does the industry's responsibility for regulation and control of the hazards and hazardous substances fit into your scheme of assumption of risk? I know you gave us a simplified explanation, but I am confused as to how much liability would be removed from the industry by informing the workers and then allowing them to assume the risk regardless of the nature of the hazards.

MR. PROVOST: Under the current law and virtually in every state in the union, a worker does not have a cause of action against his own employer for the employer's negligence. And negligence can include those circumstances in which information that should have been given to the worker was not given and, as a result of not having such information, a worker suffers injury. What I am suggesting is simply that under the doctrine of assumption of risk, as with the access to medical records regulations that we were just talking about, the employer is not likely to keep fewer records simply because he fears they will be used against him.

If you return the assumption of risk defense to the employer and give the worker the right to file a lawsuit against the employer, and if a worker is injured as a result of being denied information, all you have done is legislate that the employer do those things that would have been necessary to protect himself from liability anyway in case of injury. This proposal is not going to clean up the workplace. That is not what I am suggesting. I am simply addressing the issue of the worker's right to know. I am defining it simply as that. This would be a way of guaranteeing that the worker would be well-informed, not because of the altruistic or humanitarian motives on the part of the employer, but simply because it could mean a substantial dollar loss to the company if workers were not informed and a worker was injured as a result.

MS. KIMBALL: If you inform them, are you taking away their liability? Are you saying that if an employee *knowingly* handles toxic substances or exposes himself to noise and gets ill or loses his hearing 20 years later, the employer can say, "I told you so and you can't sue me because you knew"?

MR. PROVOST: It is not simply knowledge that is required in order for the defense of assumption of risk to be raised. The worker has to be informed of the existence of the hazard, have the knowledge of the hazard, and he must be given enough information that he appreciates the degree of the danger inherent in the hazard. For example, a worker would have to be told not only that he is being exposed to 98 or 100 decibels or whatever, but he must also be told that if he is continually exposed to that amount of noise over a period of 10 years, 20 years, or whatever, he may suffer a substantial hearing loss. The employee then has to decide whether or not he wants to wear the ear plugs, whether or not he wants to continue employment there, etc. But if the employer does not tell him, then the worker would have a cause of action against the employer. Right now, he does not have a cause of action, even if he is not informed and is injured as a result.

MR. JENNINGS: Would that be in addition to the workers' compensation remedy, or would it be a replacement for it?

MR. PROVOST: In lieu of only with regard to the specific circumstances where the cause of action was based on failure to inform, nothing else. Gross negligence or ordinary negligence in a workplace would not be removed by the compensation situation.

MR. WALDERMAN: I want to argue for the rights of government researchers to gain access to company medical records of employees without employee consent. Traditionally, in epidemiology the researcher does not seek individual subject consent, and we have to remember that there are investigations of working conditions and not of individual employees. Also, as the *Westinghouse* decision pointed out, the reason NIOSH isn't required to seek consent is that the employee may withhold consent arbitrarily, or may not understand the nature of the NIOSH investigation, or, what is probably most important, a group of employees may withhold consent because they think that is what the employer wants them to do.

I think the same thing might also be applicable to workers who write in their objections to NIOSH access to particular documents. There is a slight difference in that NIOSH doesn't have to seek the objections; it would be up to the employee to furnish his or her objections to NIOSH in writing.

DR. LIEBERMAN: What is the position of NIOSH? Are they appealing the *Westinghouse* case?

MR. WALDERMAN: There was substantial discussion as to whether or not the agency should appeal *Westinghouse,* and it wasn't solely up to the agency. There was no appeal. We will have to see how it works out in the Third Circuit. The better procedure is set forth in the *General Motors* decision.

CHAPTER 21

LEGAL AND ETHICAL DILEMMAS
OF WORKER NOTIFICATION

Merl Coon, PhD, MPH and Phillip L. Polakoff, MD, MPH
Western Institute for Occupational
and Environmental Sciences, Inc.
Berkeley, California

The legal basis for notifying workers about exposure to carcinogens and other toxic substances is clearly established in the 1970 Occupational Safety and Health Act (OSHAct), the 1976 Toxic Substances Control Act, and the Department of Labor (DOL) standards requiring that employees be informed specifically about known hazardous work conditions. Underlying these legal requirements is the ethical principle that "an employee must possess enough information to enable an intelligent choice" [1]. In spite of the technical detail spelled out in the DOL Standards and, moreover, in spite of a number of "worker notification" and other educational programs intended to inform workers of their rights to protection and to services, it is apparent that the underlying ethical precepts of worker notification are controversial when translated into practice.

The purpose of this chapter is to discuss several of the legal and ethical dilemmas underlying decisions to notify workers about hazardous work conditions. The chapter will give special consideration to the problem of how a hazardous work situation is identified, noting particularly that a work hazard may be social and psychological, as well as physical in nature, and that scientific data alone are insufficient for defining risk. The authors do not claim to have definitive answers to the many questions that are raised about worker notification. Rather, the intent of the chapter is to stimulate discussion that may eventually lead to resolution of some of the dilemmas inherent to worker notification. The chapter concludes with a discussion of how traditional methods of decision-making in law and medicine apply to worker notification situations.

BACKGROUND

The National Institute for Occupational Safety and Health (NIOSH) estimated that, in 1977, 880,000 American workers were being exposed to carcinogens regulated by the Occupational Safety and Health Administration (OSHA); an estimated 21.4 million workers were exposed full time or part time to all OSHA-regulated substances [2]. No estimates can be made of the number of persons who are exposed to hazardous work situations for which there are no governmental regulations. Overall, it seems reasonable to conclude that the number of workers in need of information about hazardous work conditions is quite possibly limited only by the number of people in our current workforce.

Programs and strategies to notify persons about hazardous work exposures can take a number of forms and come from a variety of sources. Employers are required by the OSHAct to notify their employees about specific hazards, which, of course, assumes that employers themselves are knowledgeable of such hazards. Worker notification has also been a primary or secondary purpose of surveillance and screening programs and of research. It can consist of notification of exposure, as well as notification of findings of a screening. It can take the form of indirect and impersonal notification as in the National Cancer Institute's (NCI) recent "asbestos alert" media campaign, or it can be done in the form of direct personal notification, as in the vinyl chloride surveillance and treatment program developed by the University of Louisville or the Tyler Asbestos Workers Program conducted by the University of Texas Health Center at Tyler. Notification can also come from "grass-roots" self-help programs organized by the exposed workers themselves, such as those of the Brown Lung Association, the National Association of Atomic Veterans and the several local Committee on Occupational Safety and Health (COSH) groups around the country.

This paper draws on two notification projects and some followup research conducted by the Western Institute for Occupational and Environmental Sciences, Inc. (WIOES). These projects included a NIOSH-sponsored screening program for asbestos-related disease in which nearly 2300 persons from the San Francisco Bay Area participated, a two-year contract with NCI to develop a notification program for auto repair persons and municipal firefighters working cooperatively with their local unions, and a National Institute for Mental Health (NIMH) funded research project intended to learn how families have responded to the WIOES-conducted asbestos surveillance project. In addition, the issues of worker notification have been discussed extensively with staff of NIOSH, NCI and NIMH, as well as the Workers' Institute for Safety and Health and the Institute for Southern Studies. While it is not possible to acknowledge the source of each individual idea, the many discussions with persons representing these organizations have contributed immensely to the development of this paper.

ETHICAL PREMISES OF WORKER NOTIFICATION

A WIOES staff member recently sent a questionnaire to approximately 50 people who have been engaged in some form of worker notification, asking them to identify the important ethical issues. Several different perspectives emerged from these questionnaires: (1) persons with scientific backgrounds stated a number of reservations about making ethical judgments until they see data supporting the need to do so; (2) other questionnaire respondents were apparently willing to act on their judgments that workers be notified, even with the risk of undesirable consequences; (3) a third category advocated close adherence to the rules for notification specified by the federal legislation and the DOL standards, recommending particularly that notification take place only when supported by site-monitoring data; and (4) a fourth group was very analytical, offering arguments for and against various ethical points without taking a stand. Clearly, practical application of any ethical precepts guiding worker notification depends on how the issue is viewed.

Ethical guidelines suggested for notification procedure include consideration of the following: that the content of notification information be scientifically and technically sound, that complete information be provided, that the information be delivered in a humanly sensitive and supportive fashion, that the procedure recognize the fundamental intelligence and self-sufficiency of workers and their families and avoid patronizing styles and attitudes, that there be some anticipation of the possibility of arousing undue fears and anxiety, and that the notification avoid adding to possible "hysterias" that may already exist among the public about occupational and environmental hazards and the dreaded diseases that may result from exposure to them. Underlying these guidelines are such constitutionally guaranteed precepts as the "right to know," "access to information," "freedom of choice" and "right to self-determination." In their lofty simplicity, no one would contest the righteousness of such precepts. It is in the translation of these guiding precepts into actions that one swiftly encounters dilemmas of an ethical nature. Good intentions become difficult problems when confronted by the realities of such factors as conflicting interest groups, inadequacies in epidemiological data to support a decision to notify, fears of unanticipated and undesirable consequences, and lack of consensus as to what are appropriate followup services.

Discussion of these ethical dilemmas of worker notification will proceed within a framework intended to address three practical questions:

- Who should be notified?
- How should notification take place?
- What information should notification include?

The intent of this chapter is to present the issues and viewpoints clearly. At this time, we will carefully avoid offering any of the ethical judgments that we may make personally or as representatives of any organization. Accordingly, a discussion of worker notification policy will be reserved for another paper.

ISSUES IN WHOM SHOULD BE NOTIFIED

Based on a combination of the exposure or risk involved and the type of worker notification action taken, there are two desirable, as well as two undesirable, worker notification decisions. A model for these decision categories is presented schematically in Figure 1.

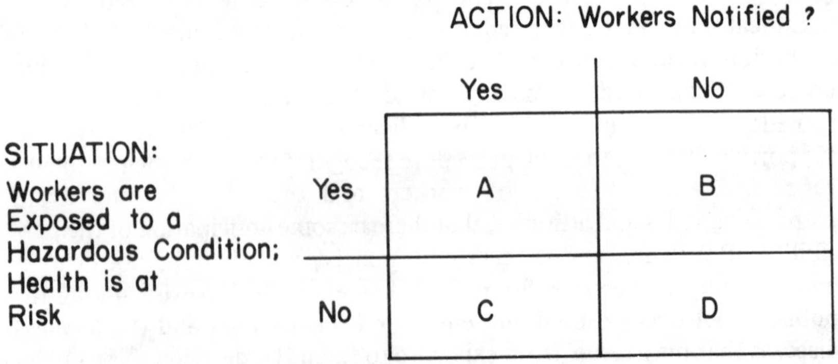

Figure 1.   Model of desirable and undesirable personal notification based on a matrix of risk situation and action choices.

Combinations A and D in the figure are desirable; notification decisions B and C are to be avoided. In other words, it is desirable both to notify workers who are exposed and at risk *and* to avoid notifying persons who are not exposed or not at risk. Contrariwise, it is desirable to *avoid* notifying persons who are neither exposed nor at risk; it is also desirable to *avoid excluding* any exposed and at-risk workers in notification. Failure to notify persons who are at risk means that they are not given appropriate information from which they can make personal decisions. For example, on the basis of notification, a worker may take no action, or he may discontinue employment in a hazardous situation, change health-related behaviors such as smoking, seek appropriate medical care, or participate in collective action intended to reduce the hazards of his work. Notification when it is appropriate, on the other hand, risks arousing undue fear and anxiety; it may lead to unwarranted and excessive use of medical services; it may lead to excessive and inappropriate litigation; and it may stir up political issues that might be better left to rest.

The intention is not to belie the complexity of the issue with this simple four-way model. Even in a situation where a health threat is clearly indicated by available data, there are important ethical considerations about whether, indeed, persons should actually be notified. In the case of notification of the cancer risks related to asbestos exposure, for example, where medical intervention can only alleviate the symptoms and do very little to slow the progress of disease, it can be argued on ethical grounds that the fears and anxieties that notification may arouse are more harmful than the medical benefits it may precipitate. Several papers on ethical dilemmas of worker notification could be written on this complex issue alone.

In any event, at the practical level the problem is remarkably simple in concept: what must be determined in making a decision to notify are both who is *exposed* and who is *sufficiently* exposed to be at risk of developing an associated disease; it is also necessary to determine who is not sufficiently exposed to be at risk in order to justify, ethically and legally, excluding them from notification. In other words, to notify about exposure to a work-related hazard, one must know who is, in fact, at risk. Though simple in concept, risk is very difficult to determine in practice.

DETERMINATION OF RISK

There is only one situation where risk seems clear and apparent. That is when a known cohort of workers has been exposed repeatedly and intensively to a specific known carcinogen or another substance that clearly puts physical health at risk. In other situations of exposure, where indications of risk are more ambiguous, notification is not so clearly appropriate. Thus, notification is not clearly prescribed when exposures are smaller and less concentrated, where exposure is less documented, or where the causal relationship between agent and disease has less scientific verification.

The most ambiguous notification situations are where the hazards of exposure are psychological, e.g., when they produce psychological stress, rather than "physical", and where the disease "agents" are social rather than physical. The dotted line in Figure 2 is intended to show that, as the

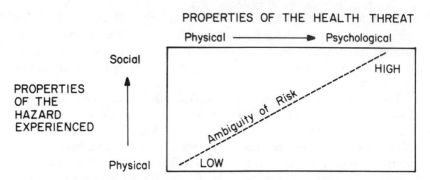

Figure 2.   Schematic representation of increased ambiguity of health risk as determined by changing properties of the hazard experienced and the health it threatens.

nature of the work exposure hazard changes from physical to social and as the threat to personal health changes from physical to psychological, the problem of whether there is sufficient risk to warrant notification increases in ambiguity. Accordingly, it is a matter of ethical judgment whether hazardous work situations other than those involving carcinogenicity or toxicogenicity might also call for worker notification.

There are further problems in determining health risk even when both the disease agent and the health threat are physical and "hard" documentation data are available. These problems have to do with inherent limitations in science. Known exposure to a specific carcinogen is neither a necessary nor a sufficient condition for the development of cancer. Monocausality cannot be assumed; allowances for error in sampling and in measurement must be made, and control of extraneous variance is only partial in any given research design. For these reasons, it can be shown *only* that increases in incidence rates of cancer for any statistical cohort are probable; they are never absolute. The level of probability that warrants a decision to notify a cohort of its increased risk is a matter of ethical rather than scientific judgment. Some data problems in WIOES' recent asbestos screening program help to illustrate this point.

In this program, severe discrepancies were observed among the X-ray readings made independently by three certified B-reader radiologists. This observation of reading discrepancies raised scientific as well as ethical dilemmas: (1) one problem was in deciding how to interpret the data, (2) a second problem was in deciding how the discrepancies should be discussed when reporting an overall prevalence rate for the cohort, especially when it was clear that policy and program decisions would be made based on this information.

The X-rays were taken in a field setting and, admittedly, there were some problems in quality. These problems notwithstanding, the purpose here is to discuss how inherent limited reliability in diagnostic testing procedures, as exemplified in the discrepancies that appeared in these X-rays, make it very difficult to determine risk, even when "hard" data supposedly are available.

Theories that guide scientific measurement do not alone give resolution to the problems. The first question to be raised is whether the discrepancies are indicative of reading inaccuracies, that is, whether they merely reflect such factors as reader fatigue or inadequate X-ray exposures. This did contribute to the confusion. It is important to realize, however, that differences among the readers' interpretations of any given X-ray reflect more than a simple matter of "right" or "wrong," valid or invalid. Rather, each reader actually may see important indications of disease which, though different from what another reader sees, may be quite valid. This is especially so when the X-ray readers are looking for indications of interstitial disease for which there are no precise clinical definitions. In any given clinical measurement, there is no way to be fully confident of the degree of accuracy it represents, especially when no precise criteria or corroborating measurement data are available. The measurement may be actually indicative of disease, or it may reflect an inaccurate reading. In medical terminology, such errors are called false-positives and false-negatives. Consequently, the summary results for any study cohort reflect many state-of-the-art weaknesses in the diagnostic processes employed. An ethical judgment is required to decide when the diagnostic data are valid enough, consistent enough, and significant enough to warrant notification of the cohort members that they are collectively or individually at risk.

Epidemiological studies have frequently handled discrepancies in X-ray readings by reporting statistical averages. For example, if reader A identified abnormalities in 50% of the X-rays, and reader B finds only 30% abnormalities, this may be reported as an average abnormal incidence rate of 40%. This averaging of the discrepancies in the readings is statistically incorrect. As such, it may lead to an inappropriate notification decision. Figure 3 is intended as a simple illustration of this statistical problem.

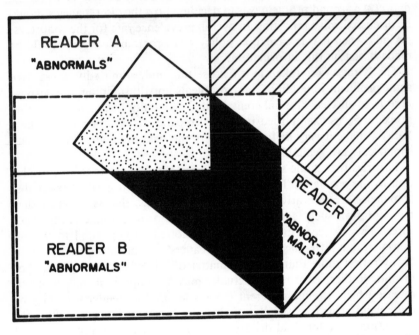

Figure 3. Schematic of the problem of interpreting discrepancies in X-ray readings in a study cohort.

In the illustration, the outer line represents the entire study cohort. The box in the upper left-hand corner represents the subset of the cohort for which reader A reported abnormal readings; the area outlined by the dotted line represents the abnormalities reported by reader B; and the diagonally placed box in the center represents the abnormalities reported by reader C. For discussion, assume that the cohort is 100 persons. Reader A reports 25 abnormalities; reader B reports 50; reader C reports 40. The numerical average for the three readers is 38.3. However, the dotted and the slanted line areas represent the only two subsets of the cohort on which all readers agree; the dotted area being the consensus "abnormal," and the slanted line area representing the consensus "normal." (The black area shows one of the three areas on which two of the three readers agree.) Thus, based on the measurement principle of interobserver reliability, the data might be interpreted as saying that a minimum of 15 (the dotted area) persons in the cohort have a reading of "abnormal," and that a maximum of 30 (the slanted line area) have a reading of "normal." The other 55% is problematical, depending on what one is willing to assume or not assume about the accuracy of the diagnostic readings. If it is assumed that one reader picks up what another misses, all identified abnormal cases of the three readers could be totaled to develop an accurate incidence rate for the cohort, in which case the incidence rate would be 70%. However, if it is assumed that each of the readers has a certain but unknown proportion of false positives, the incidence rate might be somewhere between reader A's low rate of 25% and reader B's high of 50%. Consequently, it is suggested that the data from this cohort can be interpreted accurately only by indicating that the incidence rate is between 25 and 70%. Given this lack of precision in the data, it will be an ethical judgment rather than a scientific judgment that will determine policy decisions about notification of the cohort.

If epidemiological data are used to assign measures of risk to an occupational group, single measures, such as X-ray readings, a pulmonary function test or blood pressure test, are insufficient. In epidemiological theory, disease outcome is predicated on a combination of potency of the agent and susceptibility of the host. Accordingly, the development of the disease can be predicted only by using combinations of data indicative of both of these factors. In the case of asbestos exposure, and depending on how one chooses to categorize the data, measures of "potency" might include such factors as type of asbestos used in the work situation, proximity of the worker to the asbestos, availability and use of protective equipment, and others. Measures of "susceptibility" might include a number of factors that are generally known to influence overall disease vulnerability. Age at exposure, general health history and constitution, and smoking history are some "susceptibility" factors that come to mind. Again, it is a matter of judgment as to when the data are complete enough to give an assessment of risk that is accurate enough to warrant notification.

A final problem in the determination of risk has to do with the concept of *exposure limits* or, as it is referred to in the industrial hygiene literature, *threshold limits*. According to Stokinger [3], the concept of threshold refers to the idea that exposure up to a certain limit will not produce a response, but above that limit there will be a direct and linear

relationship between the amount of the "dose" and the magnitude of response in the exposed organism.

> The premise on which the concept of thresholds rests is that,
> although all chemical substances produce a response (toxicity, irrita-
> tion, sensitization, narcosis, etc.), at some concentration, if exper-
> ienced for a sufficient period of time, it is equally true that a concen-
> tration exists for all substances from which no response of any kind
> may be expected, no matter how long the exposure, on an eight-hour
> daily, 40-hour work week basis [3].

Based on this concept of threshold limits, supported by animal research and, supposedly, with "generous safety factors" added, threshold
limit values (TLV) are recommended as good practice guidelines for air
contaminant concentrations and physical hazard levels below which "it is
believed that nearly all workers may be repeatedly exposed day after day
without adverse effect" [4]. The threshold limit concept suggests that
there are certain levels of exposure, above which worker notification is
warranted or required. That is, it suggests that workers should be notified in any situation where there are data to document "unsafe" exposures. From an ethical standpoint, however, there are some problems in
relying on these data solely as indicative of the appropriateness of notification: (1) not all scientists agree on the validity of the concept of safe exposure levels; (2) monitoring data to document excessive exposures are
not always available for both technical and sociological reasons; and (3)
monitoring measurements are frequently "crude" and, thus, the data
have limited validity.

Epstein is among the respected scientists who are critical of the concept
of safe exposure levels. Epstein questions the validity of the concept,
stating specifically that:

> It is the concensus of the informed, independent scientific com-
> munity that we do not know of any way of setting safe levels. . . . The
> greater the exposure of a population or work force or consumer group
> to a chemical carcinogen, the greater the risk. The less the exposure,
> the less the risk. But there is *no such thing* as a safe level [5] (empha-
> sis added).

And quoting the same author further:

> The arguments (about safe exposure levels) come from groups that
> have two distinct qualifications: 1) they work for industry either di-
> rectly or indirectly; and 2) they lack national or international recog-
> nition in the scientific community [5].

While risk can be defined as exposure above recommended levels, and notification could be required for all workers who have documented exposures exceeding these levels, there remain some ethical problems. If Epstein's position is correct, and if we accept the ethical precept of workers' rights to know, it can be argued that all workers with *any* known exposure should be informed of such. Possibly, the workers with low exposures should just be told the truth—that they have been exposed at levels that are below administratively defined standards but that reputable scientists disagree among themselves as to whether these low exposures are safe.

Beyond the controversy about safe exposure levels is a further problem in that data for documenting exposures often are not available. Development of suitable monitoring technology is relatively recent. Even with the availability of the technology, however, not enough worksites are monitored. Monitoring is guaranteed, apparently, only where the danger has been most publicized or where the workers are most organized; yet some of the most hazardous work conditions are found in situations where the workers are most isolated, most economically dependent and least organized. Particular examples of this would be in agricultural work, work in small businesses such as family-owned auto repair shops, and "freelance" work in such areas as building construction.

From the standpoint of the ethics of worker notification, what can be concluded about whether a documentation of risk is a necessary prerequisite for notification? The discussion here has shown in a number of ways that the concept of risk is scientifically ambiguous. Clearly, the fundamental ethical choice that must be made in every notification situation is to decide what ethical judgments we are willing to superimpose on scant and limited data in order to make a decision to notify.

ETHICAL ISSUES IN HOW TO NOTIFY

In raising the question of how to notify, the intent is not to discuss procedural detail, although certainly ethical judgments must be made in deciding whether notification should be in the form of direct personal mailings, general news articles or other means. The overriding ethical question to be addressed here concerns the amount of support and followup that should be offered with notification.

A list of minimal followup and support services to accompany notification are suggested in the literature on notification procedure. Health professionals, union officials, and the rank and file workers themselves generally agree on the desirability of accompanying notification with a

well organized mobilization of followup activities and services. The minimal services that have been suggested include: local meetings for the notified workers and their families at which various supplementary information is provided and questions are answered, a telephone "hot line"
for direct personal inquiries, medical diagnostic and treatment services,
and personal as well as family counseling [2]. General legal information
and referral have also been suggested, but it should be pointed out that
this is controversial. While there is general agreement that followup services are desirable, an ethical question arises in a situation where notification is warranted but where there are few resources for mobilizing the
services. That is to say, there is an ethical question as to whether notification should proceed or be abandoned or delayed because followup and
support services are not in place. To delay notification is to deprive an
exposed group of information that may lead individuals to take appropriate preventive or remedial action, such as seeking advice from their
personal physicians. On the other hand, to notify without support services and followup would be likely only to frustrate and raise anxieties
among those persons who do not have access to additional information
and services through their own families and communities.

This dilemma might be resolved in part by finding out what resources
are already available to the cohort. Information about the social, family
and communitiy characteristics of the cohort to be notified can be obtained through a few simple conversations with some of the workers
themselves or with some of their leaders, such as union officials, min-
isters and other persons who know the workers, their families and the
community. In situations where the workers are relatively stable in their
own communities, have an adequate health plan, are on the higher level
of blue-collar incomes, and are accustomed to dealing with problems
through their own family and community resources, it appears that the
ethical thing to do is to notify them at once, acting with some confidence
that notified persons have their own resources on which to draw. In this
case, minimal services and followup might be all that is necessary, possibly limited to an information and educational session on the problem
for local physicians and other health professionals, and a telephone service where the exposed workers may call for additional information and
referrals.

Earlier in this paper it was suggested that some of the most severe work
hazards are found in situations where the workers are most mobile, least
secure in their incomes, most lacking in employee benefits, and most dis-
enfranchised in their own communities. This appears to be the situation
where followup and support services need the most attention. In cases
where professional and financial resources to support notification are

limited, self-help methods of disseminating information and organizing screening programs, such as those used in the formation of the textile workers' Brown Lung Association, can be applied [6]. While many health professionals may consider the use of "grass-roots" methods of disseminating information about work exposures to be very risky, it is another matter of ethical judgment as to whether the benefits of an informed, self-helping work force are worth some of the risks involved.

NOTIFICATION CONTENT

Ethical precepts guiding what should be said in a worker notification seem rather simple. Clearly, a presentation of information should be straightforward and easy to understand; it should be scientifically accurate and complete; it should be sensitive to the possibility of arousing undue fear and anxiety; and it should suggest appropriate problem-solving actions the worker might take. In practice, however, many of these principles begin to contradict each other. Moreover, there are also certain legal requirements for notification that cannot be communicated in a simple way.

One need only begin drafting a worker notification statement to see the contradictions. Describing accurately what it means to be exposed and at risk is, itself, very difficult. As discussed earlier, the concept of risk is ambiguous, and if attempts are made to include all the important scientific caveats about risk, this begins to make the information very complicated. Accordingly, attempts at being scientifically accurate and complete may lead more to confusion and befuddlement than to worker enlightenment.

It is also a matter of ethical judgment as to whether fear-inducing concepts and wordings should be used. We live in an era in which the word "cancer" is very anxiety-provoking. Accordingly, there are some differences in opinion about whether a group with carcinogenic exposure should be told directly that they are at risk of developing cancer, especially if it is an older group that is probably already aware that cancer risk increases with age. Notifying a retiree who has worked hard, saved his money, and just begun to enjoy his retirement that he is at risk of developing cancer may not be the most appropriate thing to do, ethically.

There are also problems with vague and incomplete terminologies. In the Bay Area Asbestos Surveillance Project, for example, the screened individuals were given information that their X-rays were interpreted to be in one of three categories: "no abnormality," "potential serious abnormality" and "serious abnormality." Although this was acceptable

NIOSH-recommended terminology, it is needless to say that we received some very critical feedback on the sensitivity of our information and the confusion it created. It led to some rather serious evaluation of whether a large-scale screening program, where the number of participants is so large as to preclude a face-to-face method of notification, is itself ethically justified.

In any worker notification program, the notifying organization or agency may need to include some statements whose primary purpose is to protect itself legally or even politically. The personal physicians of participants in a screening program may feel very threatened by the screening information. Positive findings, for example, may be suggestive of negligence on their part. In this case, a notification statement might also want to include some caveats designed to lessen this threat. Legally, the notifying organization might want to include some wording intended to protect itself from liabilities resulting from the error of inappropriate notification, that is, from notifying and, hence, creating fear or anxiety among workers who, as it eventually turns out, may not be at risk. From an ethical standpoint, the inclusion of this information intended to protect anyone other than the notified workers themselves may be confusing and misleading and, accordingly, it detracts from the main purposes of the notification itself. Serious ethical judgments are required to decide whose interests are to be served and protected.

The WIOES Worker Notification Project staff has consulted with several attorneys about legal aspects of worker notification information, particularly in regard to how the information provided in a worker notification affects the workers' rights to file a compensation claim. In California compensation law, there is a one-year statute of limitations, which begins to run from the time a worker experiences a symptom of disability and knows the work connection to it. The receipt of notification may be interpreted as one piece of information that triggers the beginning date for the allotted compensation filing period. Although there were some differences of legal opinion, it was generally agreed that, in order to protect the notifying agency from liability and to protect the workers' compensation benefits, notification about carcinogenic exposures should include advice to consult with both a physician and a lawyer. Since there are a number of reasons to argue against such advice — that it risks burdening the worker with unnecessary fees for legal and medical advice, and that it risks flooding the medical and legal system with unnecessary requests for services and claims — it is again a matter of ethical judgment as to whether this advice should be included as a part of notification.

ETHICAL LEGITIMACY

The observations that the concept of work-exposure risk is scientifically ambiguous and that data to document risk are interpretable only in a context of ethical judgment give little direction to practical worker notification decisions. A problem is left in determining *ethical legitimacy,* that is, in knowing or justifying legitimate ethical positions that can be taken. While it can be argued that this reduces to a matter of individual or personal preference, a comparison by Scheff [7] of the ethical premises that guide law and medicine is instructive. While one possibility for a universal ethical premise to guide worker notification is suggested, the intent is not to argue for the validity of any one point of view.

As discussed by Scheff, in English and American criminal law, there is a fundamental and explicit rule for arriving at decisions when court evidence is uncertain. Innocence rather than guilt is presumed and, consequently, the evidence of guilt must be clear "beyond a reasonable doubt." An erroneous conviction is more important to avoid than an erroneous acquittal. "This concept is expressed in the maxim, 'Better a thousand guilty men go free, than one innocent man be convicted'" [7]. Where an erroneous conviction damages the individual by depriving him of citizenship rights and, moreover, where an erroneous acquittal risks harm to society by diminishing the deterrent effect of punishment or by freeing a potential sociopath, an uncertain choice is required. Direction for choice in such an ambiguous situation, however, is explicitly given in the law. Where individual rights and social benefits are weighed and compared, the individual is preferred.

The decision-making rules are not so explicit in medicine, but an implicit rule is still operative. While it is only infrequently stated as such in textbooks or in other formal instruction, Scheff points out that most physicians learn early in their training that it is far more important to avoid dismissing a sick patient than to retain a well one. This rule is often expressed by advocates of "conservative medicine" where the concept is to presume there is a medical problem until there is clear evidence to show otherwise. A missed diagnosis or delayed treatment, especially one that results in a patient death or permanent disability, is a much more serious medical error than an unnecessary application of medical procedure except, of course, when the procedure itself results in death or injury. Even considering the social costs of excessive installations of medical technologies, conservative hospital admissions or retentions, or lost work productivity, the ethical premise is the same in medicine as in law. Protection of the individual from personal risk is preferred.

If this principle is compared to the decision-choice situations diagrammed in Figure 1, further ethical direction is given to worker notification decisions. This suggests that, where there is ambiguity in information supporting a decision to notify, and where social costs are compared with individual risks, it is better to risk a Type C error than to risk an error of Type B. In other words, based on the ethical premise of protecting individual worker health even at the risk of social costs, it is better to overnotify than to undernotify.

At a practical level, this suggests a number of guiding premises for worker notification, most particularly: (1) the protection of individual worker health is worth the risk of inappropriate medical utilization or inappropriate litigation; (2) notification takes precedence over threat of economic loss to an industry; and (3) notification of hazardous exposures is preferred even at the risk of social contagions such as the creation of a "cancer hysteria."

This is not so naive as to suggest that even these directives are simple and free of contradictions. If the law protects a sociopath who then violates another person's rights by committing a crime against him, a discussion of ethics is in order. Whose individual rights were protected, in this case, is a very legitimate ethical question. By the same token, an inappropriate notification may condemn or destroy the economic underpinnings of an entire industry, cost the workers their jobs, and risk their family members of their rights to an economic livelihood. We offer no solution to the ethical dilemma that develops when it can be shown that one individual right conflicts with another. Rather, it is hoped that this may be a point from which further discussion can proceed.

CONCLUSIONS

This chapter has discussed a wide range of complexities and ambiguities in the ethical underpinnings of worker notification. It is expected that few of these issues will be resolved in the near future if, indeed, they are resolvable at all. If anything, the issues are likely to become more complex and even more controversial as the numbers of potentially toxic substances used in industrial processes increase, and as the work force itself becomes more aware of occupational health problems.

It is incumbent on all actors — whether they be labor, management, government, academic, or most important, the worker himself or herself — to participate in the development of different protocols, with the appropriate field testing thereof, to see what are the appropriate ways to notify workers of possible adverse effects of workplace exposures, whether they be physical or psychological in nature. Only through a community of effort will we be able to resolve the question of what is a just and compassionate way of carrying out worker notification. It is imperative that wherever possible worker notification be taken out of the adversary, advocacy arena so that acceptable and effective forms of communication delivery can develop.

It is important to develop good research strategies for field testing different notification protocols. Various cohort groups, not necessarily large in number, who have a variety of physical and psychological exposures seem very willing to participate in research if they understand that the intent is to provide information that will enable them to protect their personal health. Possibly, the research could be evaluated in a tripartite manner, including representatives from employers, the work force, and the community at large. Consensus meetings might then be held during which conflicting ethical positions are resolved and practical recommendations for notification protocols recommended.

One final point is to say that it is important to continually remind ourselves that worker notification itself is only a secondary consideration. From an ethical standpoint, a continuation of the work of developing appropriate control technologies, safe substitute products, and worker as well as employer education is much more important than developing protocols for notifying workers after the fact of hazardous exposures.

ACKNOWLEDGMENTS

This research was supported in part by contract N01-CH-95438-05 from the Division of Resources, Centers, and Community Agencies, National Cancer Institute; and by Grant RO1 MH34168-01 from the Center for Work and Mental Health, National Institute of Mental Health.

The authors wish to acknowledge and thank Katherine Brown-Keister, Roger Balt, Joy Carlson, Karrin Klotz and Michael Muldavin for their contribution to the ideas discussed in this chapter.

REFERENCES

1. National Academy of Sciences: Informing workers and employers about occupational cancer. Washington, D.C.: U.S. Department of Labor, Occupational Safety and Health Administration, 1978, p. 5.
2. National Institute of Occupational Safety and Health, Center for Disease Control, USDHEW: The right to know: Practical problems and policy issues arising from exposure to hazardous chemical and physical agents in the workplace, 1977, pp. 17–18; 20–23.

3.   Stokinger HE: Concepts of thresholds in standards setting. Arch Environ Health 25:153–57.
4.   American Conference of Governmental Industrial Hygienists: Threshold limit values for workroom air adopted by ACGIH for 1980, p. 2.
5.   Epstein SS: Cancer prevention. In: Work and health: inseparable in the 80's, conference proceedings. Bethesda, MD: National Cancer Institute, 1980, p. 29; p. 24. (NIH Publication Number 81-2293).
6.   Hughes C: Brown lung blues. Community Jobs: 6–7, October, 1980.
7.   Scheff TJ: Decisions in medicine. In: Scheff TJ, Ed. Being mentally ill; a sociological theory. Chicago, IL: Aldene, 1966, pp. 105–27.

CHAPTER 22

SOCIETAL IMPLICATIONS
OF OCCUPATIONAL RESEARCH

W. Clark Cooper, MD
Berkeley, California

I have chosen to construe societal implications as "costs and benefits" in the broadest sense. As Schneiderman illustrated in Chapter 17, tracking down indirect benefits and costs can carry one a long, long way from those that are direct and immediate.

There is no way to perform acceptable quantitative cost-benefit analyses in occupational health—the exercise is futile. Nevertheless, all who are engaged in the practice of occupational health base their recommendations and actions on a personal estimate of the costs and benefits.

TYPES OF OCCUPATIONAL HEALTH RESEARCH

For the purposes of this chapter, occupational research will be construed as research relating directly to the health and safety of workers. Such research draws on all the sciences, basic and applied, and is by definition goal-oriented.

Toxicological research in lower animals supplies the largest volume of information to be assimilated by the occupational health practitioner. Because of the almost infinite number of possible combinations of species, strains, dosages, routes of administration and durations of exposures, translating results into accurate predictions of human experience is as much an art as a science. One should not be surprised by conflicting evidence.

Research in nonhuman systems includes studies conducted on systems in vitro. Animal research involves not only observation and measurement of harmful effects of materials on systems, but also includes collection of information on absorption, excretion, metabolism and storage of agents, singly and in combination. This body of information forms the scientific underpinning of occupational medicine.

Whatever the purely scientific value of studies in vitro and in lower animals, the objective of the research as viewed by the practitioner of occupational health is the prediction of effects in man. The starting point for the recognition of occupational diseases has traditionally been clinical observations by discerning physicians, but isolated case reports from poorly defined populations are rarely convincing. Verification begins with the clinical observation of several similar cases in individuals with a common exposure. This is basically intuitive, nonquantified epidemiology. Still another valuable type of clinical observation is the report of cases that agree with unusual findings in experimental animals, as for example, the first cases of hepatic angiosarcoma found associated with vinyl chloride exposures [1].

Formal epidemiological studies provide the ultimate tests, both for acute toxic effects and for chronic effects, of agents that have been in use for a prescribed period of time. It is the epidemiological studies that have demonstrated conclusively the associations between a number of different types of work experience and disease, including recognition of the carcinogenic effect of hexavalent chromates, underground uranium mining, exposure to commercial asbestos and coke oven exposure. While the list is not a long one, it is convincing.

Despite the power of epidemiology to prove cause-and-effect associations where the risk is great, it is a much less certain science when attempting to ascertain low-level risks, or as proof of no effect. Related to this is the problem of adequate dose-response relationships. Those who look to epidemiologists for finely tuned answers to such questions in these areas will almost certainly be disappointed.

Epidemiology in occupational health is a fairly soft science in which the sophisticated tools of the biostatistician are often used on raw data of very dubious quality. Those who have tried to classify retrospectively the multiple exposures of workers over many job changes during a lifetime sometimes marvel that any useful information emerges. When reviewing epidemiologic reports, one should concentrate on how the information was collected, how groups were defined, how effects were identified and measured, and how exposures were quantified.

As previous authors have pointed out, there is much more to occupational health research than toxicology and epidemiology. Major contributions to our understanding of how to predict and prevent harmful effects come from studies of the absorption and metabolism of agents, the types of physiological and immune responses, potential causes of hypersusceptibility, and the scientific and technical aspects of industrial hygiene, to mention only a few.

GENERAL SOCIETAL IMPLICATIONS

What are the societal implications related to the broad areas of occupational health research that have been summarized? There are implications for society in research that is not done, in research that is done well and carefully reported, and in research that is done poorly or that is improperly interpreted. There are implications in the way it is presented to the scientific community, the general public, regulatory agencies, industrial management, workers or attorneys. There are vested interests with financial, political or emotional stakes in nearly every conceivable position or interpretation.

CHOICES FOR STUDY

In these days of diminished expectations, priorities for research become increasingly important. Society and its attitudes have an impact upon the volume and the nature of occupational health research, as well as the converse. Occupational health research definitely tends to follow fads. As an example, there have been over 2000 articles published on asbestos and health since 1965 [2]. The average number per year since 1970 has been over 150. In the preceding five years, the number averaged about 70 per year; in the five years before that, there were about 30 per year. When one goes back to 1945, the average was about 20 per year. Another example is vinyl chloride. Prior to 1973, I have been able to find only 15 or 20 published articles; there have been between 150 and 200 since 1973. Therefore, choices for investigation and funding very often follow events and utilize major resources. There remain large areas where research has been neglected. Some of these will be summarized in a later section.

FUNDING OF RESEARCH

I would like to turn now to some of the social implications that relate to the funding of occupational health research. While there should be alternative sources, in my opinion the federal government should play a considerable role, and by this I refer particularly to the National Institute for Occupational Safety and Health (NIOSH). Problems that need to be studied cross industry, union and state lines. The power to initiate studies is extremely important, so I am a wholehearted advocate of an expanded NIOSH research role. Nevertheless, much research (especially in the near future) will have to be and should be funded by industry or jointly by industry and labor.

Industry must carry out toxicological evaluation. It has access to information on exposures and exposed populations. The political changes that have altered the occupational health scene in the United States during recent months provide industry a splendid opportunity to plan and carry out studies without the pressures of regulatory deadlines. Responsible spokesmen for industry make it plain that they want to provide safe and healthful workplaces. Most agree that regulations are needed to provide a common framework for operations and that what is desired are regulations that are reasonable and based on facts.

Occupational health problems will not disappear because of a popular vote, and the wheel of fortune will turn again, you can be sure. So I think at the present time it is clearly in the best interest of industry not to lose momentum but to organize the information it has systematically collected, define hazards if they exist, document the effects of control measures and evaluate dispassionately the benefits and costs of medical surveillance. In other words, it should take a very active role in ongoing occupational health research.

As to the societal implications of the source of funds, it has been my experience that the source does not necessarily determine quality or objectivity. I have served in government, major universities and private consulting practice. There are governmental scientists with vested interests in their budgets, previous public positions and relations with essential constituencies. There are those in academia with a vested interest in anxiety, publicity and continued funding. And there are certainly those in industry who have been overoptimistic in the face of mounting evidence and who are justifiably fearful of inviting financial liability. While questions about possible bias are always relevant, the source of funds does not provide a quick answer. In the long run, one must demand enough detailed information for independent evaluation and interpretation and depend on the general integrity of the scientific community.

PERFORMANCE AND
INTERPRETATION OF RESEARCH

There are societal implications and major ethical problems involved in research operations, their interpretation and the dissemination of resultant information. The whole question of informed consent and its impact on the conduct of research is a separate question that will be left to other authors. Epidemiologic studies of mortality are usually not done with the knowledge of the individual workers involved, since many are no longer employed and must become the subject of a search to determine their vital status. Studies in which individual workers are examined require not only "informed consent," but also well thought out plans for providing workers and their physicians with information they can understand and on which they can act.

DISSEMINATION OF INFORMATION

The obligation of an investigator to make widely available information that might be important to workers' health is clearly an ethical one, but it is now also a legal obligation that one ignores at grave peril. The problem today is not so much one of concealment of positive evidence as it is the premature release of preliminary study results long before data are available for serious independent analysis.

Publication in so-called peer review journals is the accepted process by which scientific information is made generally available. However, I wish to caution against regarding peer review as a system that verifies quality. It is, of course, no such thing. Scientific journals vary tremendously in the rigor of their review processes. Anyone who has received a 30-page manuscript and been told to return it within a week knows how difficult it is to get beyond nit-picking and criticisms of statistician techniques. It is very difficult to consider basic study design and possible biases in data selection. These are difficult to determine from information in the paper.

The problem of evaluation, in part, arises from the space demands of most journals, which usually cannot afford compilations in the detail needed for independent analysis. There are, on the other hand, many detailed scientific reports widely distributed, which have not been published in the formal sense, but which provide adequate details so that any interested person can check out the analyses and conclusions. Nevertheless, these are dismissed by some because they have not been subjected to peer review while, at the same time, thoroughly discredited studies are quoted as references because they once appeared in a major journal.

The societal implications of media coverage of occupational health research is another major topic. Everyone has an opinion on this one, and attacking the media as a major villain is easy. But the fact is that bad news is more interesting than good news, and positive findings make better headlines than negative studies. A saving feature has been that a plethora of reports has made the public fairly blasé. While this can distort priorities, it is apparently unavoidable.

It is unfortunate, however, when limited conclusions are released and not followed up with a full story for critical review. The preposterous predictions of occupational cancer incidence released by the Secretary of Health, Education, and Welfare in September, 1978 [3,4] are still quoted all over the world, even though the final report was never completed, much less published.

SPECIAL CONCERNS

When considering societal implications, I have several areas that give me special concern. One is the occasional failure to consider the changes that have occurred in recent decades in the manner in which society looks at the environment and the responsibilities of government, industry and labor. The physician's view of what he should tell patients has evolved, just as has society's view of what he should tell them. It is grossly unfair to take statements and actions from the past and judge them out of context, using today's freer and less paternalistic approaches. As someone remarked in the meeting yesterday, how sure can we be that our current standards will be accepted a generation from now?

RESEARCH AREAS WITH MAJOR IMPLICATIONS

There are a number of areas in which occupational health research has been initiated that have particularly broad social implications. One is the detection, understanding and prevention of possible reproductive hazards associated with occupational exposures. Bingham emphasized this in a keynote speech at the excellent workshop on assessing reproductive hazards in the workplace sponsored by NIOSH in 1978 [5]. The problems are not simple ones, and there is need to discriminate between mutagenicity and teratogenicity, between effects resulting from exposures to the male and those resulting from exposures to the female. Individuals are now being forced to make decisions with even less information than usual. For example, even with a substance with a history as long as that of lead, it is not known whether levels of absorption resulting in blood lead concentrations between 30 and 80 μg/100 ml are hazardous to the germ cell, embryo or fetus.

Another area of needed research that has important social implications is that of variances in susceptibility resulting from genetic differences, particularly those that are associated with recognizable deviations in metabolism [6]. This is particularly sensitive because some of the best known metabolic abnormalities, such as glucose-6-phospho-dehydrogenase (G6PD) deficiency and sickle cell trait, occur nearly always in minority groups. It is essential to determine if these various genetic variants have any practical consequences for individuals exposed to otherwise acceptable levels of stress.

Still another area of needed research, which has societal implications because it relates on the one hand to wise use of medical resources and on the other hand to employability of individuals, is that of the efficacy of various types of medical surveillance. It would be very illuminating to review critically periodic chest X-rays or pulmonary function tests using protocols designed to find out whether or not there were any measurable benefits.

Another neglected area of practical occupational health research with societal implications because it tests our process of setting standards is the systematic evaluation of the efficacy of control measures. Plans for documenting effects should be built into plans for implementing control procedures.

CONCLUSIONS

My conclusions are not too controversial. Continuing research on the effects of the work environment on health is essential, and its volume or direction should not be a function of the zeal of the regulator. Financial support should come from a variety of sources. Industry, which has the basic information, should play a major role in epidemiological studies of workers. The government should be prepared to carry out or coordinate studies that cross industry and union lines and that might otherwise present great organizational difficulties. NIOSH should have a major role in studies. In this function, it should be independent of OSHA and other regulatory agencies. NIOSH should be neutral in its research role, even if not neutral in its concern for worker health. It should not regard the world as divided between an industrial complex oblivious to worker health and responsible citizens concerned with worker welfare. Conspiratorial fantasies impede an attack on the real problems. There is an important task ahead in preserving the essential functions of OSHA and in rebuilding an independent NIOSH under strong scientific leadership. OSHA should regulate and NIOSH should be responsible for scientific research and training, and they should coexist in a state of friendly tension. If they agreed on everything, it would be unhealthy. There is no question but that the interpretation of research results and their application in terms of regulations will continue to be adversarial. The most one can hope for is that data will be honestly collected, accurately processed, appropriately analyzed and fairly interpreted, with results of studies published in sufficient detail for a discerning eye to conduct a critical evaluation. The societal implications are so great that we deserve no less.

REFERENCES

1.   Creech JL, Johnson MN: Angiosarcoma of the liver in the manufacture of polyvinyl chloride. J Occup Med 16:150–51, 1974.
2.   Peters GA, Peters BJ: Sourcebook on asbestos diseases: medical, legal, and engineering aspects. New York: Garland STPM Press, 1980.
3.   Bridbord K, et al: Estimates of the fraction of cancer in the United States related to occupational factors. Draft report prepared by the National Cancer Institute, National Institute of Environmental Health Sciences, and National Institute for Occupational Safety and Health, September 15, 1978.
4.   Higginson J: Proportion of cancers due to occupation. Prev Med 9:180–88, March, 1980.
5.   Bingham E: Some scientific and social issues of identifying reproductive hazards in the workplace. Proceedings of a Workshop on Methodology for Assessing Reproductive Hazards in the Workplace. DHHS Publication No. (NIOSH) 81–100, October, 1980. Pp. 3–6.
6.   Cooper WC: Indicators of susceptibility to industrial chemicals. J Occup Med 15:355–59, 1973.

CHAPTER 23

WORKERS' COMPENSATION FOR OCCUPATIONAL ILLNESS: A CASE STUDY

L. A. Sagan, MD
Electric Power Research Institute
Palo Alto, California

The particular dilemma I wish to address is that which exists with respect to our desire to compensate occupational illness and our inability to identify the cause of illness in the individual. The system that has been developed to achieve this purpose is workers' compensation insurance, a system first developed to provide income to workers during periods of disability due to injury. There are crucial differences between injury and illness that make the compensation system work for the former. In cases of illness, however, it is my opinion that the difficulty in assessing causality and in identifying the responsible employer in the work history make the compensation system both inefficient and arbitrary. These problems and some remedies will be discussed in this chapter.

An opinion that appears desirable, although radical, is to abandon the distinction between occupational illness and nonoccupational illness and to adopt a "no-fault" insurance system.

WHAT IS WORKERS' COMPENSATION?

Workers' compensation is a mandatory form of insurance required of all employers. The first compensation law was passed at the turn of the century, and was followed by all other states by the next two decades. These laws had certain common features:

1. They removed the requirement that the injured employee prove negligence, creating an obvious advantage to the injured employee.
2. They require the employer, by law, to provide proof of his ability to pay damages — generally through insurance arrangements.
3. They require that the employee agree to give up his right to legal redress and to accept as his sole remedy the payments, limited by law, that are provided for according to a fixed schedule. The schedules vary from state to state.

The employee's loss of access to the courts was the source of considerable controversy and was at first resisted by some unions. The Maryland law of 1902 was found unconstitutional on the basis that it jeopardized the right to trial by jury. Finally, in 1917 the U.S. Supreme Court upheld the constitutionality of compensation law.

Compensation law is generally considered to have been successful in providing for compensation of medical costs and other losses associated with traumatic accidents. It reduced the heavy costs associated with adjudicating fault. By mandating experience-rated insurance, it created an incentive for the employer to improve safety. By provision of *limited* compensation, incentive was maintained for the employee to exercise personal control over safety.

It is important to understand that workers' compensation is not designed to remedy injuries. Rather, the intention is to provide the employee with some measure of income insurance. It insures against the loss of earnings from the injury or illness. The compensation model, when functioning properly, contributes significantly toward alleviating the loss of wages due to injury or illness, benefits the employer by reducing litigation costs and benefits society, which would otherwise assume the support of the worker and family.

During recent years, as the recognition of occupational illness has increased, workers' compensation has been expanded to include illness as well as injury. However, a legal fiction was created in which illness was viewed as the result of multiple, repeated injuries. Since the ill employee may have been employed by many employers over a period of many years and may have had a number of exposures, it becomes extremely difficult to identify which employer is to be held liable for compensating the illness.

Still another difficult problem arises from our inability to distinguish occupational illness from that which occurs spontaneously, i.e., how long should the employment persist before their is a presumption of causality? For example, if an employee has been employed for a short period before becoming ill, how does that affect the notion of causation?

Should the exposure level be considered as well as the period of employment? Both duration of employment and exposure level? Neither?

Cumulative injury differs considerably from the accidental injury for which workers' compensation was designed. The largest source of claims for cumulative injury comes from back injury, followed in California by heart disease, hearing loss and extremity injury. Unlike acute injuries, cumulative injury claims are filed by older workers and almost always involve litigation. The median age of the worker who files in California is above 50. Although as a portion of total claims awards, cumulative injury claims are still small, the numbers are increasing rapidly. For example, in California claims are growing at an annual rate of more than 30% [1].

The workers' compensation system frustrates ill employees with long delays. In 60% of the cases, final disposition requires more than a year. Furthermore, capricious judgments aggravate employers with high legal costs and work against the public interest by maintaining the myth that some persons do and other persons do not have occupational illness.

IS THERE A PROBLEM
WITH WORKERS' COMPENSATION?

Workers' compensation law as a management element for controlling occupational illness (as opposed to occupational injury) is considered by many to be seriously flawed. Fundamentally efficient operation of compensation law requires that there be a reasonable presumption of causality if overinclusiveness or underinclusiveness is to be avoided. For reasons to be discussed, the sorting out of occupational illness from other illness often has no scientific basis and is, therefore, rapidly creating widespread discontent with the system, both among management and employees.

In addition, I feel that the workers' compensation system has an undesirable effect in that compensation becomes an issue in labor-management relations, often resulting in management's refusal to engage in occupational health research. When such research is undertaken, the conclusions of that work become highly controversial. Clearly it would be preferable if research in the quantification of occupational risks could be carried out in a dispassionate manner protected from these self-interested challenges. I believe that the current workers' compensation system often prevents adequate research from being done.

In this chapter I shall focus on occupational radiation exposure as an example of the problems inherent in operating the compensation system. There is a certain advantage in choosing radiation for this purpose, namely that the risks associated with radiation exposure are fairly well understood. Therefore, if the compensation system is operating efficiently, it should perform well for such injuries. If the system does not function well in cases of radiation exposure, then it is unlikely to operate well for other chronic chemical or physical exposures where causal relationships are not so well understood.

I should like to make it clear at the outset that my discussion of workers' compensation pertains only to chronic occupational exposures to chemical and physical agents and not to acute or accidental exposures. With acute injuries, causal relationships are usually clear, and conventional compensation systems work well. With chronic occupational illness, causal relationships are often unclear and compensation systems operate poorly. Acute radiation injuries are rare in modern industry, whereas low-dose exposures and a risk of latent disease are increasing. This chapter is directed to the latter problem.

WHY WORKERS' COMPENSATION CANNOT WORK AS WELL FOR OCCUPATIONAL ILLNESS AS IT DOES FOR TRAUMATIC INJURIES

There are several characteristics of chronic occupational illness that enormously complicate the recognition of occupational illness and the provision for an equitable system of post-facto compensation. They are:

1. Chronic occupational illness is generally indistinguishable from non-occupationally induced disease. For example, the leukemia that follows radiation exposure is no different from and cannot be distinguished from disease that occurs spontaneously among people who are not exposed to radiation.
2. It is becoming increasingly clear that all chronic illness is "multifactorial" in nature. By that I mean that evidence is rapidly increasing to demonstrate that multiple external agents and personal biological factors may interact to affect the development of disease. Even in the most hazardous of environments, only a minority of individuals develop a bad outcome. From this observation, I conclude that any one environmental agent is weak; and, therefore, the appearance of disease will depend on simultaneous exposure to other agents or will occur in a particularly susceptible individual.

3. Disease typically appears many years after the occupational experience that contributed to its development, perhaps at a time when the employee has moved to new employment, thus making it difficult both medically and legally to attribute the illness to the responsible employer.

4. As a direct consequence of the other observations noted above, our current understanding of the relationship between environmental exposures and the development of disease is in a very primitive state. Whether some, most, or little human illness is a result of work experience can still be argued. The consequence of this state of affairs is that we must start out on our examination of social policy deeply sensitive to the knowledge that much of what may properly be called occupational illness is now unrecognized and that much of occupational illness has elements that are personal in nature, i.e, genetically determined or related to nonoccupational factors.

IS THERE A PROBLEM?

There is a problem, but it is not exclusive to the workers' compensation system. The problem arises wherever courts, agencies or other authorities are charged with adjudicating causation in chronic illness. Partly this springs from a new awareness that environmental agents may play some role in disease induction.

Recognition of an environmental element in the causation of chronic illness has led to a rapid rise in claims and litigation directed toward employers and governments. These claims follow exposure from a variety of agents including ionizing radiation and chemicals such as vinyl chloride and dioxins. The magnitude of these claims sometimes reaches staggering proportions, e.g., some 3000 veterans have brought suit against five chemical companies for $44 billion based on a spectrum of illness ranging from cancer to skin disorders, loss of sex drive and psychological illnesses. In Utah, billions of dollars in suits have been brought on the basis of claims of cancer following exposure to fallout radioactivity. Partly as a result, the cost of workers' compensation rose from $4.9 billion in 1970 to an estimated $15.8 billion in 1978. The Social Security Disability Insurance Program established in 1956 was making payments totalling $1.7 billion to 1 million disabled workers in 1960. Many of these disabilities were based on claims of occupational injury and illness. Since then, the numbers have doubled, with more than 2.8 million disabled workers drawing about $13 billion in 1978. The Department of Health and Human Services (DHHS) projected a budget of $16.6 billion for 1980.

This paper does not address itself to the issue of whether these claims are justified. There are, in fact, no valid estimates of the extent of environmental or occupational illness in the United States, a reflection of the fact that exposure histories are rarely available, dose-response relationships are poorly understood, the number of toxic agents in the environment is rapidly expanding, and there are poorly understood interactions between toxic agents and personal habits such as cigarette smoking. The more important conclusion to be drawn here is that the requirement that causation be demonstrated or excluded is becoming an increasingly important issue.

SHOULD WORKERS' COMPENSATION FOR OCCUPATIONAL ILLNESS BE ABOLISHED?

One option, of course, is to ignore occupationally induced disease and assume that all disease is the responsibility of the individual and the problem of sorting out these issues is insuperable.

This is clearly an unacceptable option. As shown in Table I, taken from a recent survey, all segments of the population agree that the employee should not have to bear the financial burden of occupational illness, even when the risk is known to the employee. In the latter case, the majority seem to prefer a sharing of costs.

It is not acceptable to ignore occupational illness, but workers' compensation may not be the best way to solve the compensation problem, as we shall see later.

WHAT DO WE WANT TO ACCOMPLISH WITH A COMPENSATION SYSTEM?

Before moving on to consider alternatives to our current workers' compensation system, at least as currently constituted, it is necessary to consider our objectives. No claim is made that the following list of guiding principles is definitive, nor shall I attempt to rank order these values among which there is a dynamic tension. Certainly, individuals will vary in the weight they would place on each of these, and weightings might well change with time. The purpose is to stimulate a more thoughtful consideration of that mixture of goals we wish to achieve in attempting to construct a policy directed toward management of occupational risk.

Table I.   Who Should Have Primary Financial Responsibility in Situations Involving Industrial Risk and Worker Safety? [2][a]

Who is Responsible for:	Type of Respondent				
	Top Corporate Executives [402][b] (%)	Investors/ Lenders [104] (%)	Congress [47] (%)	Federal Regulators [47] (%)	Public [1488] (%)
A disease caused by exposure on the job to dangerous substances years earlier?					
Employee	2	3		2	5
Employer	70	74	89	85	83
Shared Responsibility	22	19	9	11	10
Not Sure	6	4	2	2	3
An employee injury incurred on a job known by the employee to involve high risk?					
Employee	16	27	11	11	26
Employer	32	16	38	43	32
Shared Responsibility	50	55	45	45	39
Not Sure	3	2	6	2	3

[a] Respondents were asked the question, "The following situations concern industrial risks and worker safety. For each situation, please tell me who you feel should have the primary financial responsibility for the injury or disease — the employee, the employer or shared responsibility?
[b] Numbers in brackets indicate number of respondents.

Efficiency

By "efficiency" I mean the operation of industry, agriculture, and commerce in such a manner as to be optimally productive of goods and services and to allocate resources in conformity with the wishes of consumers. The underlying view is that economic goods are always limited and that greater productivity is, therefore, a good. Assuming that most will agree that more is better than less, the following statements are logical consequences:

- The illness compensation system should provide incentives to both employer and employee to optimize risks, whether of radiation exposure or other chemicals. The word "optimize" deserves some discussion. Implicit to its use is the notion that zero radiation exposure is, if not impossible to achieve, prohibitively expensive and, if achieved, would result in such costs as the denial to society of goods that would benefit the general welfare and the denial to workers of desirable jobs. What is the optimal level of radiation exposure? The "optimal" level of exposure cannot be specified since it may well vary from industry to industry; indeed, it may vary from task to task. For example, reduction of occupational exposures in medical therapy may be easily, i.e., cheaply, accomplished compared with reduction in certain maintenance procedures in nuclear power facilities. Therefore, optimal exposures under one condition may be excessive under another.
- An ideal compensation system would not generate excessive management or transaction costs. Practices that require extensive costs in managing the system or adjudicating grievances, disputes, or claims are to be avoided if efficiency is to be maintained. Legal costs or other costs of arbitration or administration are examples.
- An occupational risk system that operates efficiently will ensure that employees have access to full information regarding hazards. Herein lies a very significant problem for the traditional compensation system. Information regarding occupational illness is only now starting to accumulate. If an ideal compensation operates effectively only when full information is available, what are the options when full information is unavailable? This question will be addressed more fully later.

Equity

Justice requires that illness resulting from an industrial process be *fully* reimbursed. Although it is not always clear precisely what price compensates an individual for pain, suffering and shortening of life span, it is nevertheless clear that the vast majority of awards for chronic and disabling illnesses are purposely inadequate, since these low awards are implicitly thought to represent a deterrent to unworthy claims. But it is not only a sense of justice that mandates full compensation. If a market is to operate efficiently, all social costs should be internalized so as to allow resources to be optimally allocated. In other words, under circumstances where health costs are not fully compensated, products of hazardous industries will be overused. They are being subsidized by the injured workers.

OCCUPATIONAL RADIATION EXPOSURE: A CASE STUDY

The general point to be made in this chapter is that the excess cancer risk imposed by occupational radiation exposure is small compared with the risk to such individuals from all other causes. The incremental risk is so small that it would not be expected to be statistically detectable (Table II).

Table II.   Estimate of Annual Fatal Cancers among Persons
Exposed to Ionizing Radiation [3]

Exposure Source	Population's Dose	Radiation-Related[a]	Total Number[b]
Occupational			
Healing Arts	60,000	6	1,000
Manufacturing and Industrial	50,000	5	13,000
Nuclear Energy	50,000	5	140
Research	12,000	1	180
Naval Nuclear Propulsion	8,000	1[c]	60
Nuclear Weapons Development	800	1[c]	?
Atmospheric Nuclear Tests[d]	5,000	1[c]	25
Other Occupations	50,000	5	?

[a]The estimate for radiation-induced fatal cancers is derived by multiplying the ratio of one excess fatal cancer per 10,000 person-rem by the annual collective dose for each group.
[b]The total annual number of fatal cancers is estimated by multiplying the "natural" rate of fatal cancer by the approximate number of persons involved in each activity.
[c]The estimated number of radiation-related fatal cancers per year is less than one.
[d]Available data for the nuclear test program is cumulative for the 18-yr atmospheric test period (1945–1962). The annual amounts shown are based on prorating this cumulative information over an 18-yr period.

Standards permit average annual exposures of 5 rem. Actual average exposures are about 10% of permissible, or 0.6 rem, in the nuclear power industry. Throughout the United States only a rare individual will accumulate a lifetime exposure of 50 rem, an exposure that, on odds, has less than 1 chance in 100 of producing fatal cancer.

Since roughly one person in six will die of cancer, the problem for the compensation system is to select the radiation-related cancer from the much larger number that occur spontaneously. At the moment, the number of compensation claims for occupational radiation-induced illness is small but rapidly increasing, just as are all occupational illness claims. During the 1960s and 1970s, the Atomic Energy Commission and the Department of Labor sponsored a series of studies of workers' compensation and radiation injuries. By 1965 claims had been identified in nine states with a total of 50 delayed injury cases. Approximately 16% of the claims were awarded. Examination of these cases reveals that there were no consistent patterns in the exposure histories of these individuals that would explain judgments for the plaintiff or the defendant. It would appear that judgments are almost made on a random basis. Regardless of the outcome, legal fees are high to both parties and both sides are often left with a sense of injustice.

ARE THERE OTHER OPTIONS AVAILABLE TO EMPLOYEES WITH CLAIMS OF OCCUPATIONAL ILLNESS?

Barred from suing his employer, a worker with an occupational disease can sue the manufacturer or seller of a product used in the workplace if that product caused the illness. For instance, a worker exposed to asbestos while employed by an installer of insulation can sue the company that supplied the asbestos to his employer. Products liability actions filed by those contracting occupational diseases have proliferated in recent years thereby providing a potential source of relief for those who can locate third parties to blame for their illness [4].

To qualify for recovery in such suits, an ill employee must show a "defect" in the product. In disease cases, most of which involve potentially hazardous substances like asbestos, a manufacturer's failure to give an adequate warning of the risks of a product makes it defective. The duty to warn extends to all risks known or reasonably foreseeable at the time of sale. While some courts have extended this duty to both buyers and users of the product, others have limited the manufacturer's duty to providing cautionary information only to the purchaser of the product (the employer). Under the latter view, an employee who is injured because he was never warned about a hazard may be prevented from recovering against a manufacturer. This rule might make sense in industrial accident cases since the employer has good reason to warn employees of the hazards of a product that poses the risk of an accident. After all, he will probably have to pay an injured employee's benefits and, more importantly, an injury to a worker will slow production and require a trained replacement. With products that cause disease, however, an employer's incentive to warn employees diminishes. The disease may be one of long latency that does not affect production; the employer will rarely have to worry about liability for the illness; and the employer may not wish to cause employees concern.

If an employee can establish a prima facie case, some potent affirmative defenses can absolve third parties of liability. For instance, an employee may be held to have assumed the risk of illness if he has exposed himself to workplace health hazards, even if he feared that by not doing so he would lose his job. The longer the period of employment, the more likely it is that courts will deem a particular hazard obvious and, therefore, hold that the worker assumed the risk. These holdings may make some sense for accidental injuries since the risk of a mishap may be immediately evident; it does not take a worker long to realize he may get his hand caught in a drill press. The assumption of risk defense, however, loses logical force in the context of occupational diseases. To take one important class of cases, a worker can hardly be presumed to know that a particular invisible vapor or microscopic particle causes cancer, nor would experience necessarily increase his sensitivity to the risk.

The most serious drawback to products liability actions for occupational disease victims may simply be the expense of litigation. One study suggests that only 37.5¢ of each premium dollar of products liability insurance ever reaches the pockets of claimants. Lawyers, insurance companies, expert witnesses, investigators and courts retain the remaining 62.5¢. By any standard, these tort actions are inordinately inefficient.

WHAT TO DO?

Workers' compensation law is considered by many to be seriously flawed as a management element for controlling occupational illness (as opposed to occupational injury). Fundamentally, efficient operation of compensation law requires that there be a reasonable presumption of causality if overinclusiveness or underinclusivenss is to be avoided. For the reasons noted in the introduction, the sorting out of occupational illness from other illness often has no scientific basis and is, therefore, rapidly creating widespread discontent with the system. A federal interagency task force has recently reviewed the benefits available to persons exposed to ionizing radiation—both occupational and nonoccupational sources [3]. For reasons discussed above, the panel concluded that the benefits system, including both workers' compensation and liability, is seriously defective. One of the alternatives that has been discussed would require elimination of the requirement for demonstrating causality and would create a presumption of occupational cause. This approach would ensure coverage of all insured persons for all disease but would be enormously costly in that it is overly inclusive. Variants on this would create certain requirements for coverage. For example, one proposal would encourage or require industry to provide third-party hospital and medical insurance that would cover the expense of diseases linked to radiation for all employees who have been exposed to radiation. The right to health benefits would vest only after the worker completed a certain number of years in the industry (with one or more companies) and after exposure to a certain level of occupational radiation.

In my view, this latter plan only exchanges one set of problems for another. It requires that two separate arbitrary thresholds be set to establish eligibility: (1) the number of years of employment, and (2) a certain level of radiation exposure. Furthermore, it requires the identification of radiation-associated disease. Certain cancers, such as leukemia and breast cancer, are known to be induced by radiation. In other cases, such as multiple myeloma and gastrointestinal cancer, the causal relationship is uncertain and, therefore, likely to lead to contentiousness among all parties.

HOW ABOUT CREATING "PRESUMPTIONS" THAT WOULD PROVIDE CLEAR GUIDELINES FOR ADJUDICATING CLAIMS?

In an attempt to implement such a strategy, Senate Bill 1865 proposes that a presumption of radiation causation be established for any lung cancer appearing in a uranium miner. Based on the experience of the federal program for black lung disease, it can be expected that a federal compensation program for radiation-induced illness will gradually expand as eligibility is widened and criteria are liberalized. The black lung legislation passed as Title IV of the Federal Coal Mine Health and Safety Act of 1969 was the first federal law providing workers' compensation for a single occupational disease. The law provided legal presumptions of a work-related disease as well as a means for financing benefits. Miners were presumed to be totally disabled from black lung disease if they had worked in a mine for ten years or more and had medical evidence of complicated pneumoconiosis based on objective criteria. Program costs grew from $150 million in 1970 to almost $500 million two years later. In 1972 payments were made to 102,000 miners and 197,000 dependents and survivors.

In 1972 amendments to the law extended benefits to surface miners and liberalized the test for disability from "unable to engage in any gainful employment" to "unable to engage in regular coal mining work." In addition, the presumption of total disability (15 years of coal mining experience plus any evidence of respiratory or pulmonary impairment) was added. These are criteria that are very likely to be overinclusive; and, as a result, black lung expenditures rose to $1.0 billion in 1978 and were estimated to be $1.6 billion in 1979.

There is a second example of rapid expansion and liberalization of federal compensation programs. Recently, the Comptroller General of the United States completed a study of the Department of Labor's management of workers' compensation programs for federal employees [5]. Costs of that program are rising rapidly and will reach $1.0 billion in 1980. These authors conclude that the inability to determine with certainty the causation of occupational disease is one of the major factors contributing to this increase. As a result of this uncertainty, 90–98% of all claims for disease are awarded.

One of the reasons for liberalization of awards for disease claims is the gradual acceptance of the concept of "aggravation" of disease in defining eligibility. The law does not weigh the relative importance of cause or contributing factors; it merely inquires whether employment contributed to or aggravated existing disease. If employment was a factor, benefits can be awarded. Nor is there any effort to apportion an award on the basis of relative contribution of employment to illness. The award is all or none.

Why are definitions of occupational disease increasingly being broadened? I believe there are at least two reasons:

1. There is growing evidence of a causal relationship between exposure to industrial agents and human disease. These agents include certain metals, such as arsenic, organic compounds, such as vinyl chloride and benzene, and inorganic compounds, such as asbestos and silica.
2. The inability to identify specific cause with certainty in a particular case creates an atmosphere in which the compensation referee or judge feels obliged to resort to a doctrine of "fairness" — "the benefit of the doubt." Since in a compensation claim a sick or disabled individual confronts an anonymous corporation, governmental body, or insurance company, the individual gains considerable sympathy; "They" have ample resources; he or she has none.

Collins [6] has collected evidence that there is a change in judicial practices that is moving toward a greater inclusiveness in judging occupational causality. Two cases covering a thirty-year period and abstracted by him from a review of litigation growing out of occupational radiation exposure illustrate the point: The first case is clearly an example of underinclusiveness, and the second case one of overinclusiveness.

CASE STUDIES

Vallat v. Radium Dial Co. (Illinois, 1935)
La Porte v. U.S. Radium Corp. (New Jersey, 1935)

In 1929 Inez Vallat was employed by the defendant company to paint dials with luminous paint containing radium and, in the course of her employment, she "inhaled, swallowed or otherwise took into her system" dust containing particles of radium. The complaint charged that the defendant's employer failed to provide methods of prevention of occupational diseases and the plaintiff contracted "anemia, rarefaction of the bones, alveoli of the jaws, and other bone complications and disorders."

Miss Vallat asked $50,000 damages for her illness resulting from a violation of the Occupational Diseases Act of 1911, which required provision of "reasonable and approved devices...for the prevention of... occupational diseases...incident to such work or process." The requirement was found "unconstitutional" because it was "uncertain, vague and indefinite." Further, the plaintiff failed to show that she had suffered any disablement during her employment, she was not an employee when the disease affected her, and the suit was not filed within two years after her illness was discovered.

Justice Farthing affirmed the lower court's judgment, dismissing the plaintiff's claim.

Besner v. Kidde Nuclear Laboratory (New York, 1965)

Besner was a physicist who had been employed by Kidde for one year when he developed acute myeloblastic leukemia. He died one year later. Mr. Besner did not handle radioactive materials, although there was a shielded ion-exchange column, a 201-mCi cobalt-60 source, and a 1-mCi cobalt-60 source in the laboratory. His exposure was said to be 2250 mR during the year.

Workmens' compensation death benefits were granted to his widow and an appeal was taken. New York workmens' compensation law requires proof of an exposure that could cause the injury in question. A presumption then existed that the exposure did cause the injury unless disproven, as, for example, another nonwork-related cause could be shown. Expert witnesses testified that there was no threshold dose for radiation injury, no other cause of leukemia could be shown and the judgment for the plaintiff was affirmed.

ARE THERE OTHER POSSIBILITIES THAT WOULD STRENGTHEN WORKERS' COMPENSATION?

One reform proposed as an alternative to all-or-nothing decisions would award prorated benefits. This would avoid the problem of making full awards when the probability of occupational exposure contributing to the illness is extremely small.

A variant of this proposal would be the establishment of a threshold probability level of causation, say 25%, which would trigger the payment of full compensation. A schedule would have to be constructed to implement such a system that would project the probability of each organ type of cancer at each level of lifetime exposure. Even if it were possible to determine lifetime occupational exposures with precision, such a schedule would almost certainly generate controversy as would the threshold trigger percentage itself, particularly in cases where the exposure was just below the established threshold. Furthermore, the schedule would have to be constantly updated as new information regarding dose-response relationships and sensitivity of organs to radiation carcinogenesis becomes available. Finally, while conceivably some agreement could be reached on both the threshold percentage and the schedule for radiation-induced cancers, sufficient information to construct such schedules for other occupational illness does not exist. Therefore, an inevitable criticism of this methodology relates to the narrowness of its use and the unfairness to other employment groups where hazards might be greater but unquantified and, therefore, ineligible for compensation. In other words, a system developed for radiation compensation should be ultimately available for all industries.

WHAT ABOUT ADOPTING THE "NO FAULT" CONCEPT USED IN OTHER INSURANCE?

Workers' compensation is not unique as an insurance mechanism that has met with dissatisfaction because of inefficiencies and frustrations. There is an alternative that has worked well: pay the damages and ignore causation, or "no fault." Translating that concept into terms of occupational illness would mean simply that people who are ill and unable to work would receive disability payments without the necessity of determining the origins of their illness, a determination frequently impossible to make.

Such a system, in fact, already exists, i.e., Social Security Disability Insurance (SSDI). Although SSDI is not intended solely to compensate occupational illness or injury disability, it does provide care for injured workers. Because SSDI does not require demonstration of work-related disability, it deserves discussion here as an example of "no fault" insurance against disability.

Congress established in 1956 a program within the social security system that provides income support for persons with long-term disability regardless of cause. To qualify for SSDI, a worker must be fully insured under the system, must have worked five years out of the last ten, and must meet disability requirements. The latter imply consideration of both medical (physical and mental) impairment and vocational characteristics of the individual: age, training, education and experience. A worker must wait five months after applying to collect benefits, which are calculated on a wage replacement formula. Earnings of disabled workers are limited. No work connection is required.

The program began slowly after its inception, growing to one million disabled workers during the early 1960s with costs of $1.7 billion in 1966. By 1978 there were 2.8 million disabled workers and 2.0 million dependents. In 1980 DHHS projected costs of $16.6 billion, which would be financed by a trust fund with equal contributions by both employer and employee at a level of 1.5% of gross earnings.

The proportion of disabled workers enrolled in the SSDI program whose disability results from occupational illness is not known. Based on an unpublished survey of disabled persons carried out in 1974, the percentage is probably small. In that survey, disabled persons were asked about the cause of their disability: 17% cited job-related disabilities. Approximately two-thirds of these were illnesses, the other third the result of injuries. Of those who considered themselves occupationally disabled, about 2% of their income was from compensation sources, about one-third of their income was from social security.

A major criticism of this program is that it fails to provide incentives to employers to reduce occupational hazards to health, i.e., the employer pays 1.5% of gross wages into the fund regardless of the disability rate experienced. This contrasts with private insurance for workers' compensation, which is experience-rated and which, therefore, not only creates some incentive to reduce hazards but also more effectively influences price so as to reflect social cost.

Although SSDI is not experience rated, this could be done for the employer and for the employee. Insurance rating for individuals is, of course, already marketed for nonsmokers, joggers, etc. To "experience rate" both employer and employee could provide important incentives for a preventive health program.

It should be pointed out that workers' compensation and SSDI are not mutually exclusive. Individuals eligible for both may collect both; however, either the state or the Social Security Administration may elect to reduce benefits accordingly. Eleven states have now reduced compensation payments to take advantage of this opportunity.

SUMMARY

Although recognition of industrial illness is growing, there is essentially no reliable data on the extent of disability or illness in the United States as a result of exposure to environmental agents. Although exposures to known toxic agents such as asbestos and silica are being increasingly restricted, disease could still be increasing as a result of exposures that occurred decades ago. Exposures to agents, the effects of which are still unrecognized, may be swelling the reservoir of occupational illness to some further but unknown degree.

The same difficulties that make it impossible to estimate the magnitude of occupational illness make it even more difficult to distinguish between occupational and nonoccupational disease in the individual. In fact, evidence is accumulating that both personal and environmental factors contribute to all diseases, i.e., there is no pure "occupational" illness or "nonoccupational" illness.

In spite of the vague and poorly defined nature of occupational illness, a larger and larger portion of illness is being paid by some third party, whether it be an employer or government. The need to demonstrate probable cause is being rapidly eroded and "remote possibility" is being rapidly substituted for "probable cause" as sufficient justification for an award. The result is contentiousness, very high transaction costs, and a hodgepodge of programs and mechanisms that are poorly adapted to the social objectives for which they are designed.

Review of the insurance system, including both workers' compensation and social security, seems in order. Consideration might be given to removing the need to demonstrate fault or causality—a "no fault" system. The Social Security Disability Insurance program could provide such a mechanism, as could the private insurance industry.

An important issue to be resolved under a no-fault system relates to incentives. One proposal is to experience rate the social security system but, under such a system, should both the employer and the employee be experience rated?

REFERENCES

1. California Workers' Compensation Institute: Cumulative injury in California: a report to the industry. San Francisco, CA, 1977.
2. Risk in a complex society. Louis Harris and Associates, Inc., 1980.
3. Interagency Task Force on the Health Effects of Ionizing Radiation: Report of the work group on care and benefits. Washington, DC: Office of the Secretary, USDHEW, June, 1979.
4. Compensating victims of occupational disease. Harvard Law Rev 93:196, 1980.
5. Comptroller General of the United States: Compensation for federal employee injuries: It's time to rethink the rules. U.S. General Accounting Office, August 22, 1979.
6. Collins, VP: Summary of experience in radiation litigation. In: Brodsky, A, Ed. CRC handbook of radiation measurement and protection.

DISCUSSION: CHAPTERS 21-23

Larry C. Drapkin, W. Clark Cooper, T. Lamont Guidotti,
Leonard A. Sagan, Kenneth S. Cohen, Phillip A. Polakoff,
Leslie I. Boden and Joseph T. Hughes, Jr.

MR. DRAPKIN: Dr. Cooper, you mentioned that we have inadequate data on reproduction hazards. In light of that statement, how do you feel about the use of exclusionary policies? Maybe you could focus on lead as an example because it has been a controversial one.

DR. COOPER: I think that there is evidence from animal studies and from other studies that lead does potentially affect the reproductive process. I do not think we know how to translate these into human experiences. If one presupposes an effect, one is going to be further confronted with the argument that it may be a no-threshold effect, which places the medical advisor to an industry in an impossible position.

Therefore, I have consistently recommended that women of reproductive age not work in situations where lead absorption could produce blood lead levels over 50 μg, even though I have no knowledge that this would be harmful. Fortunately, I have never had the direct responsibility for an exposed group. I know that in any group of 100 or 1000 pregnancies there will be 3% or more birth defects, depending on definition. Such an occurrence is an extremely emotional and traumatic experience for a family. Therefore, I prefer — in a state of lack of knowledge — to go on the side of preventing a possible birth defect that might or might not be associated with lead.

I think that creating a situation in which people are sterilized in order to get a job is very bad and is antisocial; I would say unethical. But, if I had my choice, I would very much prefer that women in childbearing age not be exposed to lead in those uncertain concentrations.

DR. GUIDOTTI: There are some epistemological underpinnings to some of the remarks Dr. Cooper made, and a point of view that I think lends some perspective to our discussions. When Dr. Hine presented his discussion, he gave fairly conventional definitions for the two dominant modes of science—basic science and applied science. I think he missed a concept that was developed about ten years ago by a British professor of the history of science at the University of Leeds by the name of Jerry Ravetz—that of "critical science."

Dr. Ravetz defined critical science as the emergence of a new mode of science similar in its social and epistemological features to basic and applied science but directed toward monitoring or providing feedback on the direction of society in technological innovation. In the new view of Ravetz, critical science would develop a new scientific basis for technology assessment, which at this point is a discipline that has borrowed and, in many cases, misapplied its techniques from other fields and has not been terribly successful.

A large part of what we do in occupational medicine and research is what Ravetz would consider critical science. I think this paradigm is quite interesting and useful. It is also true that critical science is liable to emotionality because it has a tendency, once it gains momentum, to become an advocacy science. But I think it is also true on the eve of the launching of the space shuttle that applied science can be no less arbitrary in its selection of methods and its selection of strategy to achieve a goal. I submit that perhaps the manned space program is not entirely rational in itself. Without question, the strongest case for protecting the health and welfare of the worker rests on the best and most accurate data. But it is also true that complete and accurate information is not usually available; all things eventually come down to a matter of judgment and, really, the question becomes, "Who will be the judge?"

I would like to make some specific suggestions. One is the expansion of publication opportunities for papers that raise issues as opposed to settling them. Perhaps what we need is a "journal of tentative findings" in which we recognize the tentative nature of these things. On a more serious note, investigators can use the services of the National Technical Information Service to provide basic documentation of their raw data to a much greater degree than they are now doing.

Finally, we need to increase recognition of the merits of activity related to this new mode of critical science in hiring and promotion and to insist on the same rigor and methodology, recognizing that the data are not always the best.

DR. COOPER: I think you have given us an excellent talk on this subject, and I don't see anything that is less than admirable in what you have said. I find no area of disagreement with you as to who makes these decisions. We already have OSHA; the Occupational Safety and Health Act has already stipulated who makes the regulatory decisions. It is the input and the variety of people and the sources of information that are critical in decision-making.

I don't believe in conspiratorial theories. There really is no such thing as "industry," although there are powerful people in industry and powerful trade associations. Let's not believe in a monolithic entity called industry. Nevertheless, if "industry" had planned a scenario to discredit OSHA, it couldn't have done much better than OSHA did on its own. An example of this has been the blowing up out of all proportion a drop in the beryllium standard from 2 μg to 1 μg. Another was emphasizing out of all proportion the need for a new arsenic standard. These diverted attention by unnecessary contention. There are hundreds of standards in OSHA regulations right now. You don't have to come out with a 20- or 30-page document to have a standard. There are numerical values for acceptable concentrations available to OSHA for hundreds of currently unregulated substances.

I think that the scenario was such that OSHA let itself be discredited in the eyes of the public and legislators unnecessarily. OSHA's enemies couldn't have planned it better.

DR. SAGAN: I understood the question to be, "Who should make decisions regarding the adequacy of occupational health and safety standards?" I understood the answer from Dr. Cooper to be that OSHA logically should be the agency to make decisions with respect to risk acceptance. Then he added that OSHA hadn't done a very good job of that.

I would like to intervene on behalf of the worker himself or herself. I believe that, in risk acceptance, the role of the individual should be maintained as large as possible; it is the worker whose body is at stake. In discussions of who it is who should decide, it is often assumed that the worker is unable to make that decision. There is an implicit and patronizing judgment that the worker doesn't have the competence to understand the risk or to make judgments with respect to his or her own benefits. There always seem to be plenty of moralists around who feel that they know best how much risk is good for others.

In order that a worker be in a position to make rational risk-benefit decisions, that worker must have access to information. That was the reason I felt Dr. Polakoff's presentation this afternoon was so terribly important. Many of us are beginning to consider how to make information available to the worker in a comprehensible form so that reasonable decisions can be made. There will never be perfect information about health hazards, and there will always be uncertainty. It is the nature of science that you can never be certain that a hazard does not exist; you can only prove the existence of one.

Dr. Cooper talked about lead exposure, for example. I disagreed with him on that. He said that any lead exposure could be considered potentially harmful to an unborn child. Well, I would point out that there is no concrete evidence that anything to which a pregnant woman is exposed, whether in her diet, in the medications that her obstetrician offers her, in the materials and the air around her, might not be a teratogenic agent. So, if you were to apply Dr. Cooper's prescription consistently, a pregnant woman should not eat or breathe for nine months. She can never be certain that any agent is not harmful.

I have a great deal of respect for Dr. Cooper and I would like to press him to respond to that. In my opinion, the answer is that you have to use common sense. You have to use some judgment about what is harmful and what is not. Finally, I think that it is the individual who must make the judgment, not the expert. The expert does have an obligation to provide information, whether it be to his patient or his employee.

DR. COOPER: First, with regard to decision-making, of course the worker should be involved. I mean decisions made by the Secretary of Labor or his representative should certainly include informed labor's input. These are not mutually exclusive. With regard to birth defects or reproductive hazards, there are odds or degrees of probability to deal with. The reason that I would be hesitant in this range of lead is because of the supportive evidence that in higher doses it has an effect. Clearly, the same strictures would apply to medication where there was evidence of a high probability.

DR. COHEN: I have a quick comment for Dr. Sagan. A point for everybody's information is that, in the state of California and several other states, a *one-year rule* in workers' compensation action is becoming popular. An illustration of the one-year rule, for those unfamiliar with this area of compensation law, is the exclusion of prior compensation carriers extending back through the workplace history of an employee, once disability has been demonstrated. The regulation limits the liability to the insurer of record at the time of injury identification. This cumulative trauma type of injury would not reflect back on prior compensation insurers unless the situation of microtrauma or reexposure can be excluded from the current employer. This concept is becoming very popular in many states.

But, to answer Dr. Polakoff, I would like to pose the following hypothetical situation. There are six women accountants who have worked over a bond paper copier for three years. The purchasing department forwards an alert from the manufacturer identifying a carcinogen that has been shown to cause cancer of the breast in laboratory animals. There is no established threshold limit value (TLV) or human reports in the literature. When a dilemma such as this exists, would you: (a) Apply your concept to the facts of this hypothetical situation? (b) If you were the plant physician, would you do the same? (c) How would you notify the women working in that situation?

DR. POLAKOFF: First of all, I think what is most important is that dialog and communication be open as soon as data are available and that discussion of the relative merits of the data take place. If I were the plant physician, I would try to include as many different types of expertise in the discussion as possible. I would go into the legal community and find out what they're doing. I would develop a background scientific report to provide support of the statements. I would get in contact with the work-force early on to inform them where my efforts were leading. Having had a good deal of exposure in media, I can see the pros and cons of utilizing the media.

DR. BODEN: I have a question for Dr. Sagan. In your suggested revolution of the problem of occupational disease and worker compensation, you mentioned a program like Social Security Disability Insurance (SSDI). SSDI is only for long-term disability. Would you also include partial disability in your proposal?

DR. SAGAN: I see no problem with existing workers' compensation law for trauma cases. I believe the system has worked well for such cases by providing for medical expenses and lost income.

I think that a person should not have any loss of income because of an occupationally incurred illness or injury. I'm not yet certain as to how the mechanisms might be brought to bear on these issues, but my instinct is that there should not be loss of income because of illness and that the worker should not be subsidizing the clients of that industry, whatever it is.

DR. BODEN: Suppose your system is in place and employers are rated on the basis of general illness history of their employees' occupation and nonoccupation. I have a concern that a number of employers may then become very careful about whom they hire, about previous health history, about smoking history, about genetics (Did your parents have cancer of the pancreas?), and so forth, and then, employers who are otherwise good employers and pay well might become quite concerned about the health of workers, thus ultimately leading to large groups of workers becoming unemployable because of their health histories. I wonder if you thought about that problem and what your comments might be?

DR. SAGAN: Workmen's compensation is and has been experience rated and that has not led to the kinds of discriminatory practices that you fear.

DR. BODEN: It has to some extent. For example, lots of employers will not hire people who have had cancer and are in remission. I'm worried that the problem will become even greater as financial risks of employers under such a system become greater.

DR. SAGAN: The costs of insuring against occupational illness and injury are generally a very small cost to employers compared to other production costs; therefore, I find it difficult to believe that employers will ever act on the basis of these costs to the extent you fear.

For the same reason, I am also skeptical that experienced-rated insurance costs will ever become an important incentive for improving workplace safety. My own hunch is that employee attitudes are more important in achieving safety but, in the final analysis, we really know very little about the factors that contribute to workplace safety.

Professor Wildavsky, who is at the School of Public Administration at Berkeley, writes about the "silent hand" of occupational health and safety that has been operating to provide the historical reduction in occupational risk. He, too, does not understand how to explain it. So perhaps I gave too much emphasis to disincentives when I talked about experience-rated systems. Maybe we don't need economic disincentives for either employers or employees.

MR. HUGHES: Dr. Polakoff's comments about worker notification made me think about the number of cohort studies that have been done in this country over the last decade and my own byssinosis cases, where the results have been published in numerous medical journals, but the people who are the statistics never knew about the publication or were aware that they had a disease.

I wonder, number one, if in epidemiology studies, it is not possible to include in the protocol a mechanism to ensure workers that those found to be disabled would be provided some kind of services—certainly something more than just being part of a study that is written up and then forgotten about.

Number two is in answer to Dr. Sagan's comments about Social Security Disability. It is my feeling that you have described basically a situation in which society must bear the burden. Financial burden for occupational disease has been completely shifted onto the federal government, and it was my feeling that there needs to be some way to shift that burden back the other way—toward the employer. I don't think that your system would do that without setting disease standards and without establishing industry-based trust funds that would directly apportion liability back to industry.

And just one last thing, I don't think that knowledge on the part of the worker that there is a risk should somehow make him a co-sharer in the responsibility for his disability because, under OSHA, the employer has the responsibility to provide a safe and healthful workplace.

DR. POLAKOFF: We all have technical expertise; however, it is defined and manifested differently. I think that is really a very good statement in the sense that it might put the whole definition of professionalism in a slightly different context. In designing any protocol where human subjects are involved, I believe the people who are being studied should be consulted in the development of a human subject protocol. In any large-scale epidemiological cohort studies, those people who represent the workers should be a part of the development of the study early on. This will increase the acceptance of the study and decrease workers' anxiety.

SECTION 3

WORKERS' COMPENSATION

CHAPTER 24

BYSSINOSIS COMPENSATION IN NORTH CAROLINA: A TEN-YEAR POLICY REVIEW (1971–1980)

Joseph T. Hughes, Jr.
Institute for Southern Studies
Durham, North Carolina

For the past decade, the state workers' compensation program in North Carolina, where the bulk of the nation's cotton textile production is located, has been the focus of an ongoing controversy as disabled victims of byssinosis, a chronic respiratory disease, have attempted to recover disability benefits from their former employers.

During the 1970s, byssinotics became aware of the occupational connection of their disease and of the system that had been set up to compensate them, and they applied in increasing numbers for relief. Over the decade, the state compensation system's stated legal mandate "to provide swift and adequate relief to injured employees" [1] came to be viewed as a broken legislative promise and a medicolegal illusion. In spite of the existence of state statutes that would seem to permit compensation of work-related diseases such as byssinosis, the state system has functioned more as an administrative barrier than as a quasilegal agency charged with making reasonable and speedy determinations and, in the words of the North Carolina Act, thereby "compelling industry to take care of its own human wreckage" [2].

At the turn of the century, workers' compensation systems were set up to replace the costly and uncertain common law liability system and to ensure that workers would pay the costs of work-related disability. Over the past 80 years, however, the state workers' compensation system has come to mirror the worst aspects of the system it was designed to improve. There are excessive delays in reaching administrative decisions; benefits are uncertain; litigation is excessive; and, as was the case in the early part of the century, there is a continued dependence by work-injury victims on public funds for their survival [3].

The problems of occupational disease claims have provided the toughest challenge to state compensation systems in their 80-year-old history. Although the underlying principle of workers' compensation is "liability without fault," the costs of work injuries and illnesses are allocated to the employer, not because of personal blame but because of the inherently hazardous nature of industrial employment. This has not been the case for occupational diseases [4].

Originally, the system was set up to deal with injuries where the causation of physical impairment was clear-cut and the occurrence of the disability immediate. Many cancers and occupational respiratory diseases, however, are characterized by long latency periods that increase the difficulty of linking a specific disease process with a prior hazardous work exposure. In addition, many work-related disases, such as byssinosis, have symptoms and pathology that are indistinguishable in their chronic stages from other nonwork-related conditions, such as emphysema and chronic bronchitis.

Because of the preponderance of difficult-to-answer medicolegal questions, the byssinosis compensation dilemma in North Carolina represents an important microcosm of legal and ethical problems that arise in attempting to deal with any occupationally related health condition. In this chapter, the major medical, legal and administrative issues that underlie the byssinosis controversy will be delineated; data on byssinosis cases in North Carolina will be compared with a similar national survey; and ethical implications will be drawn concerning the importance of workers' compensation reforms in an era of rollback in occupational health standards.

The question of who is now paying *and* who should pay the costs of occupational diseases will be the critical policy issue of the 1980s.

LEGAL ISSUES AND PROGRAM ADMINISTRATION

With the passage of the Occupational Safety and Health Act (OSHAct) in 1970, and the recognition of black lung as a compensable condition, state legislatures came under pressure to specifically recognize and cover work-related diseases and illnesses. North Carolina, like most other states, endorsed broad general clauses that included occupational disease coverage to replace the existing procedure of listing new diseases on schedules as they were recognized by medical science. The North Carolina disease statute, passed in 1971, reads as follows:

> Any disease... which is proven to be due to causes and conditions which are characteristic of and peculiar to a particular trade, occupation or employment, but excluding all ordinary diseases of life to which the general public is equally exposed to outside of employment [5].

With the passage of these general disease clauses by many legislatures, state workers' compensation agencies were given broad discretion in interpreting their application and administering their implementation. After the 1971 amendments were passed in North Carolina, the State Industrial Commission enlisted the aid of Dr. James Merchant and a number of other byssinosis reasearchers in setting up a special medical panel to evaluate claimants' disability and to determine the occupational connection of any impairment. With a flood of claims expected, the panel's role was to screen out invalid applications. Since by North Carolina law the employer was required to pay for medical exams, the employer chose the examining physician.

To enter the system, a claimant would file a form signed by his local doctor indicating that he had possible byssinosis. If the State Board of Health and the employer thought the claim was valid, a panel exam would be ordered. If the panel doctor thought there was a bona fide case of byssinosis, the claim would either be settled by a clincher agreement or would be scheduled for further hearings [6].

Although the system appeared reasonable and simple at the time, it did not consider that most disabled workers lived in isolated mill villages spread across the state. None of the workers had ever heard of byssinosis or even knew about workers' compensation for diseases. Most local doctors knew little about byssinosis and none would risk putting a diagnosis in writing, as required by Commission procedures.

As a result of the difficulties encountered in getting into the compensation system, only 63 claims were filed during the first five years (1971–1975), although an estimated 10,000 disabled millworkers in the state were potentially eligible for benefits. During this first five years, 36 claims were settled, averaging $9000, 14 cases were denied, and none were awarded according to the Commission.

The controversy over byssinosis cases has refocused attention on the inadequacies and inconsistencies of workers' compensation agencies' administrative procedures. For example, the North Carolina Commission is exempt from the state's Administrative Procedures Act, which requires public notification of changes in rule-making procedures. In addition, none of the agencies' files are open to public scrutiny, making it all but impossible to learn what precedents may have been established by previous decisions [7].

Because of the lack of internal monitoring systems on pending cases, organized case management is almost nonexistent in the North Carolina Industrial Commission. The lack of case management results in interminable delays at each step in the claims process. In a study we conducted of the North Carolina System, prepared for the U.S. Department of Labor, a sample of pending cases revealed that it took almost six months from filing data to receive an initial medical evaluation, over a year to receive an initial informal hearing, more than 1.5 yr for a second hearing, and 2 yr to receive an initial opinion from a deputy commission, which is subject to later review by the full commission and the courts [8, p. 25]. Similar experience was found in our study of South Carolina cases.

In addition to the lack of case management or efficient claims administration, excessive litigation by employers and their insurance carriers have further exacerbated the byssinosis compensation situation. Larson [9] notes that, from the beginning, early court interpretations of compensation statutes sought to block claims administration from escaping common law procedures, although the system was set up as a less litigious alternative. "Interpreting 'due process' liberally for the defendants has permitted endless hearings and appeals of commission decisions and an increased volume of litigation" [9].

Using estimates from the Governor's Brown Lung Study Commission, only 38 of the 1657 cases filed have received an award through the Industrial Commission after 10 years [10, p. 32]. Of these 38 awards, 16 cases are still pending in appellate courts, after waiting two to four years for an Industrial Commission opinion [11, p. 1]. A number of these cases may set important new precedents for occupational disease victims. In *Taylor v. J. P. Stevens,* the North Carolina Supreme Court affirmed that the time for filing and notice on an occupational disease claim runs, not from the date of last exposure, but from "the date that the employee has been advised by competent medical authority that she has an occupational disease." *Taylor* also struck down a North Carolina requirement that disablement must occur within one year from last exposure, an impossibility in many chronic degenerative diseases [11, p. 4].

Two other critical national issues are also currently being reviewed by the North Carolina Supreme Court in pending byssinosis cases: prior liability and apportionment of disability. In *Wood v. J. P. Stevens* (1979), the court held that a claim could be filed and considered, even though the date of last exposure came prior to the passage of occupational disease coverage. In North Carolina, prior to 1963, damage to internal organs was not covered by workers' compensation statutes [12, p. 1].

The most important case to date, *Morrison v. Burlington Industries,* is currently before the North Carolina Supreme Court. This case concerns the issue of apportioning work- and nonwork-related disability. Over the past 80 years, this concept has been continually rejected by appellate courts across the country once a causal connection has been made with employment. In *Morrison*, the textile industry is contending that compensation payments should be reduced if a claimant has smoked or has other causes of disability [12, p. 5].

This case is expected to have a significant impact on settlement negotiations and clincher agreements, which account for 93% of all compensation payments made over the last ten years [10, p. 32]. For example, compromise agreements average $10,000 less than Industrial Commission awards ($15,000 v. $25,000) and are usually reduced because of other causes of disability. In addition, clincher agreements have no provision for future medical coverage while awards leave employers liable for future medical coverage [8, p. 38].

MEDICAL ISSUES AND OCCUPATIONAL DISEASE PANELS

In addition to the problems involved in legal interpretation and claims administration, developing a reliable medical determination stands as the largest roadblock for claimants who are left to carry the entire burden of proof in occupational disease cases. In 1977 the Federal Interdepartmental Task Force on Workers' Compensation recommended that all states adopt expert medical panels to dispose of disease cases. In spite of the good intentions of its founders, the Textile Occupational Disease Panel in North Carolina has become a new battleground for exacerbating medical disputes about byssinosis rather than an effective mechanism for fairly resolving them.

Because the Industrial Commission and the panel have been unwilling to develop standards for diagnosis, an arbitrary and inconsistent set of medical criteria has been evolving by default. Under these criteria, the success or failure of a particular claim is dependent on the preconceptions or misconceptions of an individual physician rather than on the merits or weaknesses of a claimant's case. With no guidelines for diagnosis and with the burden of proof resting on the claimant's shoulders, the system has become "a medical roulette game" with "doctor shopping" accounting for most of the administrative delays.

In our policy study of 200 North Carolina cases, 85% of the claimants received exams from the panel, but only 49% received an opinion on the critical question of cotton dust causation, and only 68% were given an opinion on disability. In more than 90% of the cases in which a physician did venture an opinion, cotton dust was thought to be a causative factor. In 58% of the cases with a positive opinion on work-relatedness, cotton dust caused over half the disability [8, p. 32].

Although the burden of proof is on each individual claimant to prove both the work-relatedness of his condition and that he is at increased risk compared with nonexposed populations, the results of studies by epidemiologists and other scientists of exposed populations is considered irrelevant in evaluating individual cases. Thus, it becomes all but impossible to satisfy the law's requirements.

TEN YEARS IN RETROSPECT

In January 1971, the North Carolina Industrial Commission released statistics covering the past ten years of byssinosis compensation experience. These are the only data that the Commission compiles on filed cases. While 1657 cases were filed during the ten-year period, 821 (50%) are still pending at the Commission. As Table I indicates, there were almost no cases filed before 1975, which was when the Carolina Brown Lung Association was organized and began holding screening clinics and filing valid claims. Each year since then, the number of claims filed has jumped substantially, with the largest increase occurring during 1980 following Governor Hunt's appointment of a study commission to investigate claim delays and inequity in the byssinosis compensation procedure.

As the backlog of cases has increased, so have the number of cases settled and the average amount of the settlements. Although 543 claimants have received some payment, less than 50 of those payments have come as a result of a Commission award. In addition, of the 264 cases that have been dismissed or denied only 25 were denied by official Commission opinions [11, p. 1].

Based on extrapolations from epidemiologic data, less than 20% of the potentially disabled cotton mill workers in North Carolina with byssinosis have filed claims with the Commission, leaving approximately 8000 possible claimants outside of the workers' compensation system.

Table I.   Statistics on Byssinosis Claims in North Carolina[a]

Year	Number of Cases Filed	Awarded or Settled		Compensation Paid ($)	Average Award or Settlement Amount ($)	Cases Dismissed or Denied		Cases Heard and Pending	
		Number	% of Filed			Number	% of Filed	Number	% of Filed
1971–1975	63	36	57	324,000	9,000	14	22	13	21
1976	112	42	37	424,000	10,000	17	15	53	48
1977	175	49	28	549,098	11,206	29	17	97	55
1978	297	65	22	725,675	11,164	41	14	191	64
1979	655	172	26	2,053,362	11,938	103	16	382	58
1980	913	286	31	3,760,359	13,148	186	20	437	48
1981	1,657	543	33	8,009,258	14,750	264	17	821	50

[a]Data supplied by the North Carolina Industrial Commission.

In no way is this situation unique to the workers' compensation system in North Carolina. In fact, North Carolina has done a much better job than all other southern textile states (Georgia, Virginia, Alabama and Mississippi) where awards for byssinosis have yet to be made.

In the major national study of occupational disease compensation completed by Barth and Hunt [4] during 1976, very similar conclusions were reached (Table II). Data from the Barth and Hunt study on disease cases in workers' compensation indicated that occupational disease compensation claims represented the most seriously disabling, the most vigorously contested, the most frequently compromised, and the most difficult medical and legal claims to dispose of [4, pp. 160-185].

Barth concluded that "controversion and formal compromise are the rule and not the exception in occupational disease cases. For this group of cases, it would not be unfair to question whether the workers' compensation system is handling these cases at all. It is difficult to use the term "no fault" to apply to a group of cases that have, in the overwhelming majority of cases, been controverted on the question of compensability. For these claimants, the typical experience bears little resemblance to the theoretical or textbook description of the operation of the workers compensation system" [4, p. 187].

In reviewing North Carolina's experience with byssinosis compensation, it is difficult to dispute the conclusions of Barth and Hunt. Our state's problem is an integral part of a larger national dilemma regarding both occupational and environmental compensation for victims of illness and disease.

PROSPECTS FOR REFORM

Ever since the National Commission on State Workman's Compensation was set up as part of the original OSHAct, state compensation programs have been encouraged to voluntarily reform and upgrade their inadequate programs. If the experience of North Carolina is any guide, this process is continuing at a snail's pace. The lack of progress in North Carolina is particularly instructive since the Industrial Commission has been under constant pressure from victim advocacy groups, has been the subject of hearings in both Houses of Congress, and has been reviewed by the Governor's office, the press and a maverick Commissioner of Insurance.

Obviously, there are limits to the possibility of innovation on the state level. In North Carolina the Industrial Commission, which administers workers' compensation programs, is a division of the state Commerce

Table II.   Comparison of North Carolina Byssinosis Cases with a National Survey of Work Injury and Occupational Disease Cases[a]

	Cases Controverted (Percent of Total Filed)	Cases Compromised (Percent of Total Paid)	Average Age of Claimant at Filing (yr)	Total Compensation Amount Paid ($)	Elapsed Time, Claim Filing to Payment (mo)	Percent of Permanent Total Disability Cases
North Carolina Byssinosis Cases[b]	98	93	57.5	14,750	18	85
National Survey of Occupational Respiratory Disease Cases	88	82	52.5	11,426	13	76
National Survey of Work-Related Injury Cases	9.8	15	46.5	2,822	1.5	20

[a]National data from the Interdepartmental Task Force on Workers' Compensation (1977), analyzed by Barth and Hunt [4].
[b]North Carolina data from the Governor's Brown Lung Study Commission [7,10] and Hughes' [8] study for the U.S. Department of Labor.

Department, whose main task is to recruit new industry to the state. Ironically, the paucity of workers' compensation benefits in the state is touted as a benefit by industry recruiters; and the understaffed Industrial Commission is always at the bottom of the Commerce Department's legislative requests for funding.

On the national level, outlooks for workers' compensation reform are even bleaker, in spite of a recent report by the U.S. Department of Labor that less than 5% of the three million occupational disease victims in this country ever receive any workers' compensation benefits [13]. With the Reagan Administration's calls to weaken occupational health standards and for application of cost-benefit analyses to work-related diseases, such as the current cotton dust controversy at the U.S. Supreme Court, there will be another opportunity to raise the question of who *is* and who *should* be paying the costs of occupational disease.

In the cotton dust situation, the U.S. Department of Labor found that the costs of the disease — $7.51 billion — were more than ten times the estimated cost of clean up, which was $655 million. Even more important than the cost of disease and disability is answering the question of who is now paying these costs [14].

In my study of 270 byssinosis victims in the Carolinas, I found that 82% of the study cohort's income maintenance benefits came from the federal government, almost entirely through the Social Security Administration, while workers' compensation payments from the employer accounted for less than 7% of the benefit total. Even with Social Security funds in crisis, it is apparent that a massive shift of costs has been made from the employers to the general taxpaying public, due to an almost nonexistent workers' compensation system — a form of public subsidy to disease-producing industries like textiles (Figure 1).

Byssinosis is not a newly discovered disease. It is an affliction that came to us as a by-product of the original Industrial Revolution. If we cannot first design programs to deal with diseases that we have known about for centuries, how are we as a society going to come to grips with more recent modern diseases such as cancer, where the frontiers of medical understanding have yet to be explored? How are we ever going to orient our society to understand that it is much easier in the long run, both in terms of dollars made and lives saved, to clean up hazards in the workplace now, rather than to take the risks now in the hopes that someone else will pay the consequences later?

"Time," many brown lung victims always say, "is something that we just don't have." And time, for many of the 35,000 textile workers who are already victimized, is running out.

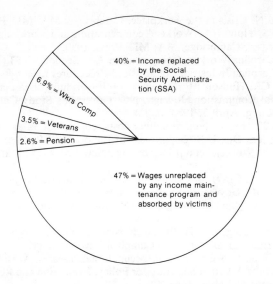

Income Replacement Over
Seven Years Of Disability

$20,916 = SSA Ave. Benefit
$ 1,819 = VA Ave. Benefit
$ 1,398 = Pension Benefit
$ 3,574 = Wkrs Comp Benefit

$27,636 = Total Average
Wage Replacement

Average Expected Earnings
Over Seven Years
Of Work Disability[1]

1973[2] =	$6033[3]
1974 =	$6284
1975 =	$6664
1976 =	$7358
1977 =	$8019
1978 =	$8500
1979 =	$9000 (est.)

$51,858 = Total Expected
Earnings

(1)– Average length of disability
     before 65 years old for study population = 7 years
(2)– Average date of leaving the mill disabled
     for the study population = 1973
(3)– Average annual textile wage levels
     according to the Bureau of Labor Statistics (1979)

Figure 1.   Income replacement of lost wages due to disability.

REFERENCES

1. *Barnhardt v. Yellow Cab Company,* 266 N.C. 419; 146 S.E. (2d) 479, 1966.
2. *Barber v. Minges,* 233 N.C. 213; 25 S.E. (2d) 837, 1943.

3.   Ashford N: Crisis in the workplace. Cambridge, MA: MIT Press, 1976.
4.   Barth PS, Hunt HA: Workers' compensation and work-related illnesses and diseases. Cambridge, MA: MIT Press, 1980.
5.   North Carolina Workers Compensation Act, Section 97-53 (13).
6.   North Carolina Industrial Commission. Procedures for a byssinosis claim.
7.   Hassell CR, Hudson RE: Statement on brown lung disease and the N.C. Industrial Commission. Raleigh, NC: Governor's Study Commission on Brown Lung, April 3, 1980 (p. 32).
8.   Hughes JT: Brown lung disability: costs, compensation and controversy. Washington, DC: U.S. Department of Labor/ASPER, 1980.
9.   Larson A: Workers' compensation law. New York, NY: Mathew Bender, 1980.
10. Final report. Governor's study commission on brown lung compensation. Raleigh, NC: Governor's Study Commission on Brown Lung, April, 1980.
11. Case summary report. Raleigh, NC: NC Industrial Commission, January 5, 1980.
12. Hammer S, Singletary H: Recent decisions clarify law on brown lung disability. Buies Creek, NC: The Campbell Law Observer, July 25, 1980.
13. Interim report to Congress on occupational diseases. Washington, DC: Office of the Assistant Secretary for Policy, Evaluation and Research, U.S. Department of Labor, June, 1980 (p. 5).
14. Report to Congress—cotton dust: a review of alternative technical standards and control technologies. Washington, DC: U.S. Department of Labor, May, 1979 (pp. 98-99).

CHAPTER 25

PRESUMPTIVE STANDARDS: CAN THEY IMPROVE OCCUPATIONAL DISEASE COMPENSATION?

Leslie I. Boden, PhD

Occupational Health Program
Harvard School of Public Health
Boston, Massachusetts

> In 1899 the lower house of the Indiana Legislature finally stepped in where effete academics had been pussyfooting around for centuries: the Hoosiers passed a bill setting the value of pi exactly equal to four [1].

At the end of the nineteenth century, there was no workers' compensation in the United States. The only recourse for an injured worker was to sue his employer, claiming liability based on negligence. The employer's legal defenses were so strong that the injured worker was rarely successful in obtaining compensation for his injuries, no matter how serious they were. Workers' compensation laws substituted a no-fault system and did away with the three major defenses of the employer (contributory negligence, the fellow-servant doctrine and the assumption of risk). The major goals of the new system were sure and speedy compensation for workers and a minimal amount of litigation. While its benefits may be inadequate and its coverage of employees incomplete, workers' compensation has succeeded in providing direct and speedy payments to most injured workers.

In contrast, workers' compensation has generally not met these goals for victims of chronic occupational disease, who often never receive compensation. Those who apply are commonly faced with litigation, years of delay, and relatively small awards.

Recently, the possibility that presumptive standards could improve the functioning of workers' compensation for occupational disease has been suggested [2-4]. Such standards specify conditions under which it is presumed that claimants have satisfied the burden of proof necessary to receive workers' compensation payments. Generally, presumptions are designed to clarify and lighten a claimant's burden of proof, thereby reducing the amount of litigation and increasing the number of workers whose injuries or illnesses are compensated. Following is a discussion of the nature of the problems facing occupational disease compensation and an assessment of the possible role that presumptions might play in solving these problems.

PROBLEMS IN COMPENSATING OCCUPATIONAL DISEASES

There are many roadblocks to receiving compensation for occupational *disease* as opposed to occupational *injury*. First, the workers' compensation system expects a testifying medical expert to say whether or not a worker's illness was *caused* or *aggravated* by work. Physicians are asked, "Was this illness caused by workplace conditions?" This is a question for which medical science does not have a simple answer. Some common characteristics of occupational diseases that make the disabled worker's burden of proof difficult to sustain are:

1. Physicians may not realize that their patients have become ill as a result of workplace exposures. Many physicians are not able to identify occupational diseases because their medical training in this area was inadequate. Many have not even been trained in taking occupational histories. As a result, occupational disease victims often do not suspect that their disease is job-related and therefore never apply for compensation.

2. Signs and symptoms of the disease are rarely uniquely related to an occupational exposure. Medical and epidemiological knowledge may be insufficient to distinguish clearly a disease of occupational origin from one of nonoccupational exposure. For example, shortness of breath, an important symptom of occupational lung disease, is also associated with other chronic lung diseases.

3. A disease may have several causes, only one of which is occupational exposure. A worker who smokes and is exposed to ionizing radiation at work may develop lung cancer. Since both cigarette smoke and

ionizing radiation are well-established risk factors for lung cancer, it may be impossible to say which of these two factors "caused" the disease.

4. The disease may develop many years after exposure began, and perhaps many years after exposure ceased. This means that memories of events and exposures may be unclear, and that records of employment may not be available. Some occupational injuries, like chronic illnesses, occur as a result of extended exposure to a hazard. These are "cumulative trauma" injuries, such as carpal tunnel syndrome, noise-induced hearing loss, and chronic low-back pain. As with chronic occupational diseases, it may be difficult to prove the work-relatedness of these injuries. Whereas, some illnesses, such as dermatitis, are like most typical injuries because they occur during or soon after exposure.

5. Records of exposure to occupational hazards may never have existed. Even when a worker knows of the type, level, and duration of exposure, he/she may have no written evidence of this.

All these reasons explain why so few claims for compensation for occupational *disease* are filed. A study of occupational disease in Washington and California found that, of the 51 "probable cases of occupational respiratory conditions," only one was reported as a workers' compensation claim [5].

When claims for chronic occupational disease are filed, many are contested (Table I). For these claims, payments to disabled workers are delayed and uncertain. Workers with chronic occupational diseases wait an average of more than a year to receive compensation payments (Table II). In addition, administrative and legal costs absorb many of the resources devoted to compensating occupational diseases.

Table I.   Percent of Cases Contested by Category
of Disease [4]

Category	Percent
Dust Diseases	88
Disorders Due to Repeated Traumas	86
Respiratory Conditions Due to Toxic Agents	79
Cancers and Tumors	46
Poisoning	37
Skin Diseases	14
Disorders Due to Physical Agents	10
Other	54
All Diseases	63
All Accidents	10

Table II.   Delays in Compensation for Occupational Disease,
by Category of Disease [4]

Category	Mean Number of Days From Notice to Insurer to First Payment
Skin Diseases	59
Dust Diseases	390
Respiratory Conditions Due to Toxic Agents	389
Poisoning	111
Disorders Due to Physical Agents	79
Disorders Due to Repeated Trauma	362
Cancers and Tumors	260
Other Illnesses	180
All Accidents	43

Litigation results not only from the difficulty of deciding whether illnesses are occupational in origin, but also because occupational disease claims tend to be costly. Chronic occupational illnesses are, on average, much more severe than occupational injuries. They are both more disabling and have higher medical costs. More than 60% of occupational disease claims are for permanent disability or death. The comparable figure for occupational injury claims is only 20% (Table III).

Because the monetary stakes are high, insurers have incentive to scrutinize potentially costly claims and to deny them. If a claim is contested, the insurer generally keeps the settlement, and is able to receive investment income until the case is closed and the claimant is paid. Since a contest delays payment by an average of over a year, this investment income is an incentive to contest even in those cases that the insurer is sure to lose.

In the face of protracted litigation, an injured worker may accept the offer of a compromise settlement. The claimant may have no wage

Table III.   Highest Level of Disability, by Case Type [4]

	Occupational Disease (%)	Occupational Injury (%)
Temporary Total	35.0	71.9
Permanent Partial	53.8	19.5
Permanent Total	6.1	0.5
Death	1.1	0.3

income and large medical bills. Rather than waiting one or two years for a larger settlement, the disabled worker may be willing to agree to a much smaller settlement today. For the insurer, the investment income and potential smaller settlement resulting from a contest must be weighed against its administrative and legal costs. The higher the uncontested award would be, the more incentive there is to contest. Table IV shows that claims for permanent disability and death are contested much more frequently than those for temporary disability. This is true for both injury and disease claims.

Table IV.   Percent of All Claims Contested by
Highest Degree of Disability [4]

Temporary Total	3.3
Permanent Partial	39.1
Permanent Total	51.9
Death	41.8

As insurers have incentives to contest expensive occupational disease claims, workers have incentives to file claims for nonoccupational illnesses if potential settlements are high. They are also more likely to be able to find lawyers to represent them if there is a large potential settlement involved since workers' compensation lawyers generally are paid on a contingency basis.

HOW WOULD A PRESUMPTIVE STANDARD WORK?

A presumptive standard attempts to address the above problems in three ways:

1.   It can establish a set of conditions under which the claimant's burden of proof is considered to have been met. For example, a presumption can define a level of evidence sufficient to demonstrate a causal relationship between occupation and disease (e.g., ten years of work in a listed occupation combined with specific medical tests). Note that the existence of a presumption does not necessarily prevent the employer from rebuttal. If a presumption is rebuttable, the employer may attempt to demonstrate that his firm controlled exposure unusually well. Such evidence may sway a judge to deny compensation even though the cri-

teria of the presumption are satisfied. On the other side of the coin, if the criteria of presumption are not met, claimants may still present evidence to convince the judge that they should receive compensation.

2. It is also possible to create negative presumptions. A negative presumption creates a minimum standard of evidence for compensation to be awarded. Claimants not able to produce this evidence are barred from receiving compensation. In this case, workers who otherwise might have brought claims are discouraged from doing so. Fewer cases enter the system, and less litigation takes place.

3. A presumption can also be used to provide payments to disabled workers during the period between the filing of a claim and its settlement. Since claimants can often perform little or no work, they may be under great financial pressure. This can lead them to accept small settlements for their contested claims. Providing interim payments to all workers with contested claims would probably be financially expensive and politically unfeasible, because such payments would provide an incentive to file questionable claims. If, instead, payments require approval by a workers' compensation examiner, a set of minimum criteria for approval ought to exist. A presumptive standard can offer such criteria.

Most observers of workers' compensation agree that there is too much litigation over occupational disease claims. There is, however, a great deal of disagreement about whether too few victims of occupational disease or too many workers with nonoccupational disease receive compensation. There are two reasons for this disagreement. It is impossible, on the basis of current evidence, to determine the true extent of disabling occupational disease. The epidemiological literature gives us some information about the extent of specific occupational diseases in study populations, but without appropriate exposure information it is not possible to extrapolate from these studies to the general working population. We also do not know how many people with occupational disease did not bring claims, nor what proportion of claimants actually were suffering from occupational disease.

In the knowledge vacuum that I have described, the personal experiences, ethical beliefs, and self-interest of observers are often the basis for opinions. In particular, there are widely differing beliefs about how bad it is when (1) workers with occupational diseases are *not* compensated, and (2) workers without occupational disease *are* compensated.

Since methods of increasing compensation generally increase compensation for *both* groups, authors of occupational disease presumptions are faced with deciding how much additional "inappropriate" compensation they are willing to pay in order to compensate for "legitimate" claims.

The examples that follow display a wide range of responses to this trade-off and demonstrate the variety of ways that presumptions can specify the conditions under which claimants can satisfy their burden of proof.

EXAMPLES OF WORKERS' COMPENSATION PRESUMPTIONS

To receive workers' compensation, a claimant must demonstrate a causal link between workplace exposure and disease. Presumptions can specify evidence legally sufficient to establish this link. They can:

1. Specify a level or duration of exposure that is deemed sufficient to cause an occupational disease. The claimant must still demonstrate this exposure in order to demonstrate work-relatedness.
2. Deem that working in a specific occupation implies exposure to a harmful substance. The claimant must then demonstrate existence of the associated occupational disease to demonstrate work-relatedness.
3. Link working in a specific occupation directly to an occupational disease.
4. Require specific evidence, e.g., a minimum of five years of workplace exposure, to establish causality. Workers without such evidence would be denied compensation.

This section provides examples of the different kinds of occupational disease presumptions. Note that the examples are not model presumptions and that they may or may not conform to clinical and epidemiological understanding of disease processes.

Link Between Exposure and Illness

The first type of presumption links an exposure to an illness. The illness is presumed to be work-related if the worker can demonstrate that an exposure to a listed hazard has occurred. For example, New York law (Section 47) provides that any exposure to harmful dust for a period of 60 days or longer or any exposure to compressed air is presumed to be harmful in the absence of substantial evidence to the contrary. Thus, a worker with lung disease who was exposed to silica dust for longer than 60 days would be presumed to have silicosis, unless the insurance carrier or employer could demonstrate otherwise. Note that, without such a presumption, a causal link between exposure and disease must be demonstrated by the claimant. With this presumption, only exposure and disease must be demonstrated.

Kentucky has a similar presumption, which states (Section 342.316(5)): "Where the occupational disease is compensable pneumonoconiosis and there has been employment exposure for ten years or more to an industrial hazard sufficient to cause the disability of pneumonoconiosis there shall be a rebuttable presumption that the disability or death was due to the compensable pneumonoconiosis." If an employee demonstrates a compensable disease and exposure to a hazardous substance that could have caused the disease, the law presumes the causation has been demonstrated. Several aspects of the Kentucky statute are interesting to note. First, New York requires only a 60-day exposure, while Kentucky requires a 10-year exposure to activate the presumption. Both are legally valid presumptions but, obviously, the New York requirement is more favorable to claimants than is the Kentucky requirement. Moreover, the Kentucky presumption only covers exposure to hazards *sufficient to cause* the disability of pneumonoconiosis. In practice, this may imply not only exposure but some (unstated) level of exposure must be demonstrated before the claimant is covered by the presumption. Both the New York and Kentucky presumptions are rebuttable. That is, evidence can be offered to refute the presumption that a given exposure was causally related to the claimed disease. Obviously, in both states there is ample room for laywers to interpret the presumption and to challenge its application.

Link Between Occupation and Exposure

A second type of presumption helps to establish the link between occupation and exposure. In several states, employees in specified occupations or processes are presumed to be exposed to hazards associated with those occupations, even if there is no evidence to support this assertion. For example, in New York, any workers who develop anthrax while working with, or immediately after handling, wool, hair, bristles, hides or skins, are deemed to have anthrax caused by their occupation. Note that a presumption like this would not be very useful to workers with diseases of long latency.

In addition to presumptions that link occupation and exposure and those that link exposure and disease, some states have presumptions that directly link occupation and disease. For example, some states have presumptions that apply to firefighters, police officers, or both who have died or become disabled as the result of circulatory diseases. In this case, the claimant's occupation is considered sufficient evidence of harmful exposure and of the relationships between this exposure and the worker's disabling disease.

Finally, about twenty states have *negative* presumptions for some diseases. The typical negative presumption states that there must be minimum exposure to the relevant hazard for compensation to be paid. About half of these negative presumptions are rebuttable, while in ten states there is no opportunity for workers with less than the mandated exposure to receive compensation.

The occupation disease presumption which has received the most publicity over the past decade was established by Title IV of the Coal Mine Health and Safety Act of 1969 to provide benefits for coal miners totally disabled by pneumoconiosis. As part of this law, coal miners were presumed to be totally disabled if they worked in a coal mine for ten or more years and had medical evidence of complicated pneumoconiosis. When this legislation was passed, it was realized that liberalizing compensation for coal workers' pneumoconiosis (CWP) would create an enormous liability for coal mine operators, who would suddenly become responsible for all prevalent cases of CWP which met the criteria of the presumptive standard. Because of this, the original legislation empowered the federal government to pay benefits for all claims filed before December 31, 1972. By that date, annual program expenditures had already grown to almost $500 million. In 1972 the act was amended to extend until the end of 1973 the federal responsibility for payment of benefits. The 1972 amendments also added a presumption that miners with complicated pneumoconiosis would irrebuttably be presumed to be totally disabled. Federal expenditures on this program continued to grow, rising to about $1 billion in 1977. Beginning in 1974 the responsibility for paying claims switched to the coal mine operators. Of those claims filed from 1974 through 1978 for which the Department of Labor identified a responsible mine operator and awarded compensation, 96% were contested. In 1977 almost half of every dollar spent on the Black Lung Program was spent to determine work-relatedness and the level of disability of claimants.

This aspect of the history of Black Lung legislation suggests that presumptions may play only a limited role in reducing litigation, and that perhaps the most important factor in litigation is, indeed, financial responsibility. The federal government seems to have been led to this conclusion, since the 1977 Black Lung Benefits Revenue Act moved away from the identification of responsible coal mine operators to the creation of a federally administered trust fund to pay disability benefits for most claimants. The trust fund is financed by an industrywide tax on coal of

$0.50 per ton for underground mines and $0.25 per ton for surface mines. When individual employers have nothing to gain from contesting claims, they do not contest. As a result, non-benefit expenses have dropped from 50% to 10% and the payment rate has shot up. Federal payments of Black Lung benefits now are at a rate of almost $2 billion annually.

PROBLEMS IN SETTING PRESUMPTIVE STANDARDS

In this section, I would like to suggest some criteria for evaluating presumptions, and to explore some of the problems involved in writing them.

1. Reasonableness: note that a presumptive standard establishes a *legal* basis for paying workers' compensation. A legal standard may have a close, or a tenuous relationship to medical and epidemiological knowledge. A presumption should have a reasonable relationship to our knowledge about the disease which it addresses. Criteria for presuming work-related disability should be generally consistent with current understanding of the causes of disabling occupational diseases.
2. Fairness: within the class of at-risk workers, the population falling under the presumption should have a greater risk of disabling occupational illness than those who do not meet its criteria. In addition, to the extent feasible, presumptions about different diseases should not create classes of occupational disease victims, some of whom receive compensation a great deal more easily than others. Workers with asbestosis should not be much more likely to receive compensation than those with silicosis.
3. Reducing litigation: reducing litigation should help greatly to improve the effective delivery of those resources which enter the workers' compensation system.
4. Increasing compensation of occupational disease victims: any presumption should attempt to be as inclusive as possible of occupational disease victims, within the constraints imposed by other goals of workers' compensation.
5. Limiting compensation of workers without occupational disease: workers' compensation was not designed to pay for nonwork-related injuries and illnesses.
6. Political acceptability: obviously, in order to become enacted, presumption must receive political support.

Even with well-defined goals, it may be very difficult to forecast how effectively those goals will be achieved — or for that matter, to evaluate effectiveness after a presumptive standard has been implemented. We do now know how much exposure there is to occupational hazards, nor, for most substances, do we know the risks of exposure. The same uncertainties that motivate the implementation of presumptions make their evaluation almost impossible.

The paucity of information about occupational disease leaves a great deal of room for political maneuvering to determine the final form of a presumptive standard. When only a few facts exist, prejudices and self-interest often take their place. As a result, the formulation of presumptive standards may largely be based on the political power of the groups affected. Strong unions could get permissive standards for diseases which are common among their members, while weak unions or unorganized workers could perhaps get nothing at all.

Even when we know the shape of the relevant dose-response relationships and the exposure levels of the working populations, any actual presumptions can be more or less stringent in their criteria. More stringent criteria would mean both that fewer claims would be paid to people without occupational disease and that more people with occupational disease would not meet the criteria of the presumptive standard. This underlines the fact that the design of a presumptive standard is to a significant extent a policy decision. The underlying policy question is: how much excess compensation are we willing to allow in order to serve deserving, disabled workers for whom workers' compensation was designed?

The arbitrary nature of presumptive standards is most obvious in cases where diseases have multiple, perhaps synergistic causes. How, for example, do we "fairly" decide which cigarette-smoking asbestos workers with lung cancer should be compensated? Do we disregard the fact that smoking and asbestos have a multiplicative effect on the risk of developing lung cancer? If we do not disregard this fact, do we apportion the cause of lung cancer between asbestos and cigarette smoking? If we do this, on what do we base apportionment? Again, this is a political decision as much as anything else.

EFFECTS OF PRESUMPTIVE STANDARDS ON WORKERS' COMPENSATION

Some speculation about the probable impacts of presumptive standards on workers' compensation for occupational diseases is warranted.

One of the major benefits of presumptive standards is the potential educational impact on workers, lawyers and physicians. Knowledge about occupational disease is not widespread and, therefore, many workers do not even realize that they may be suffering from an occupational disease. Moreover, lawyers who may represent workers in occupational disease compensation cases are, themselves, often not adequately informed. The impact of presumptive standards on workers would probably be to increase the number of claims brought for occupational disease. A well designed disease presumption can bolster the willingness of claimants' lawyers to take on claims that meet the criteria under the standard and to refuse those claims that do not seem to be covered by the standard. The net impact would probably be to increase the number of claims brought into the system.

The second benefit of occupational disease presumptions is the clarification of the claimant's burden of proof. This would reduce the uncertainty about whether a claim could be upheld by a workers' compensation hearing examiner. In contested cases for which occupational and nonoccupational diseases cannot be differentiated medically, legal rules could clarify the likely results of a hearing. They could also impose consistency on decisions about compensability. Suppose, for example, that a worker who has been exposed to ionizing radiation at work develops leukemia. It should be easier for the lawyers of those workers with lifetime exposures of greater than 50 rems to satisfy the burden of proof of work-relatedness. However, workers who develop leukemia with less than 50 rems of lifetime exposure might find it more difficult to pursue successfully a claim for workers' compensation under this presumptive standard than they would without such a standard. Even if the standard does not contain a negative presumption, both lawyers and judges are likely to be influenced by the fact that the worker does not meet the criteria of the presumption and may reject the claim on this basis.

The use of interim payments, as suggested above, could increase settlements and decrease contests by insurers. Even if interim payments did not change the probability of a claimant's victory, the receipt of income before the case was closed would improve the bargaining position of the claimant, especially in cases of total disability or death. In the 60% of all claims for occupational disease that are compromised [6], the average settlement should increase as a result. In turn, both the earlier payments to claimants and their improved bargaining position will mean that fewer cases will be contested by insurers, decreasing the amount of compensation money paid to claimants' lawyers and insurers' lawyers.

In summary, we can expect that occupational disease presumptions would increase the number of workers claiming compensation, since it would be easier to win such claims. While for any specific claim employers would be somewhat less likely to contest, the general increase in the number of claims might lead to more legal contests. As a result, more occupational disease victims would be compensated—and more victims of nonwork-related disease would also be compensated. Finally, if a smaller proportion of occupational disease claims is litigated, the proportion of premiums going to legal and administrative costs will probably fall.

Let me add a note of caution. We don't know whether the Black Lung compensation experience is the rule or the exception. If it is a portent of what could happen in other areas, then we must be very cautious about predicting the impact of future occupational disease presumptions.

First, if the cost of individual cases is high, insurance carriers and self-insured employers will generally litigate. As noted above, of those Black Lung cases in which a responsible mine operator was identified, more than 95% were contested, in spite of the fact that the Department of Labor had decided that these miners met the criteria of the presumptive standard. In fact, it is easy to imagine a case in which presumptions would lead to a higher proportion of litigated cases than would otherwise occur. If a standard leads to a large increase in the number of claimants, if the standard is not considered reasonable by the affected industry, and if claimants can impose very large costs on insurance carriers and employers, then a great deal of litigation may follow.

Second, presumptive standards for occupational disease may be very expensive. In such cases, Congress seems ready to transfer the burden of workers' compensation payments from individual employers to taxpayers and consumers. This may be a reasonable policy decision, but those who oppose it as a bail-out of employers with unhealthy workplaces should understand that it is a likely outcome of an expensive presumptive standard.

Finally, when the government sets standards they are often quite political in nature. The decisions about which standards to set and how to set them may not flow from a careful attempt to assess their benefits. Given the scientific uncertainties in this area, there is a great deal of leeway for the setting of a restrictive or an expansive standard. Thus, outcomes are by no means guaranteed.

REFERENCES

1. Shodel M: Rights of the cell. Science 81 2:26–8, 1981.
2. Larson LW: Analysis of current laws reflecting worker benefits for occupational disease. National Technical Information Service Report No. ASPER/PUR-78/4385/A, 1979.

3. Solomons ME: Workers' compensation for occupational disease victims: federal standards and threshold problems. Albany Law Rev 41:195–259, 1977.
4. U.S. Department of Labor, Assistant Secretary for Policy, Evaluation and Research: An interim report to Congress on occupational diseases, 1980.
5. Pischer DP, Kleinman GD, Foster FS: Pilot study for development of an occupational disease surveillance method. Washington, DC: U.S. Government Printing Office, 1975 (NIOSH Publication No. 75-162).
6. Barth PS, Hunt HA: Workers' compensation and work-related illnesses. Cambridge, MA: MIT Press, 1980.
7. Cooper and Company: Survey of closed workers' compensation cases in the United States. Washington, DC: Intergovernmental Workers' Compensation Task Force, 1975.

CHAPTER 26

DESIGN DEFECTS
AND WORKERS' COMPENSATION

Willie Hammer
 Cerritos, California

In terms of correcting design defects to minimize accidents in industrial plants, workers' compensation laws are possibly the greatest fraud ever imposed on the industrial workers of this country. They have probably done more to deter improvements for safety in the workplace than they have to encourage them, as it was originally believed they would.

Before the workers' compensation laws were passed, both employers and workers hoped they would provide specific benefits. Employers wanted a cheap means to end costly litigation resulting from accidents due to unsafe conditions in their industrial plants. This is what they got. The resulting legislation actually should have been entitled "Employers' Injury Litigation Relief Laws."

The workers did not get the two principal benefits they wanted: (1) the laws do not ensure an adequate income for injured persons or their surviving dependents; and (2) the laws have not resulted in improvements in workplace safety. The improvements that have taken place have generally been due to other causes.

It might be noted that improvements in workplace safety would result in fewer accidents and reduce the payments to workers injured on the job. At the same time, these same improvements would reduce the costly litigation, which was a primary objective of employers. At the time they were enacted, workers' compensation laws were viewed by employers as a cheap solution. The legislators assumed that the costs of insuring unsafe workplaces would force employers to make improvements. The workers'

331

compensation laws are not only unmitigated failures, but they have been counterproductive.

The failure of workers' compensation to force improvements in industrial plant safety is due to a complex intermix of economic, legal, judicial and technical factors. To understand what happened and is still happening, it is necessary to review some history.

During the Industrial Revolution, a prevalent philosophy in business was that a productive enterprise should not be hampered as long as it did not subject anyone to an "unreasonable risk." Managers in control of these enterprises were interested only in performance, costs and profits. Job safety was the responsibility of the worker. An "unreasonable" or "reasonable" risk has never definitely been established. Managers believed that it was not reasonable to provide safeguards that could not be justified economically (an argument still heard today). Since most safeguards cannot be economically justified before an accident, few safeguards were provided.

In England, where most of our early legal precedents originated, the courts supported the philosophy that a productive enterprise was not to be hampered. The justices were probably inherently biased: almost all came from the same upper-middle class that provided the industrial enterprise managers. Some of the legal precedents these justices set affected industrial working conditions for years. These decisions included:

1. The doctrine of assumption of risk (*Priestly vs. Fowler,* 1837) declared that a person who accepted a job or undertook a task known to be dangerous voluntarily accepted the hazards and possibilities of injury that the job entailed. Therefore, no claim could be made against the employer for injuries suffered.

2. The Rule of Privity (*Winterbottom vs. Wright,* 1842) required that there be a direct contractual relationship between two parties if one wanted to sue another. Winterbottom was the driver of a mail coach that had been made by Wright and sold to the Postmaster General of England, Winterbottom's employer. While he was driving the coach, a wheel failed, the coach overturned, and Winterbottom was injured. He attempted to sue Wright, but the court declared there could be no suit since no contractual relationship existed. Wright had sold the coach to the Postmaster General, who had not been injured. Winterbottom could not sue the Postmaster General since he had not made the defective coach. The Rule of Privity held in most cases for almost 75 years. It is still cited as a precedent in Great Britain.

3. In *Brown vs. Kendall* (1850), the courts ruled that in order to hold a party liable for an accidental injury, the plaintiff had to prove negligence. This rule held in product liability cases until 1963, 113 years later.

4. In various cases, courts held that if a plaintiff was negligent to *any* degree, however minor, in avoiding or causing an accident or injury, the defendant, no matter how large his share of the blame, could not be held liable for damages. This rule of contributory negligence served as precedent until the 1970s when it was superseded in the courts of California and several other states by the concept of comparative negligence. Under this concept, blame is apportioned to all parties concerned, and the plaintiff collects only that portion of any damages not due to his own negligence.

5. Historically, injured workers could not expect to collect damages from an employer if the accident resulted from the negligence of a fellow employee.

6. If a worker was killed instantly or died subsequently, his surviving dependents could not sue for damages. Only the injured person could sue. This situation was later rectified by "wrongful death" laws.

These are only a few of the legal principles that for long periods of time were used as precedents in the courts. Although most have been superseded, either entirely or in part, many managers and engineers, as well as many in the general public, still believe they hold as rules of law.

Prior to the passage of workers' compensation laws, an injured worker who brought suit against an employer following injury generally lost the case. It often took years for the case to be heard in court, during which time the worker often had little or no income. It was extremely difficult to prove that neither the worker nor a fellow employee had been negligent, that the worker did not know the risks of a job or task, that the employer had been negligent, and so on. A great number of accidents were called unavoidable "Acts of God" (who appeared to be on the side of employers) for which no one could be blamed.

When claims of employer negligence were filed, fellow workers refused to testify out of fear of reprisals. The injured worker often refrained from bringing suit, afraid he would destroy any opportunity of reinstatement in his job or of finding work with another employer. In addition, insurance company lawyers were very clever in demonstrating that the plaintiff had known the job was dangerous but had chosen to assume the risk and that action might have been taken by the plaintiff to avoid the accident or injury, or that negligence of a fellow employee (who was sometimes miles away) had caused the accident.

Plaintiffs won few cases and, when awards were made, they were generally so small as to provide little incentive to employers to improve workplace safety. Permanently injured workers or their dependents often became public wards.

In attempts to reduce this problem, Massachusetts passed a law in 1867 requiring factory inspections and, ten years later, another instituting guarding of dangerous machinery. A few years later, Alabama, Massachusetts and other states passed employer liability laws that voided the employers' right to use Assumption of Risk and certain other usual defenses in legal suits. Generally, these new laws were ineffective; there were too many ways by which they could be circumvented.

The cost of insurance for employers was high, not because of the few and meager damage awards, but because of the high legal fees charged to defend employers in the great number of suits. Employers sought a system to reduce legal and insurance costs. They demanded a low-cost system that would eliminate suits and lawyers, and provide predictable payments to workers for injuries. Workers' compensation laws were passed, and employers are the ones who have benefited most.

Historically, engineers were as little concerned with worker safety as were managers. Their loyalties were to the employers, a situation that still exists. They were interested only in avoiding permanent failures of equipment and in minimizing construction and operating costs. The American Society of Mechanical Engineers (ASME) was formed in 1880 to discuss problems of mutual interest; safety was not on the agenda. The principal items of interest were standardization of pipe and bolt sizes and thread configurations and dimensions. In 1910, 30 years later, when plants were experiencing 1300–1400 boiler explosions per year (a boiler exploded every 36 hours), ASME finally moved for an investigation. It was apparent that designs had to be improved to reduce the occurrence of explosions. It was only then that the famous ASME "Boiler Code" was developed.

It can be noted that the National Electric Code was not written by an electrical engineering society. It was developed by the National Fire Protection Association to reduce the number of fires in industrial plants and structures from electrical causes. There is very little in the code written specifically to protect persons from electrocution.

Engineers had little knowledge about accident prevention and generally ascribed accidents to worker negligence. The engineering schools taught future designers almost nothing of safety, and with two or three exceptions, still do not. Schools teach safety to safety engineers, industrial hygienists and industrial engineers, but not to the undergraduates who will become equipment and facility designers. The assumption is "if a piece of equipment works as designed without failure, it will be safe unless some knothead screws it up."

In 1907 Wyoming passed the first state law requiring registration of engineers, supposedly to "protect the health, safety, and welfare of the public." The law was passed to end the common practice by lawyers, realtors, barbers and other persons without suitable training of preparing land survey maps whose inaccuracies led to legal disputes. However, the public, unhappy with all the railroad accidents that were occurring (railroad companies were then regarded with the same kindly and affectionate feelings that oil companies and OPEC members are today), determined that engineers (most of whom were civil engineers) should be tested for competence and registered before they practice. All other states followed Wyoming with similar laws, but most exempted engineers in industry, who were generally mechanical engineers, from the acts. Today, those registration laws are still on the books to protect the public health and safety, but the examinations contain no questions on safety principles or methods of accident prevention. There is no means to determine whether designers, uneducated in safety matters, have acquired even the slightest ability to "protect the health and safety of the public."

After the turn of the century, laws were passed to improve public and worker safety, but these laws were the ineffective engineer registration and workers' compensation laws. They were placebos. It was hoped that the workers' compensation laws would result in safe workplaces, but this did not happen. The meager compensation schedules kept insurance rates low, making it cheaper for employers to pay compensation than to provide safe equipment. In addition, from the time the first workers' compensation law was passed until 1916, workers covered by the law could sue employers for injury only in exceptional cases. The compensation laws required that workers give up the right to sue except in those cases where there was "gross and willful negligence." The Rule of Privity prevented an injured worker from suing the manufacturer of the defective equipment that had caused the injury since the equipment had been sold to the employer and not to the worker.

The Rule of Privity for products was superseded in 1916, but an injured worker still had to prove negligence in order to win a liability suit. The failure of engineering schools to teach safety principles and accident prevention methods created such widely accepted misconceptions among engineers that engineers would not testify, except in obvious cases of negligence, against their peers. In general, equipment manufacturers' negligence could only be demonstrated in those cases in which there was a very apparent production defect that caused the equipment to fail, resulting in an injury, or if there was a failure to warn of a covert hazard, which the engineer himself probably did not recognize.

In 1963 the concept of "strict liability" was adopted in California and is now observed in most state courts. With strict liability a plaintiff need no longer prove negligence by a manufacturer. The plaintiff must show that the product had a dangerous defect when it left the manufacturer or other party in the chain of commerce and that the defect caused the accident and injury. The adoption of strict liability has improved the safety of industrial equipment. Presently, 70% of the product liability claims and cases for amounts greater than $100,000 involve injuries to workers caused by industrial equipment. The average award is approximately $100,000 in an industrial product liability case. The overall average nationwide is $14,000 when nonindustrial cases are included.

In 1977 a California worker who lost eight fingers because of a defective safety switch on a press was awarded $1.1 million in a legal suit brought under the strict liability concept. If he had lost all ten fingers, he would have received $40,000 plus a small lifetime pension from workers' compensation. A few months ago a Los Angeles man was awarded $458,700 for the loss of seven fingers in a stamping press. Workers injured by industrial equipment are, therefore, highly disposed to sue the manufacturers, which they can do even if they are receiving compensation benefits. The workers' compensation insurers are eager for workers who have reasonable claims to sue. They will assist them and will bring suits on their behalf. If the plaintiff wins, the workers' compensation insurer is reimbursed for payments made and is relieved from further payments.

The observance of the strict liability concept by the courts has improved industrial plant safety far more than has the workers' compensation laws. Improvements in industrial safety have occurred for other reasons, but few improvements are attributable to workers' compensation. Many improvements were due to state safety codes and laws that required that specific safety requirements be met and to the passage and enforcement of the Occupational Safety and Health Act (OSHAct) and the standards imposed under OSHAct. The strict liability concept, however, due to the threat of very high awards, has been more effective in ensuring that manufacturers improve equipment safety than has OSHAct in improving plant safety because OSHAct fines for violations are small.

The common law regarding workers' compensation allows injured workers to sue employers when accidents are due to "gross and willful negligence." Recently, the courts have ruled that gross and willful negligence existed in cases where accidents resulted because of violations of OSHAct standards or state laws. Thus, in spite of the workers' compensation laws, an employer can be sued.

The employers pay workers' compensation insurance costs, and those costs have been increasing. Benefits to injured workers have been increased by state legislatures; types of injuries covered are being extended; and increasingly costly medical rehabilitation coverages are causing workers' compensation insurance costs to skyrocket. In addition, as equipment grows bigger, more energetic, more complex and potentially more dangerous, the number of workers who can be injured and the injury severities increase. If rising insurance costs are not to become so prohibitive that productive enterprises are hampered, the number and severities of accidents must be reduced. This can be done only through safer designs and procedures. The engineers who create these designs and who prescribe operating procedures must be knowledgeable in safety principles and accident prevention methods.

During the past 20 years, a new engineering discipline, system safety, has developed these safety principles and accident prevention methods. Developed to minimize accidental losses in the military services, system safety principles and methods are being used more and more by civilian agencies of the government. Studies by system safety engineers, by insurance companies, and of product liability cases have shown that a large percentage of accidents is due to design defects. To reduce the number of these defects, engineering schools should make courses in safety principles and accident prevention methods mandatory for all students, not to train them to be engineers, but to ensure that they are safety-minded engineers.

CHAPTER 27

ALTERNATIVE REMEDIES FOR WORKERS OTHER THAN THE OCCUPATIONAL SAFETY AND HEALTH ACT AND WORKERS' COMPENSATION

Frederick M. Baron, JD

Baron and Cowley
Dallas, Texas

The purpose of this chapter is to describe methods by which injured workers can avoid the "exclusive remedy" of most state workers' compensation acts and obtain realistic economic recompense for injuries sustained as a result of exposure to toxic chemicals.

Before I begin, however, I would like to comment on a statement made by Mr. Hammer, which I believe describes an important sociological trend. The statement was made that *courts* have been awarding "substantial" benefits to workers who file product liability third-party suits. Such is not the case. *Juries,* composed of citizens from all walks of life, are the ones who award benefits in these cases, not the courts. I believe that the real reason we are seeing large verdicts in product liability cases is not because of any changes in the law, but because the juries (i.e., the American public) are becoming outraged by the way many industrial corporations exercise an almost total disregard for the health and safety of their employees and for the consumers of their products. It is my sincere belief that what we are witnessing is an effort by citizens, through their participation in the administration of justice, to register their disapproval of the way in which many companies conduct their business affairs.

As the title of my chapter implies, the objective our office pursues when representing injured workers is to find an alternative to workers' compensation benefits. Why do we want to do that? There is one basic

339

reason. The benefits provided by most workers' compensation statutes are woefully inadequate. In fact, they are unrealistic. In my home state, Texas, we provide a maximum benefit of approximately $136 a week for a worker who has been injured, regardless of what his income was prior to the injury. There is no inflation factor whatsoever, and the benefits for total disability are limited to a maximum period of 401 weeks. Thus, when a client comes into my office who has lost both of his legs in an explosion, is 18 or 19 years old, has a wife and a young child, I have the unfortunate experience of advising him that the total benefit he will receive for the rest of his life, undiscounted for present value, is approximately $50,000, which is to last him indefinitely. Moreover, in the event that he receives social security benefits, his social security benefits will be offset by the amount of workers' compensation benefits he receives.

There is not an individual in this room who can argue that this is truly adequate compensation. Until we make state workers' compensation laws realistic, more and more lawyers are going to be filing more third-party claims, seeking to protect the rights of their clients who have been injured in industrial accidents.

As a further complicating factor, in occupational disease cases it is often very difficult to determine the origin of the disease. When an employee works for five years in a plant that uses numerous chemicals, then goes to a second chemical plant that also utilizes many toxic chemicals, and then subsequently develops a lung disease, who is to bear the blame, or who bears the risk or liability for the disease? Historically, under state workers' compensation, the answer has been that the worker himself must bear the burden, if he is unable to prove with some degree of certainty where it was that he contracted the disease. Thus, workers' compensation benefits have been entirely unavailable in many occupational disease cases.

Moreover, another problem exists in the case of an individual who, for instance, worked for an asbestos company in the 1940s, never worked for another employer that used asbestos, and then in 1980 contracts asbestosis. If the worker files a workers' compensation claim, he may be required to file it against his employer for the period of his exposure in the 1940s, if that employer is still in existence. In most states, he will receive benefits at the level provided during his years of exposure. In Texas, for instance, the maximum benefits prior to 1949 were $21 per week, or a total of approximately $8000 for permanent total incapacity.

In addition, there are provisions in state workers' compensation acts that require compensation claims be filed within six months of the date of last exposure. These provisions virtually ensure that no benefits will be paid in those cases that involve a latency period between time of exposure and development of disease. The result of these problems, which are inherent in the law, has been that the worker is the one that bears the risk of injury. Lawyers have, therefore, vigorously pursued third-party lawsuits for individuals who have not been able to protect themselves against medical, financial and family disaster caused by their job-related illness.

There are several methods by which a diligent attorney can attempt to circumvent the applicable state compensation act. In most states, there is a provision that permits an employee to pursue a "common law" remedy for intentional injury inflicted by his employer. This means that if the employer assaults his employee, the employer is no longer immune from direct liability. As stated by one court, "an employer cannot correct and punish with whips the mistakes his employees committed in the course of employment and protect himself against civil liability for the result of his assault under the coverage of workers' compensation." Yet, in other states, and in particular Georgia, courts have held that an employer who physically assaults an employee (in the Georgia case, with a baseball bat) cannot be sued by the employee because of the exclusive remedy provisions of the state workers' compensation act. Some courts have interpreted the words "intentional injury" to encompass gross wanton and willful misconduct. In those states, which include New York and Ohio, the injured or diseased claimant can recover against his employer if he can prove that gross or wanton misconduct by his employer caused his injury. This stops somewhat short of actual assault.

Recently, a California Supreme Court in the case of *Rudkin v. Johns-Manville* declared an even further exception to the exclusive remedy provisions of the California workers' compensation act: situations in which the employer has engaged in fraud, deceit and misrepresentation that cause injuries to employees. In the case in point, the employer was advised by the company physician that an employee, Mr. Rudkin, had developed asbestosis. Despite this knowledge, the employer failed to advise Mr. Rudkin of his condition and permitted him to continue his employment and exposure for several years after the diagnosis had been made. The aggravation of the disease caused by the continued exposure caused the employee to become a pulmonary cripple. The court found that the facts, which were so obviously horrendous, required it to create an exception to the compensation act. The court held that in a situation where misrepresentation or fraud on the part of the employer aggravates a condition caused by a job-related injury, the employee may sue for the aggravation of the condition but not the underlying condition itself. Thus, Mr. Rudkin was permitted to recover for any aggravation of his asbestosis but not for the underlying condition itself. The California Supreme Court specifically held, moreover, that it was the burden of the employer to prove the difference between the underlying disease and the aggravation; this seems like a very difficult task and would permit a jury to refuse to divide the injury from the aggravation.

The only other way by which the courts have allowed injured claimants to successfully pursue actions against their employers is through a device called the "Dual Capacity Doctrine." This is referred to by the courts as the "two-hat theory." If, on one hand, the employer is acting as an employer, yet also acting as a manufacturer of a product used by the employee, and the employee is injured by the product that was manufactured, the court deems that to be a product liability suit and not barred by workers' compensation law. Although only a few states have adopted this theory, there has been a substantial amount written about it in the legal periodicals, and I believe that this doctrine will gain great favor in other jurisdictions.

Another potential "third party" can be the parent company of a subsidiary employer. The famous case in this area is *Boggs v. Blue Diamond Mining Company. Boggs* was a lawsuit filed by 15 widows of coal workers who had been killed in a methane gas explosion inside a mine in Kentucky. The mine was owned by a company called Scotia Mining Company, which was a wholly owned subsidiary of Blue Diamond Mining Company. Following award of workers' compensation benefits, which were minimal, the widows filed suit against Blue Diamond, claiming that Blue Diamond was aware of the condition inside the mine and had done nothing whatsoever to correct it. This was borne out by numerous citations from government mine inspectors requiring Scotia to make changes that were never implemented. In fact, Blue Diamond spent a great deal of its time monitoring its subsidiary corporations and could have been found to be the operator of the mine. After suit was filed against Blue Diamond, the district court dismissed the case, accepting Blue Diamond's contention that the workers' compensation act provided a bar to recovery. The United States Court of Appeals for the Sixth Circuit, in a very well-written decision, reversed the lower court and said that the company itself was a separate corporate entity; moreover, Blue Diamond had created the separate entities and there was a policy of law not to set aside a corporate entity unless it could be proved that the two were completely interrelated. Although Blue Diamond owned 100% of the stock of Scotia, each company paid separate taxes, had separate auditors, and were, in fact, separate entities for all other purposes.

Pursuant to an old doctrine of law, sometimes known as the "volunteer act," Blue Diamond could be held responsible under these circumstances. One who adverts to act and does so negligently can be held responsible for the consequences. In this case, the plaintiffs were able to prove that Blue Diamond had undertaken to correct the situation inside the mine but had done so negligently and had failed to warn workers of hazards. Thus, the court permitted recovery.

Several other jurisdictions have likewise held that corporate entities cannot be integrated to provide an exclusive remedy under the workers' compensation act. In most courts, there have been decisions, particularly in the tax law area, that a subsidiary corporation will be held to be a separate and distinct entity. In the chemical industry in particular, this is not uncommon. Our office is currently handling a case in Delaware filed by employees of Amoco Chemicals Corporations, a wholly owned subsidiary of Standard Oil. Standard Oil Company provides corporate medical and industrial hygiene services to their subsidiary and, by doing so, creates potential legal exposure. Since Standard is a separate entity from Amoco, Standard is not immune from responsibility under the Delaware state workers' compensation law.

Other remedies that might be available to injured workers include suits against professional co-employees who might be operating in a nonemployer context. There have been numerous cases over the years concerning the liability of medical personnel. What is the liability of a company physician who is aware of hazardous exposures inside the plant and early X-ray changes in certain patients but does not warn these patients about the exposures and early changes? Does the doctor have an affirmative duty to the patient to advise him to wear a respirator or to protect him against the hazards to his health that are present in the plant? The moral question is much easier to resolve than the legal question. Courts have gone in all directions in attempting to define responsibility in this area.

In the state of Indiana, for instance, they have found that there are certain responsibilities inherent in the practice of medicine. If the plaintiff can show that the doctor breached one of these responsibilities by not telling him of a hazardous condition or by not showing him how to protect himself, the courts have held that the doctor himself can be liable, even though he is a retained physician paid by the employer. Needless to say, many physicians are concerned about this liability. Yet, in such situations, the company may be responsible for the injuries sustained as a result of the negligence of their retained medical employee, even though the worker is receiving workers' compensation benefits. Thus, the ability of a worker to sue a physician in this context may provide a manner in which he can recover against his employer, successfully circumventing the exclusive remedy provision of the workers' compensation act.

Other courts have not adopted that theory. In some of the more conservative jurisdictions, the rule has been that the doctor need only diagnose accurately and advise his patient. In other states, the courts have held that a physician is absolutely immune from liability because he is held to be a fellow employee and, therefore, entitled to the insulation of the workers' compensation act. Many of the cases in the latter category are old cases and have not been tested in many years. This is an area of law in which the courts have a substantial amount of leeway in drafting new doctrine. When confronted with a difficult fact situation, I feel that most courts will create a remedy for a seriously injured employee against those who caused and continued his injury.

Another potential "third party" are the safety personnel other than physicians who are employed by the employer. In this context, I am particularly referring to industrial hygienists and independent testing laboratories. In the Tyler, Texas, asbestos litigation cases, suit was filed against the United States government because its industrial hygienists had inspected an asbestos plant, found grave and immediate danger to employees, yet failed to warn the employees in any manner. In fact, in one deposition, a government employee testified that he did not wear a respirator inside the plant when he conducted the inspection because he was afraid he would alarm the workers to the hazard that existed. In my opinion, when a court is faced with this type of testimony and this type of conduct and if the area of law is hazy, the court will fashion a remedy. Thus, I believe it safe to say that such safety personnel have an independent duty to the employees in a plant to disclose to them, regardless of any contractual duties to the employer, any hazardous conditions that exist inside the plant.

The primary area in which third-party litigation has been carried on is in the area of product liability. Product liability as a doctrine has been around since 1916. This doctrine states that anyone who manufactures a hazardous product is responsible for the damage that product causes. In a product liability case, the jury assesses the efficacy of the product, not the conduct of the manufacturer. If a determination is made that the product was not a good, safe product, the manufacturer can be held responsible for injuries even though he was not negligent. There are two major reasons for this law: first is the concept of "spreading the risk." Unquestionably, it is much easier for a manufacturer or supplier of such defective products to spread the cost of injury among numerous consumers than it is for an individual who is seriously injured to bear the entire financial burden of his injury. Second, the courts have recognized that it is an extremely difficult task for an injured worker to prove the negligence of a manufacturer in the manufacturing process itself. Thus, by eliminating the requirement of proof of negligence and merely allowing the plaintiff to show that the product itself was defective, the courts permit the plaintiff to have at least a reasonable shot at recovery.

In the toxic substance area, the strict liability theory that is primarily used is that the product is defective because it is not accompanied by an adequate warning. The classic example is asbestos insulation material. In the case of *Borel v. Fibreboard,* the court found that since the asbestos manufacturers were aware of the hazards of asbestos, they had a duty to place an adequate warning on the material being used by installers. In the absence of such a warning, the product would be rendered defective even though it was properly made, and the manufacturer would be held responsible for injuries sustained by the unaware consumer.

There are defenses to this theory. The primary defense is that of "misuse." If an employee misuses a product and that misuse is the cause of his injury, he is barred from recovery. Also, in some states the courts have refused to permit recovery if the employee "assumes the risk." This means that if an employee is aware of a hazard and voluntarily chooses to encounter it, he may not recover for any injury he sustained. This is a harsh doctrine. The economic reality of things is that employees must earn a living to support their families, and they are faced with daily risks of danger in fulfilling their jobs. The general social policy behind the law is to attempt to minimize these dangers and spread the risk; in my opinion, except in the grossest of cases, the doctrine of assumed risk should not be applicable.

The rule of product liability exposure has been applied to other than the manufacturer of industrial products. In the case of *Porter v. American Optical Corporation,* the plaintiff was an employee of a company that manufactured asbestos-containing preformed boards. His employer, in accordance with the OSHAct, provided him with a respirator to use in fulfilling his job function. Unfortunately, however, the employee contracted asbestosis even though he was using his respirator in a proper manner. Suit was subsequently filed against the manufacturer of the respirator on a product liability theory. The court found that since the respirator manufacturer had advertised its product as capable of preventing such diseases, it would be responsible for the injured plaintiff's harm. This does not seem unreasonable. In the words of Ralph Nader, the law should act to instill "responsibility" in those who present themselves as experts, whether in the area of medical practice, industrial hygiene, or product supply. We cannot operate in a vacuum. I do not believe that it is stretching things to hold a company responsible for doing that which it said it would do, i.e., produce a safe, workable product.

The final area in which there has been a substantial amount of movement in liability cases is that involving government agencies. In a typical case, the mine safety inspector found a front-end loader that did not contain a roll bar, in violation of the Mine Safety Act. The inspector in question gave the employer a 60-day period in which to remedy the problem. At the end of the 60-day period, the employer sought and was given another 30-day compliance period. In all, a total of five delays were granted for no apparent reason, and the inevitable occurred; an employee was killed when a front-end loader rolled over and crushed him. The court in that case found that, under the "gratuitous undertaking" or "volunteer" theory previously discussed, the government was responsible. Noting that the government does not have the responsibility to inspect, the court found that when they do inspect, they have a duty to do it carefully. The court specifically found that it was discretionary as to whether or not the government decided to do the inspection, but that it was not discretionary in how it was carried out and implemented. Again, the court found that once the line is crossed and the inspector attempts to do something, he has a *responsibility* not only to himself but to those who rely on him to do it properly.

DISCUSSION: CHAPTER 27

Herbert K. Abrams, Frederick M. Baron,
Kenneth S. Cohen, Mark Veckman and Michael Zacks

DR. ABRAMS: It was stated by an earlier speaker that industry is attempting to nullify product liability through legislation in various states. Can you bring me up to date on that?

MR. BARON: There has been a concerted activity by manufacturers' associations and insurance companies to limit product liability claims. The most devastating offshoot of this "reform" legislation is the passage by many states of a so-called "statute of repose." These statutes provide that if more than a designated period elapses (usually 12 years) from the date of sale of the product until the injury occurs, no responsibility will be imposed on the seller or manufacturer of the product. In the case of a chemical or other product that causes latent injury, this eliminates the possibility of recovery on behalf of the injured consumer. In other words, if a user of asbestos contracts asbestosis 25 years from the date of the initial use of the asbestos, he will be barred from recovery by the statute of repose. Needless to say, in those states where this law has passed, toxic substance litigation has been all but eliminated. I should add, however, that at least one court, the Supreme Court of Florida, has held that this type of law is unconstitutional in that it violates a citizen's rights to access to courts for redress of grievances.

The purpose of the law is to cut off the "tail" on product liability claims and to protect insurance companies against exposure for things that occurred many years ago. Quite frankly, I don't believe that this "social policy" outweighs the interest of injured consumers. Moreover, I suggest that these "reform" acts are merely methods to insulate manufacturers and insurance companies from further responsibility for bad and unsafe practices.

347

DR. COHEN: Could you comment on defendants who suggest "standard of care in a community" or "state-of-the-art knowledge" as an excuse?

MR. BARON: In most states, a defense is available to the manufacturer of a hazardous product in a product liability case if the manufacturer could not have reasonably known, at the time the product was sold, of the hazardous propensities of the product. Unfortunately, some courts have held that the state-of-the-art is to be defined as "industry practices." This is both illogical and unfair. The idea of product liability is to help stimulate research to determine whether products are, in fact, hazardous and, if so, to eliminate their use. By equating state-of-the-art with industry practice, the court is merely suggesting that industry ought not find out about the hazards of its products before they are marketed.

Fortunately, several courts have recently held that state-of-the-art is not a defense in a product liability case. The Colorado Supreme Court, for instance, found that to adopt such a defensive theory would be, in effect, returning to a negligence standard. The issue should be whether the product was safe, not whether or not the manufacturer knew of the hazard. If we intend to spread the risk and promote safety through testing, I believe "state-of-the-art" should not be an issue in these cases.

DR. COHEN: What about "standard of care" as that term is used to define professional liability based on community standards?

MR. BARON: I would like to note first that the cases against an industrial hygienist or company physician are not based on malpractice. In a malpractice case, the doctor or other professional is judged by the standard of care exercised in the community in which he practices. The cases that I have been referring to against an industrial hygienist or a retained company physician are essentially "failure to warn" cases. In those situations I do not believe it appropriate to look to other industrial hygienists or physicians to determine what they would do under the circumstances. I think the only relevant inquiry is whether it is fair to impose liability based on the facts before the court in the case at hand.

MR. VECKMAN: Are you saying that the hygienist has a duty to warn the employee or merely to warn the employer?

MR. BARON: That's a very good question. This issue has not really been resolved by the courts. I take the position that the duty to warn is to those people who are in peril. Thus, a warning by the industrial hygienist to the employer will do nothing to give immediate warning to those individuals who are at risk, namely the employees. How can one fulfill his responsibility, which should be defined as protecting workers' safety, without notifying workers of industrial hazards? In this area of developing law, if I were an industrial hygienist and found a hazard in a plant,

I would want to be absolutely sure that the employees were aware of it. If there were a union in the plant, I would certainly notify the union, at a minimum, and perhaps also place notices on the bulletin board advising employees who work in the area about the nature and extent of the hazard.

MR. ZACKS: You said during your speech that the company could be held responsible for its physician's act. What if the doctor were chosen by the union? Would the liability flow back to the union?

MR. BARON: Again, this is a question that has not been tested. The cases that have been litigated are those in which the company has retained a doctor. I know of no case against a doctor who has been retained by both the company and the union. It certainly is an interesting issue. There have been, however, suits against labor unions for failure to warn their employees. As of this point, they have been unsuccessful because the courts have held that the only duty that a labor union has to its members is defined by Section 301 of the Labor Act, which states that they may not discriminate against their members. Absent discrimination, I do not think that it is likely that a union will be held responsible to its members for failing to point out a hazard. This remains to be seen.

CHAPTER 28

WORKPLACE SAFETY AND HEALTH: RECENT DEVELOPMENTS IN UNION LIABILITY

Larry C. Drapkin

Labor Occupational Health Program
Center for Labor Research and Education
Institute for Industrial Relations
University of California—Berkeley
Berkeley, California

Morris E. Davis

United States Merit Systems Protection Board
San Francisco Field Office
San Francisco, California

It has been a little over a decade since Congress enacted the 1970 Occupational Safety and Health Act (OSHAct). This legislation arose out of an unrelenting crisis—an epidemic of injury, illness and death in America's workplaces. The enactment of this legislation, together with the federal Coal Mine Safety and Health Act of 1969, signaled official recognition of a problem already well known in the industrial world.

The OSHAct signaled a new approach to the problem of workplace hazards. For the first time, Congress had enacted a comprehensive federal law, the intent of which was to prevent workplace injuries and illnesses, as opposed to compensating those already injured or ill. Under the common law an employer was often legally liable for the work-related injuries of employees. With the passage of workers' compensation legislation, employers again were recognized as being primarily responsible for those injuries and illnesses arising "out of and incidental to...employment." Likewise, under the OSHAct, it is the employer who must

> furnish to each of his employees, employment and a place of employment which are free from recognized hazards that are causing or are likely to cause death or serious physical harm to his employees [1].

Although employees have a duty to comply with OSHAct regulations, the standards are primarily geared toward, and enforceable only against, employers [2]. At no point does the OSHAct explicitly or implicitly require an employee's union to ensure or to seek safe and healthful working conditions. Further, for OSHA purposes, the safety and health obligations of employers cannot be delegated or contractually assumed by other employers or unions [3].

Organized labor has influenced, and has been influenced by, the OSHAct. Although health and safety issues had been considered "mandatory" subjects of bargaining under the National Labor Relations Act (NLRA) prior to 1970 [4], widespread union activity on these issues was lacking. However, since 1970 a significant number of bargaining agreements have incorporated specific health and safety clauses. These provisions, among other things (1) establish labor-management safety and health committees, (2) establish workplace safety and health practices, and (3) limit exposure to hazardous work.

As unions began to bargain for the right to influence and, at times, control workplace safety and health practices, they were subjected to greater legal scrutiny. Since the early 1970s, numerous lawsuits have been brought against unions by union members, employers and third-party manufacturers. These cases generally involve allegations that the union (particularly those that have negotiated safety and health provisions) inadequately utilized its power to secure safer and more healthful working conditions. Consequently, the fear of costly and time-consuming litigation hangs over many unions. In some instances it has caused unions to rethink or withdraw from more active negotiations on health and safety issues.

The issue of union liability for health and safety activities begins to take on an even greater importance as more emphasis is placed on concepts, such as "voluntary compliance," that emphasize workplace (employer, employee and union) monitoring and resolution of dangerous or unhealthy working conditions. For unions to continue to engage in active negotiation for and enforcement of meaningful health and safety provisions, we must ensure that they not be saddled with the legal duties and responsibilities that are traditionally those of the employer—the responsibility to ensure workplace health and safety. At this juncture union participation and involvement in health and safety matters is integral in ensuring employer compliance with OSHAct regulations. Further, with the expected decrease in governmental regulatory activity, unions may have to initiate and bargain for more stringent exposure standards in their collective bargaining agreements. It is for these reasons that the law should not hold unions liable for their health and safety activities. Liability, or the threat of potential litigation arising out of union health and safety activities is too great a deterrent to union health and safety activity at a time when such interest and action is more necessary than ever before.

This chapter will assess the legal duties that unions shoulder when involved in health and safety matters. Our analysis will look at the cases that have been brought against unions and assess the strengths and weaknesses of such cases. We will examine the cases that have arisen under labor law's duty of fair representation (DFR), as well as those brought under various common law negligence theories (i.e., breach of contract or tort actions). We will discuss some ways in which unions can minimize their legal risks without deterring their interests in promoting healthier and safer work environments. Finally, we will explore some simple and practical legislative solutions to this problem.

DUTY OF FAIR REPRESENTATION

Origins of the Duty

Federal labor law vests a great deal of authority in the recognized bargaining representative—the union. Unions have the right exclusively to represent and bargain for their membership. With this grant of authority comes some resulting difficulties. Unions often represent bargaining units that are large in size and complicated by occasional conflicts among the membership, specific unit members and management. Because of the diversity of functions, as well as the possibility that some members' interests may wrongfully be overlooked, the Supreme Court has interpreted both section 9(a) of the NLRA [5] and section 2 of the Railway Labor Act [6] to establish a union duty of fair representation (commencing when the union becomes the exclusive bargaining representative of its unit members).

This doctrine was not specifically enumerated in the various labor legislation. In establishing or reading into the legislation this duty, the Supreme Court, in *Steele v. Louisiana & N.R. Co.,* stated:

> It is a principle of general application that the exercise of a granted power to act in behalf of others involves the assumption toward them of a duty to exercise the power in their [the unit members'] interest and behalf, and that such a grant of power will not be deemed to dispense with all duty toward those for whom it is exercised unless so expressed [7].

The Court compared the similar functions of a legislature and a union—particularly the duty of representatives to nondiscriminatorily represent their constituencies.

In *Steele,* a case involving a union's attempt to eliminate current and future jobs for black workers within the bargaining unit, the Court held that unless unions are held to a duty of fair representation, those in disfavor with the majority will be denied "equal representation." Thus, a union now has the duty to fairly represent all employees in the unit for which it is the representative. This duty applies both in contract negotiations and in the evaluation and processing of grievances [8, p. 171].

How DFR Suits Are Brought

Under current case law, suits for the breach of the DFR can be brought in both state and federal courts on an implied cause of action under the exclusive representation provision of the appropriate labor law [9]. Many courts have also entertained such suits under section 301 of the Labor Management Reform Act (LMRA), which afford employees a right to sue for breach of a union-management contract. Additionally, the NLRB can independently and concurrently determine whether a union committed an unfair labor practice by breaching its duty of fair representation.

What Constitutes a Breach of the DFR?

Although the courts vary in determining what constitutes a breach of the DFR, it is widely accepted that a breach occurs when the union's conduct toward a bargaining unit employee is "arbitrary, discriminatory, or in bad faith" [8, p. 207]. This does not imply a duty to process all grievances or to negotiate equal working conditions for all unit employees.

The union has a "wide range of reasonableness...subject always to complete good faith and honesty of purpose in the exercise of its discretion." Numerous cases now hold that unions breach the DFR by bargaining for discriminatory contract provisions or by ignoring or failing to evaluate carefully the basis of a grievance or by handling the grievance in a perfunctory or discriminatory manner, with or without bad faith. The courts still maintain that union negligence, alone, does not constitute a breach of the DFR.

Breach of the DFR over Safety and Health Issues

From a DFR perspective, a union generally has considerable flexibility in negotiating contract language. However, a union cannot use its power to bargain away those rights that, by law, are guaranteed to individual employees. Thus, it is a breach of the union's DFR to encourage or bargain for discrimination based on race, sex or union membership. From an occupational health perspective, it is arguable that a union's bargaining away of workers' OSHAct rights might constitute a breach of the DFR. Further, if a union willfully allowed certain employees to be exposed to health hazards because of race, sex or union membership status, then the DFR may be breached. Thus, when unions fail to identify and seek resolution of health hazards predominantly affecting a specific work population (e.g., women, minority workers, ethnic groups, etc.), a breach may occur. However, without a showing of discrimination, it appears doubtful that a breach of the DFR will be found in the failure to bargain for health and safety improvements.

Many unions have negotiated for specific health and safety language [10] which, for example, establishes joint labor-management safety and health committees with the right to inspect working conditions and make appropriate recommendations. Other provisions might allow a union the power to withdraw workers from hazardous job assignments. A union's nonenforcement of such provisions, or inadequate enforcement, can result in lawsuits alleging breach of the DFR. Traditionally, these cases have arisen out of accidents or workplace disasters. The DFR cases brought by workers or their survivors seek to recover those consequential damages that workers' compensation benefits do not adequately cover. The DFR suit against the union is often the worker's only available cause of action, due to the employer's protected status under the exclusive remedy provisions of most state workers' compensation laws. Further, some employers or manufacturers who are sued can join a union as a defendant, if it can be shown that the union member would have had a cause of action against the union either for the breach of the DFR or under common law theories of negligence.

The first DFR case involving occupational health and safety was *Brough v. United Steelworkers of America* [11]. In that case the Steelworkers were sued by a member who was injured while operating an allegedly faulty machine. The suit contended that the union ineffectively and negligently performed its role as a "safety advisor" to the employer. The employee alleged that the union breached its DFR and also sought tort damages under a New Hampshire state common law principle that an employer's "safety advisors" are liable for negligent inspections of faulty machinery. The federal court dismissed the suit in favor of the union on the DFR count and stated that federal labor law imposes upon the union only a duty of good faith representation, not a general duty of due care. However, the state court claim for negligence under the common law was remanded to state courts. The union impleaded the company and settled for $10,000 in order to avoid further litigation. This general standard was followed and expanded in *Bryant v. International Union, UMW of America* [12, p. 1].

The *Bryant* case was brought by the survivors and estates of coal miners killed in a mine explosion against both the employer and the union for failure to ensure the company's compliance with the standards of the federal Mine Safety Code. The collective bargaining agreement provided that the union safety committee "may" inspect any workers from an area it deems unsafe. The decision, written by First Circuit Court Judge Celebrezze, held that: (1) the union did not fail to seek correction of any known violations; and (2) the bargaining agreement provided only that the union "may," rather than "shall" or "must," inspect the mine area and, therefore, the union had no duty to inspect. The court added that without allegations of discriminatory, arbitrary, or bad faith behavior by the union (on knowledge of a violation), no DFR action could prevail. Further, and most importantly, the court expressed its doubt that the contract necessarily makes the union financially responsible for failing to demand correction of code violations, even in situations where such violations have been reported by federal mine inspectors. The court looked at the nature and purpose of bargaining agreements:

> It would be a mistake of vast proportion to read every power granted the union by management as creating a corollary contract right in the employee as against the union. Such interpretation of collective bargaining agreements would simply deter unions from engaging in the unfettered give and take negotiation which lies at the heart of the collective bargaining agreement [12, p. 5].

In what has become the main policy reason against allowing union liability, Judge Celebrezze noted that by allowing liability the courts would deter unions from including safety and health provisions in future contracts. The First Circuit recognized that such a development would obstruct the goals of national labor policy.

Other courts have been sensitive to another policy concern inherent in allowing union liability—a shifting of health and safety responsibility from the employer to the union [13]. Such a development would be in conflict with the well established common and statutory law developments that make the employer responsible for assuring workplace health and safety. A Pennsylvania court has determined that Congress did not attempt to alter the employer's responsibility for health and safety:

> By imposing on it [the union] the duty of fair representation, Congress sought to prevent it from neglecting the wishes of the minority "electorate." Congress did not seek to make the union responsible for its members' working conditions [14].

Both state and federal courts have unanimously failed to find that a union has breached its DFR when it had negligently enforced, or failed to enforce, a safety and health clause. A negligent failure to ensure safe working conditions, even when the union has the power to remove employees from any hazardous condition, will not breach the DFR [15]. Likewise, it has been ruled that, in the absence of an employee complaint, a union safety committee, established to monitor safety conditions and to make appropriate recommendations, does not breach its DFR when it is aware of a disconnected safety switch and does not seek to correct it [16]. Neither will a union's failure to search out and discover workplace health hazards constitute a violation of the DFR standard. Nor will a union breach its DFR to an injured member if it negligently referred to the worksite the violence-prone individual who maliciously attacked the injured member [17].

Although negligence alone does not constitute a breach of the DFR, a negligent act may be termed "arbitrary" conduct. Under *Ruzicka v. U.A.W.* [18], an "arbitrary" action stemming from *unintentional negligence* constituted a breach of the DFR. In *Ruzicka* a union representative forgot to process a grievance within the appropriate time limit. The "negligent" action was deemed intolerable because the union was on notice that its member wanted to continue his grievance to the next stage of processing. *Ruzicka,* however, does not mean that failure to enforce the union's inspection right is a breach of the DFR. *Ruzicka* can be distinguished from those health and safety cases previously discussed on the grounds that it involved procedural negligence, which is most easily preventable.

The *Bryant* case can also be distinguished because the specific concern with promoting union flexibility in enforcing the contract generally is quite different from unacceptable inaction on a specific member's grievance.

A union owes its unit members a duty of fair representation such that it will not refuse to process grievances arbitrarily, discriminatorily, or in bad faith. In health and safety grievances, the same standard applies. A union can evaluate a meritorious grievance and decide not to proceed with it for numerous reasons. The union can refuse to process a grievance further because, in its good faith opinion, the grievance has little chance of further success. For example, in *Powell v. Globe Industries, Inc.* [19], the union refused to take a grievance to arbitration. The union reasoned that the company's offer to reinstate the grievant, without back pay, was the best possible settlement. The grievant alleged employer retaliation for OSHAct-related activities and refused to settle. The court held that the union was motivated by a proper concern that if the reinstatement offer were rejected the grievant could be left jobless if he lost the arbitration.

It is doubtful that a union could arbitrarily or perfunctorily refuse to grieve matters arising under a particular clause in the agreement solely because it has decided not to enforce that section. Summers [20] has suggested that such a decision would constitute an undemocratic nullification of the contract.

> The union's refusal to enforce a clear provision in the collective agreement cuts at the very root of its duty of fair representation. The union as representative of the employees owes to them the duty an agent owes to his principal. How can an agent authorized to make a contract on behalf of his principal make the contract and then deprive his principal of its benefits? [20]

Summers' point is an important one in the occupational health and safety arena. It suggests that a union could not refuse to assert its contractual health and safety rights because of its leadership's noninterest in pursuing them.

The Summers proposal would not, however, change the results in cases such as *Bryant* and *House* where the union's nonexercise of its enforcement power did not breach its DFR. The distinction is simple—a union can choose not to enforce a contractual provision as long as its unit members have not sought to enforce that section.

However, a union handling a health and safety grievance that involves the interpretation of technical or scientific data may breach the DFR by failing to seek out expert witnesses on behalf of its grievants. A recent district court decision has held that the United Transportation Union violated the DFR by providing incompetent nonexpert representation for a member seeking reinstatement after surgery. The court reasoned that the union was guilty of "perfunctory representation" for failing to supply expert representation [21]. Thus, a union may have a duty to not only enforce meritorious health and safety grievances but to also provide "expert representation."

Ironically, we have come full circle in our discussion. While a union may be obligated to grieve all meritorious claims and supply expensive expert representation, it might still successfully argue that the union budget cannot afford such costly grievances.

NEGLIGENCE ACTIONS UNDER COMMON LAW

In numerous cases, plaintiffs who could not prevail under the DFR standards have brought common law tort actions against unions. In the area of occupational health and safety, most of these cases have alleged that the union breached its duty of care to its membership by negligently enforcing or failing to enforce relevant health and safety provisions or by failing to discover or warn employees about health and safety standards.

While negligence alone is not sufficient to show a union's breach of its DFR, it will support a cause of action under many state tort laws. However, the doctrine of federal preemption, which we shall examine later, has precluded most DFR cases from being brought as tort claims.

The landmark negligence case arising out of a union's health and safety duty is *Helton v. Hake* [22]. In *Helton,* an ironworker was electrocuted while working near noninsulated high-tension wires. The union was sued in tort for failing to comply with its obligation under the bargaining agreement to ensure that work not be performed around high tension lines until the safety of the employees could be assured. Significantly, the Iron Workers contract provided that the union steward "shall see that the working rules' provisions are complied with and that the employer is not held responsible for the performance of those functions by the steward." The Missouri Court of Appeals held that under the preceding contractual language the union chose

> to go far beyond a mere advisory status or representative capacity in the processing of grievances. Rather, it has taken over for itself a managerial function, namely the full independent right to enforce safety requirements [22, p. 321].

The court then held that once the union assumed the affirmative duty of ensuring a safe workplace, it could be liable in tort for a failure to perform that duty.

In finding the union and the steward liable, the *Helton* court relied heavily on the contract language expressly stating that the union had sole responsibility for assuring workplace safety. At no point did the court consider whether the union's assumption of the employer's duty to provide a safe and healthful workplace was consistent with public policy. It is arguable that the employer cannot legally delegate the responsibility to assure workplace health and safety. In OSHAct enforcement cases, the Occupational Safety and Health Review Commission (OSHRC) and the courts have consistently held that employers cannot contractually escape their primary responsibilities held under the OSHAct [23].

Subsequent to the *Helton* decision, the Idaho Supreme Court has held that survivors of a mining disaster may bring negligence actions against the union for failure to adequately function as an accident prevention representative [24]. The Idaho Court followed *Helton* and rejected the federal preemption analyses offered by many of the courts deciding similar cases.

Few other health and safety cases against unions have been successful under a common law negligence action. The only other reported case, *Nivins v. Sievers Hauling Corp.* [25], involved an action brought by an employer against a union with whom he had contracted to supply construction workers. The agreement required that competent employees be referred by the union to the employer. The court found the union had an express duty to provide .competent workers. Thus, when the union breached its obligation and an injury subsequently occurred, the union was held liable for the resulting damages. Because this case was brought by the employer and not a unit member, no DFR questions arose.

FEDERAL LABOR LAW PREEMPTION OF STATE COMMON LAW REMEDIES

As a general rule, the relationships among most private sector labor unions, union members and employers are governed solely by federal law. Traditionally, federal preemption has been found when the activity the state seeks to regulate is arguably protected or prohibited by the NLRA. Therefore, federal labor law will preempt inconsistent provisions of state law. The U.S. Supreme Court has emphasized the need for a uniform national labor policy:

> The course of events that eventuated in the enactment of a comprehensive national labor law...reveals that a primary factor in this development was the perceived incapacity of common-law courts and state legislatures, acting alone, to provide an informed and coherent basis for stabilizing labor relations conflict and for equitably and delicately structuring the balance of power among competing forces so as to further the common good [26].

However, the states are not totally precluded from regulating behavior that is *peripheral* to the policies sought to be promoted by the federal law, particularly when the regulated conduct is traditionally a matter of deeply rooted local concern.

While a breach of the DFR is an unfair labor practice, a union's negligence in the nonenforcement of a general contractual provision (absent a grievant) does not constitute a violation of its DFR. Therefore, union negligence of this sort is not prohibited or protected. Yet, a number of courts have held that a union's only duty to its members is subsumed under the DFR. The courts have held that any claim arising out of any duty allegedly owed by the union to an employee must be governed by federal labor law's DFR standards.

In *House v. Mine Safety Appliance Co.* [27], the survivors of miners killed in the 1972 Sunshine Mine disaster sued the mining company. The employer filed a third-party complaint against the union alleging that the union committed a common law tort by negligently performing its assumed duty of preventing unsafe conditions. The federal district court held that any claim of negligence arising under the enforcement of duties assumed in a bargaining agreement must be governed by the DFR and not state common law. The court concluded that a common law negligence suit alleging the union's breach of its safety duty was "inextricably intertwined and embodied in the union's duty of fair representation [27, p. 945]. Thus, the court held that the common law action was preempted. Other courts have also held that tort claims alleging a union's inadequate protection of workplace health and safety pose a significant potential of interference with federal labor law [28]. These courts have, therefore, found federal preemption of such state claims.

A few courts have not specifically held that the federal law preempts state law. However, a number of those courts have either declined to rule on the merits of the negligence issue or found that the unions had not breached any common law duty to the employee.

Two recent significant cases, *Helton v. Hake* and *Dunbar v. United Steelworkers of America,* now allow common law negligence actions regardless of the preemption doctrine. In *Helton* the court held that the contractual language created a *mandatory* duty in the union—to seek out and correct health and safety hazards. *Helton* sought to distinguish both *House* and *Bryant* by showing that the union did not fail to exercise a *permissive* right to inspect the workplace. To the contrary, the court held that under the bargaining agreement the union took over the managerial function of the employer to provide a safe and healthful workplace and thus assumed consequential common law duties arising from that contract. The language in *Bryant* declaring that all contract provisions are not enforceable against the union was held to apply only when the union has a *nonmandatory* duty.

The *Dunbar* court did not seek to base its decision on the distinction between mandatory and permissive functions. Its holding was much broader. In deciding that a common law action for wrongful death is not preempted by federal law, the court emphasized that the NLRB does not hear such cases per se. Under the *Dunbar* analysis, a union could potentially be liable for any "duty" in ensuring workplace safety and health regardless of whether that duty is mandatory or permissive.

The *Dunbar* and *Helton* courts determined that the negligent enforcement of a bargaining agreement could not constitute a breach of the DFR. These courts went on to find that a common law cause of action would not create a conflict between state and federal law because the common law duty is "peripheral" to the intent of the DFR.

The cases show distinct disagreement as to whether common law negligence actions will actually interfere with the goals and purposes of the NLRA in regard to improvement of working conditions. Successful wrongful death suits may directly deter unions from bargaining on at least some types of health and safety provisions. If the *Dunbar* case is followed in other jurisdictions, unions may well be deterred from establishing health and safety committees or from taking an active role in hazards identification and controls. Such consequences could retard the federal policy of promoting workplace cooperation and improvement. To impose more stringent standards would arguably serve to undermine union efforts to supplement and make more effective the workplace health and safety protections accorded by the OSHAct.

HOW TO AVOID UNION LIABILITY

Unions can act to lessen significantly the chances of ever being embroiled in litigation arising out of either a breach of the DFR or of a common law duty for the negligent performance of their health and safety functions. This section will look at various methods by which unions can insulate themselves from liability—both through careful drafting of their health and safety contract provisions and by careful and thorough exercise of their contractual prerogatives to encourage workplace health and safety. In addition, this section will consider a few of the legislative developments that may significantly alter and prevent union liability for union health and safety activities.

Contractual Language

The major case holding a union liable for its negligent performance of its health and safety function, *Helton v. Hake,* based its holding on the fact that:

> the union has chosen to go far beyond a mere advisory status or representative capacity in the processing of grievances. Rather it has taken over for itself a managerial function, namely the full independent right to enforce safety requirements. With its demand for and successful acquisition of that management right, it must also accept the concomitant responsibility [28, p. 321].

The court based its opinion on the fact that the Ironworkers had the unilateral power to enforce the health and safety provisions, and to therefore remove workers from hazardous work areas. This unilateral power, which is seldom found in bargaining agreement provisions at present, establishes far greater union control over workplace conditions. The *Helton* court, recognizing this fact, held that unions can contractually assume a duty of care to its members when it begins to exercise control over workplace conditions.

While some unions may not seek contractual independent control, others may perceive such power as essential in order to adequately protect unit members. Therefore, the response to the *Helton* case is not to retreat from negotiating for the unilateral power to change working conditions. The union may retain the same powers and influence over workplace conditions by negotiating clauses that give them the *permissive* power to exercise control. Thus, in *Helton* the clause providing that the union steward "shall see that the provisions of these working rules are complied with," could have read the union steward "may require that the provisions of these working rules are complied with." The permissive power would then have paralleled the union's authority in *Bryant v. United Steelworkers of America.* In that case, the court pointed out that a union does not assume a duty to enforce all of its contract rights. In addition, the court noted that the bargaining agreement stated that the union "may inspect any mine." The court further held that "the use of the permissive 'may' rather than the obligatory language in the clause clearly negates the possibility that any duty was to be created [12, p. 5].

Aside from changing the nature of the union's duty from mandatory to permissive, other contractual provisions may seek to protect the union. A union should attempt to have the contract reiterate the employer's exclusive duty to provide a safe and healthful workplace. This approach reinforces the main employer obligation under the OSHAct. Also, the contract should contain a clause that states that the union, by negotiating for and establishing its powers on health and safety matters, does not assume any of the employer's exclusive duty.

Further, unions can negotiate language that provides that the international union, local unions, union safety committees, union officers, employees, and agents will not be liable for any work-connected injuries, disabilities or diseases. Such language intends two distinct results: (1) to insulate the union from lawsuits from its unit members and their survivors, and (2) to insulate the union from third-party lawsuits brought by employers or manufacturers.

While such clauses may sound attractive to a cautious union, they may be in part both deceptive and against public policy. If the bargaining agreement is perceived as a contract between the employer and the union for the benefit of the members under a third-party beneficiary theory, then these employees are not necssarily bound by any contract clause insulating the union from lawsuits brought by unit members [29]. In addition, the courts are not inclined to uphold such disclaimers if, in their opinion, injustice will result.

However, the employer's waiver of any legal action resulting out of alleged union negligence (third-party actions) may likely be upheld by the courts. The legal doctrine of assumption of risk allows the contractual parties to contract to specifically agree in advance that one of them will not be liable for the consequences of conduct that would otherwise be negligent [30]. It is likely that the agreement exempting a union from all liability arising under its health and safety functions would be valid because: (1) it is a waiver of liability for only one area of the unions' function; (2) health and safety concerns are generally the responsibility of the employer; and (3) an agreement between two parties that affords the traditionally weaker one an advantage must be presumed to be the product not of coercion but of free and open negotiations between the parties.

Some unions have also attempted to bargain for indemnification clauses that provide that the employer will compensate the union for any damages, settlements and legal fees arising out of the union's negligent performance of its health and safety functions. While such clauses may be difficult to attain at the bargaining table, they do serve to lawfully insulate the union from its negligent performance of its health and safety powers. Such agreements have been approved by the courts [30, p. 442], and are common among manufacturers.

Preventing Liability by Exercising Health and Safety Authority

A union should not merely seek to negotiate protective contractual provisions. Lawsuits over a union's negligent performance of its health and safety functions may be prevented not only by contractual language but, more importantly, by diligent and concerned enforcement of explicit and implied powers. A union can show its good faith efforts through: (1) effective health and safety educational efforts geared toward its officers, representatives and unit members; (2) timely and effective exercise of inspection and committee membership rights; (3) a willingness to consult the employer and outside agencies such as OSHA and the National Institute for Occupational Safety and Health (NIOSH) on workplace conditions that the union or its membership perceives as dangerous; (4) soliciting, supporting and prioritizing the health and safety protests of unit members when those objections have merit under the contract; (5) exercising diligence and careful analysis on health and safety grievances; (6) a willingness to consult outside experts when safety and health problems are of a technical or specialized nature; and (7) effectively utilizing union rights of access to illness and injury data, as well as medical and industrial hygiene information.

Legislative Proposals to Limit Liability

Finally, as the courts in *Bryant* and *House* suggested, there are some compelling policy reasons why unions should not be held liable for their health and safety activities. This reason is simple and straightforward—if widespread liability is attached to negligent performance of these functions, unions may well respond by refusing to negotiate for further health and safety powers.

The policy concerns voiced in *Bryant* and *House* have prompted some significant legislative developments that seek to address the problem of union liability. One pertinent example is a provision already incorporated in the Michigan workers' compensation statutes that exempt unions, their officers, agents and committee members from liability for union activity or lack of activity in health and safety matters [31]. While the Michigan law is significant, it does not address these concerns on more than an individual state approach.

There is some national movement toward legally insulating union healthy and safety activities. Interestingly, it was the Schweiker bill (SB 2153), the proposal aimed at reducing OSHA's presence in the workplace, that promoted some protection for the health and safety activities of committee members. While the Schweiker bill sought to protect the committees and their members from liability, it was unlike the Michigan workers' compensation provision in one very significant way—it did not expressly apply its protections to unions:

> Notwithstanding any other provision of law, no claim of liability for an occupational illness or injury may be asserted against any advisory safety committee or provider of consultation services against any member or employee of such committee or consultant based on any activity, relationship, or breach of duty within the scope of functions of such committee, consultant, or individual required or authorized pursuant to this section [32].

While unions were not specifically exempted, they may likely be considered a "provider of consultation services," although the opposite argument could be made. Nevertheless, the Schweiker bill recognized that when workplace committees assume a greater role in promoting workplace health and safety, they must be protected legally. The reasons for such protections likely include the fact that these committees are voluntary compliance efforts and that participation in them should not be deterred by the threat of liability.

Further, proposed federal legislation has also provided specifically for union nonliability. The Williams-Javits National Workers' Compensation Standards bill provides that the proposed federal workers' compensation remedy,

> shall constitute the employee's exclusive remedy against the employer, the employer's insurer or any collective bargaining agent of the employer's employees...for any illness, injury, or death arising out of and in the course of his or her employment [33].

This provision is the fairly typical exclusive remedy provision found in most compensation laws and, thus, the Williams-Javits bill parallels the Michigan Workers' Compensation Law.

CONCLUSION

As organized labor bargains over the health and safety concerns of its membership, it must bear in mind that it may be exposing itself to potential legal liability. By far, the unions' greatest vulnerability does not arise under the DFR standards of federal labor law; rather, the possibility of a lawsuit alleging a union's breach of state tort law looms as the most significant legal threat. The preemption doctrine has been used to limit the state court actions but in a few cases this approach has been rejected. It is those cases, *Helton v. Hake* and *Dunbar v. United Steelworkers of America,* that make unions apprehensive.

The threat of liability can be minimized without sacrificing union influence or authority over workplace health and safety practices. Through careful drafting of contract provisions, as well as through adequate and informed enforcement, unions can effectively limit the possibility of liability for health and safety activities. Further, various legislative proposals may offer unions the legal protection they seek.

Union liability for its health and safety practices is a relatively recent phenomenon. It is likely that it will attract increased attention as unions assert more influence over the issues and as public awareness of occupational hazards increases. While the threat of liability can cause unions to be more diligent in health and safety matters, it can also cause unions to withdraw from further involvement in this crucial area. For this reason, unions argue they should be free of the legal web of common law negligence when they become involved in the health and safety arena. This is the policy issue that the courts, the Congress, and individual state legislatures must begin to resolve, particularly in light of the expected decrease in government involvement in the regulation and monitoring of workplace health and safety.

NOTE

For a more detailed and comprehensive treatment of this subject, see Drapkin LC, Davis ME: Health and safety provision in union contracts: Power or liability? Minn Law Rev 65:635, 1981.

REFERENCES

1. O.S.H. Act § 5(a)(1), U.S.C. § 654(a)(1).
2. O.S.H. Act § 5(b), 29 U.S.C. § 654(b); § 5(a)(2), 29 U.S.C. 654 (a)(2); § 2(b)(3), 29 U.S.C. § 651(b)(3).

3.   *Howard Electric Co.*, 6 O.S.H.C. 1518, 1521 (O.S.H.R.C. 1978); *Frohlick Crane Service Inc. v. Occupational Safety & Health Review Commission*, 521 F.2d 628, 631–632 (10th Cir. 1975).
4.   *NLRB v. Gulf Power*, 384 F. 2nd 822 (5th Cir. 1967).
5.   29 U.S.C. § 159(a); *Ford Motor Co. v. Huffman*, 345 US 330 (1953).
6.   45 USC § 152; *Steele v. Louisville & N.R. Co.*, 323 U.S. 192 (1944).
7.   *Steele v. Louisville & N.R. Co.*, *supra* at 202.
8.   *Vaca v. Sipes*, 386 U.S. (1967).
9.   NLRA § 9(a), 29 U.S.C. § 159(a); 29 U.S.C. § 1337; *Textile-Workers Union of America v. Lincoln Mills of Alabama*, 353 U.S. 448 (1959); *Teamsters v. Lucas Flour Co.*, 369 U.S. 95, 102–104 (1962).
10. U.S. Bureau of Labor Statistics, Department of Labor: Bulletin No. 1425-16, Major Collective Bargaining Agreements: Safety and Health Provisions 3 (1976).
11. *Brough v. United Steelworkers of America*, 437 F. 2d 748 (1st Cir. 1971).
12. *Bryant v. International Union, UMW of America*, 467 F. 2d (1st Cir. 1972).
13. *House v. Mine Safety Appliance Co.*, *supra* at 946; *Brooks v. New Jersey Manufacturers Insurance Co.*, 405A.2d 466 (N.J. Super. Ct. 1977).
14. *Carollo v. Forty-eight Insulation, Inc.*, 381 A. 2d 990 (Pa. Super. Ct. 1977).
15. *Helton v. Hake*, 386 F. Supp. 1027, 1032 (W.D.Mo. 1974).
16. *Brooks v. N.J. Manufacturers Ins. Co.*, 405 A. 2d 466 (N.J. Super. Ct. App. Div. 1979).
17. *Hartsfield v. Seafarer's International Union*, 422 F. Supp. 264, 269–270 (D.C. Ala. 1977).
18. *Ruzicka v. U.A.W.*, 523 F. 2d 306 (6th Cir. 1975).
19. *Powell v. Globe Industries, Inc.*, 431 F. Supp. 1096 (N.D. Ohio 1977).
20. Summers: The individual employee's rights under the collective bargaining agreement: what constitutes fair representation? U. Pa. Law Rev. 126:251 (1977).
21. *Curtis v. United Transportation Union*, 102 LRRM 2961 (E.D. Ark. 1979).
22. *Helton v. Hake*, 564 S.W. 2d 313 (Mo. C.A. 1978), *cert. den.* 439 U.S. 959 (1978).
23. *Howard Electric Co.*, 6 O.S.H.C. 1518, 1521 (O.S.H.R.C. 1978); *A. J. McNulty and Co.*, 4 O.S.H.C. 1097, 1101 (O.S.H.R.C. 1976); *Frohlick Crane Service, Inc. v. Occupational Safety & Health Review Commission*, 521 F. 2d 528, 631–632 (10th Cir. 1975).
24. *Dunbar v. United Steelworkers of America*, 602 P. 2d 21 (Supreme Ct. Idaho 1979).
25. *Nivins v. Sievers Hauling Corp.*, 424 F. Supp. 82 (D.N.J. 1976).
26. *Amalgamated Association of Street, Electrical Railway and Motor Coach Employees of America v. Lockridge*, 703 U.S. 286 (1971).
27. *House v. Mine Safety Appliance Co.*, 417 F. Supp. 939 (D. Idaho 1976).
28. *Carollo v. Forty-eight Insulation Inc.*, 381 A. 2d 990 (Pa. Super. Ct. 1977); *Farmer v. General Refractories Company*, No. G.D. 77-24854 (Ct. Common Pleas., Allegheny Co., Pa. 1979).
29. Feller: A general theory of the collective bargaining agreement. Calif. L. Rev. 61:663, 773–805 (1973).

30.   Prosser W: Handbook of the law of torts, 4th ed., pp. 442–443, 1971.
31.   Mich. Comp. Laws Ann. § 418.827 (8).
32.   S. 2153, 96th Cong., 1st Sess. (1979) § 4(A)(4).
33.   S. 420, 96th Cong., 1st Sess. (1979) § 10(a).

DISCUSSION: CHAPTER 28

Kenneth S. Cohen and Larry C. Drapkin

DR. COHEN: Please comment on the concept of unions negotiating in the arena of environmental or dirty pay: for example, iron workers who negotiate 10¢/hr more to work on zinc-based metals, and roofers, particularly on the East Coast, who negotiate $1/hr benefits for working with coal tar pitch rather than asphalt roofing materials. Does this, in fact, make the union a party in actions against the employer for an unsafe act resulting in disease litigation?

MR. DRAPKIN: I think, under the standard of care that I gave you, that under the "duty of fair representation" this is probably not a capricious action, but rather a matter of contract negotiation. Remember I said that in cases of contract negotiation, the courts are relatively unwilling to go very far in finding liability. I am troubled, just as you apparently are, that other approaches aren't being taken. But I think that if you were to look at the developed standards, they are probably not liable.

CHAPTER 29

IMPAIRMENT VERSUS DISABILITY

Attilio D. Renzetti, Jr., MD
> Division of Respiratory, Critical Care,
> and Occupational (Pulmonary) Medicine
> Department of Internal Medicine
> University of Utah College of Medicine
> Salt Lake City, Utah

In the course of their daily practice, physicians are often asked to pass judgment on whether or not their patients are unable to perform their usual work because of ill health. Such requests usually come from either the patient or his or her employer. Most such requests relate to temporary reversible illnesses common to the population at large and present little problem to the physician. In fact, in contemporary society, excused absence from work with pay is an accepted condition of employment.

Much more difficult for both physicians and employers are those more prolonged or even fatal medical conditions that have led to the development of social legislation aimed at providing economic support for the victims. Unfortunately, this development has failed to achieve a goal of providing equal treatment under the law for all citizens. In general, this failure can be attributed both to the medical profession, for its inability to provide criteria for the presence and degree of health impairment, and to society, as expressed through government, for the conflicting social objectives of disability legislation.

There are a number of problems that need to be addressed if a system for disability determination is to be developed that will be valid whatever its social objective may be. These include: (1) definitions of impairment; (2) assessment of impairment (Is it temporary or permanent? What is the degree? Will it be determined through clinical or laboratory methods?); and (3) the relationship of impairment to disability.

An expansion of the definition of impairment by Gaensler and Wright [1], as suggested by Richman [2], should find wide acceptance in the medical community. Namely, that impairment should be defined as a reduction from normal of body or organ function demonstrable by medically acceptable clinical and laboratory methods. Further, the definition should be made more specific for each organ system; for example, for the respiratory system the reduction would be specified as being present in one or more of the three components of respiration, i.e., ventilation, perfusion and diffusion.

In the assessment of impairment, the physician should indicate whether or not it is temporary or permanent but, in addition, it would be well to build into any system for disability determination a requirement for periodic reexamination to determine permanency or changing degrees of impairment. Degrees of impairment need to be specified, but in general it may be difficult to be more quantitative than indicating two or three degrees, such as mild, moderate and severe. Perhaps most important to the assessment of impairment is the requirement for a comprehensive evaluation that includes clinical methods, such as a thorough history and physical examination, as well as the use of objective methods for measuring organ function. For example, pulmonary function tests provide quantitative data of great utility in assessing impairment in disorders of the respiratory system. It needs to be emphasized, however, that the medical concepts of impairment relate to some reduction from "normal" and, unfortunately, for many measures of organ system function the medical community has yet to agree on how to define normal and has yet to collect adequate data on which to base such a definition.

Finally, given that we have or will have satisfactory data on which to determine degrees of impairment, there remains the question of relating the degree of impairment to the ability to perform both occupational and nonoccupational functions. Any physician with experience in this area recognizes the numerous variables that determine this relationship, such as age, physical conditioning, obesity and motivation, to mention just a few. Although it is possible to quantify energy requirements of various job functions and energy output capability of individuals with physical impairment, it seems likely that less precise means, such as a general knowledge of various daily activities, job requirements and recreational endeavors, will suffice for the physician trained in a medical specialty and experienced in disability assessment to match physical impairment and job requirements.

The social problems of disability determination constitute the major impediments to providing equal benefits under the law to disabled workers. Richman has pointed out that the purpose of any future legislation in this area should be determined before it is promulgated [2]. Are compensation benefits intended to function as income maintenance, health insurance or retirement pension? The alternative of choice should help to solve some of the most urgent questions in need of solution, namely, the source of the money with which to pay compensation benefits, and the need to define an attributable cause or causes of the impairment.

A brief review of the definitions of disability in current compensation law illustrates the conflicting social objectives. Although workers' compensation laws vary widely among states, certain principles are shared in common. "Compensable disability is inability as the result of a work-connected injury to perform or obtain work suitable to the claimant's qualifications and training." Additionally, disability benefits are payable for impairment alone, despite ability to perform usual work. Disability is judged to be total or partial on the basis of whether or not a worker can perform his former work. In short, if he continues his former work, he may be impaired but is not "disabled." If he can work at a lesser rate of pay, he is partially disabled, and if there is no available work for which he is medically and vocationally capable, his disability is total. Under certain circumstances, when there has been permanent loss of use of an impaired organ, individuals may receive benefits despite the fact that they can continue their usual work. Obviously, then, in state compensation law, "disability" is synonymous with loss of earning power, whereas "impairment" bears no necessary relationship to "disability".

In social security law disability is defined as an inability to engage in any substantial gainful activity by reason of any medically determinable physical or mental impairment or impairments. It should be noted that with multiple impairments each is given a percentage, and the sum of these is used in arriving at a decision. Clearly, the term disability here means total disability.

Finally, under federal Black Lung law, disability is equated with total disability, since any reduction in a miner's ability to perform his usual coal work will entitle him to total disability benefits. In reality, benefits under this law have become similar to the types of benefits paid for scheduled permanent bodily impairment, as in state worker's compensation law. (It is worth adding that the definition of impairment used in the regulations for this law is one with which few pulmonary specialists would find fault.)

It seems obvious from this brief review of compensation law that the terms impairment and disability are being defined differently in different applications. It is no wonder, then, that confusion often exists among physicians and lawyers and perhaps most tragicallly among laymen who are subject to these laws. In addition, there are important issues that have not yet been satisfactorily addressed in existing laws.

The first of these—the presence of disease without impairment—is well illustrated by the federal Black Lung law wherein category 1, simple coal-workers' pneumoconiosis (CWP), has led to the award of "disability" benefits in the absence of any functional deficit! Is this a proper use of the term disability, and is this a reasonable justification for compensation?

Another problem is that of a detectable impairment without decline in earning power. For example, does a 72-inch-tall, 65-year-old motorman in an underground coal mine whose FEV_1 is 2.45 liters and whose chest X-ray is normal but whose hobby is deer hunting in the mountains of Utah have disability? This FEV_1 is only 79% of predicted normal for this man, hence, by current standards of less than 80% of predicted being abnormal, he has a slight degree of functional impairment. This would probably adversely affect his ability to chase deer at high altitudes. Under workers' compensation law, the slight impairment would not interfere with his earning power and, hence, there would be no disability. Under the federal Black Lung law, there is sufficient impairment for "total disability due to pneumoconiosis."

Finally, what about the question of risk in those workers with disease without impairment? For example, does a worker with a pneumoconiosis (e.g., asbestosis, CWP or silicosis), but with no physical impairment by pulmonary function tests, have disability because of the *risk* of developing progression of his disease and deterioration of his pulmonary function if he continues his current employment? The *legal* answer to this question will not only vary among different state workers' compensation laws, but also in the adjudication of the federal Black Lung law. What is needed here, of course, is a definition of disability that takes into consideration risk of disease progression on the basis of sound medical data.

The issues that have been raised relate to the respiratory system, a field with which I am familiar. I have no doubt that there are numerous other issues related to other medical specialties. Add to this the fact that in recent years a number of bills have been introduced into the U.S. Congress that will provide disability benefits to individuals with so-called "white lung", i.e., asbestosis, "brown lung", i.e., byssinosis, and perhaps others. It seems obvious that such categorical legislation is bound to add to the existing chaos in dealing with impairment and disability.

In preparing this paper, I have borrowed very heavily from Richman [2]. He also introduced the idea, and I will take it one step further by strongly recommending that a medicolegal group be assembled and assigned "the task of arriving at definitions of impairment and of disability which even our legislators will respect"!

REFERENCES

1. Gaensler EA, Wright GW: Evaluation of respiratory impairment. Arch Environ Health 12:146–189, 1966.
2. Richman SI: Meanings of impairment and disability. Chest 78:2, 367–371, 1980.

DISCUSSION: CHAPTER 29

T. Lamont Guidotti and Attilio D. Renzetti

DR. GUIDOTTI: Is the American Thoracic Society supporting the concept that workers having occupational lung disease should have equal representation and rights under the law? Specifically, should agricultural workers be excluded under various state compensation legislation in addition to being excluded from OSHA regulations?

DR. RENZETTI: I have to admit that, in the discussions of our committee, agricultural workers were not even brought up. I presume you are referring to disability incurred from the inhalation of pesticides?

DR. GUIDOTTI: No. Specifically, farmer's lung and, now, the emerging major issue of the grain dust exposures.

DR. RENZETTI: These could all be considered in our recommendation as occupational lung diseases, including occupational asthma, which, as you know, is a rapidly growing problem.

SECTION 4

JOB DISCRIMINATION

CHAPTER 30

JOB DISCRIMINATION BASED ON EXPOSURE CONSIDERATIONS AND SMOKING

Theodor D. Sterling, PhD
Computing Science Department
Simon Fraser University
Burnaby, British Columbia

The restriction on smoking in the workplace has an old history, probably as old as smoking itself. Current research has unearthed early examples of smoking rules in the workplace, such as those in the manual of the American Central Life Insurance Company (now American United Life). That handbook for employees prohibited smoking or carrying of lighted cigarettes or cigars in laboratories, the supply division or vaults. Interestingly, the same manual informs us that the company did not ordinarily accept married women for employment and encouraged employees with spare time after lunch to go outside for fresh air and exercise. More recent practices, documented by Bennett and Levy for Massachusetts [1], probably hold for the United States and Canada as a whole. Jobs or work areas in which smoking was prohibited were designated by 64% of industries with more than a thousand employees designated. Of the companies that restricted smoking, 91% cited possible danger to products or equipment as a reason for this prohibition. The rest (hospitals, restaurants, retail outlets and providers of services such as insurance agencies) listed contacts with clients as reasons for the smoking restriction. None of the respondents to the Bennett and Levy survey listed hazards to workers' health as a reason for their nonsmoking policies.

Smoking has been restricted and/or smokers have not been hired for reasons other than safety of products or equipment or contact with clients.

For the most part, this has occurred in small, nonunionized plants. Recently, however, large employers, such as the Alexandria, Virginia, Fire Department, the Quin-T company of New Hampshire, and, perhaps the most notable, the Johns-Manville Corporation, have followed suit. These organizations have based their restrictions on health grounds.

Policies to restrict employment to nonsmokers or to exclude smokers from certain compensation and perquisites represent a development that needs careful evaluation. Such policies affect individuals who are not ill, have no criminal record and are in need of work.

It is the purpose of this chapter to examine with care the claimed reasons for and the advisability of employment restrictions on smokers. Discriminatory practices are very often based on contentions that, on closer inspection, tend not to be true or at least not to be relevant to the discriminatory practice. (While we shall briefly examine some major flaws in the argument linking smoking causally to increased disease among certain occupations, an extended analysis of the state of the scientific evidence is peripheral to our discussion. For a full discussion see Sterling [2].)

Perhaps the prime example was, and still is, the refusal of many industries to employ women by maintaining that women are not capable of performing many jobs, for physical or psychological reasons. Of interest to us will be considerations of whether employment practices regarding smoking and smokers have punitive effects and whether they fit within the system of fairness and justice to which democratic societies are devoted.

EVALUATING REASONS GIVEN FOR DISCRIMINATORY HIRING PRACTICES

Discriminatory practices of any sort are usually based on apparently rational arguments [3,4]. In the case of smokers, the "objective" arguments appear to be:

1. smoking interacts with substances to which the worker is exposed at his job so as to heighten the risk of disease;
2. smoke of others is injurious to nonsmokers; and
3. smoking workers represent an additional cost factor to industry.

We have pointed out for many years now that almost all epidemiological studies dealing with the claimed effects of smoking on health have neglected to take into account the confounding between smoking as a lifestyle characteristic and employment of smokers in occupations where workers are exposed to irritating and toxic dust and fumes. In many industries, there are higher percentages of smokers in those jobs that involve heavier exposure to dust and fumes. On the other hand, non-smokers and exsmokers are more heavily represented in those jobs where exposures to airborne industrial hazards are considerably less (Table I)

Table I.   Comparison of Current and Former Smokers
Between a Blue Collar Occupation and a Professional or
Managerial Occupation for Selected Industries [6]

Industry and Occupation	Percent Current Smokers	Former Smokers (Percent of Ever Smokers)
Agriculture		
Farm Laborers	43.3	21.1
Farmers (owners)	28.1	53.1
Construction		
Painters	70.7	19.1
Civil Engineers	31.9	50.0
Miscellaneous Machinery		
Operators	51.0	32.7
Managers	37.9	50.0
Electrical Machine Equipment Supplies		
Operators	48.4	31.3
Electrical Engineers	20.0	66.7
Educational Services		
Mechanics	53.1	26.1
Teachers	26.9	53.3

[5,6]. As a consequence, whenever groups are compared that differ with respect to their smoking habits, they also most likely differ with respect to the number of individuals who are exposed to hazardous industrial substances. Because of this confounding, the emphasis on the reported relationship between smoking and disease may hide a relationship between workplace exposures and disease. The category "smoker," in a statistical sense, may be primarily a major index for "exposure" to occupational hazards. In short, increased disease incidence, ascribed to smoking by some epidemiological studies that failed to control for occupation, may well be of occupational origin.

Recent studies of lung disease in occupationally exposed workers support this contention. A relatively large number of these studies have now compared the incidences of disease among smokers and nonsmokers. Some failed to find any significant effect due to cigarette smoking, while others found even higher disease rates among nonsmokers than among smokers.

Some Examples

In a study of zinc and lead miners, Axelson et al. [7] report a greater risk for lung cancer among nonsmokers compared to smokers. Another study of miners' mortality, reported in 1980, showed that nonsmoking miners had a higher mortality from lung cancer than smoking miners [8]. Axelson emphasized that these Swedish studies deal with a lifetime followup of miners, whereas most other mining populations have been studied by means of cohorts with a followup of not more than about 25-30 years or less [9]. In other words, the completeness of the followup leads to results with added reliability.

An inverse relationship between smoking and lung cancer was also found among workers exposed to chloromethyl methyl ether [10]. The author, who is known for his militant opposition to smoking, observed: "The data suggest that continued cigarette smoking entailed a factor which partially inhibited the carcinogenic effects of chloromethyl ethers."

Pinto [11] found elevated lung cancer mortality rates among the nonsmokers in his study population of arsenic-exposed workers when compared to the general population.

A study of chrysotile asbestos miners in Canada [12] reported that lung cancer deaths in nonsmokers showed a greater increase in incidence with increasing exposure than did lung cancer in smokers. This is one of the very few studies that estimated levels of asbestos dust exposure among the workers.

These recent studies are examples of circumstances in which smoking does not appear to interact with substances to which workers are exposed so as to heighten the risk of disease. In fact, Axelson and Weiss separately raised the possibility that smoking may have protective properties for some types of work.

Some scientists have claimed that exposure to asbestos and cigarette smoking interact to cause disease in workers. However, even here the evidence needs to be examined carefully. For example, mesothelioma does not appear to be related to smoking at all.

Although the well known studies of insulation workers by Hammond and co-workers [13-15) have reported an apparently large effect on lung cancer rates due to the claimed interaction of smoking and asbestos, that effect has become less important with successive reports as increasing numbers of lung cancers appear among nonsmoking workers. It is important to note that the latest report by these investigators has shown a fivefold relative risk of lung cancer mortality for both smoking and nonsmoking workers exposed to asbestos. There are serious problems with this study that have been discussed among others by Saracci [16] (that the large fraction of subjects with unknown smoking habits, 6000 out of the 17,800 insulation workers, makes uncertain any quantitative assessment of the joint effects of smoking and asbestos) and by Sterling [17] (that it was impossible to believe that these insulation workers died with identical proportional mortality rates from all cancers whether or not they smoked without assuming that the PMR for nonsmokers was higher from cancers other than of the lung or that serious errors existed in data acquisition and/or analysis of the Hammond and Selikoff study).

The question of whether the tobacco smoke of other workers represents a health hazard to nonsmokers on the job is far from resolution. A recent California study has reported somewhat lower lung function measures for nonsmokers working along with smokers [18]. There are a number of criticisms of the design of that study [19–21]. Also, its results disagree with other investigations [22,23].

While claims are made that smoking represents increased costs to industry, the economic reasoning used to support these claims is weak and unpersuasive. The claims about "economic losses" from smoking are mostly anecdotal or conjectural [24–26]. It is almost impossible to find reliable estimates of the so-called "economic cost" of smoking. In addition, estimates of economic cost depend on whatever data are selected to support *loss* or *gain*. For instance, there are data showing less chronic morbidity among moderate smokers than among nonsmokers (Table II) [27]. The U.S. Surgeon General's Report of 1979 indicates that some moderate smokers have fewer days of disability than nonsmokers [28]. Such data could conceivably support claims of economic gain due to smoking.

PREVENTIVE HEALTH ASPECTS AND OTHER CONSIDERATIONS

The control of contagious diseases through public health measures has made an impressive contribution to the overall health status of society. As our medical understanding of the health effects of industrial exposures and pollutants has grown, a realization that occupationally related diseases may be controlled by regulation and protective measures has emerged. However, there has been in some quarters an insistence that it is far more important to control workers' lifestyle characteristics than it is to reduce or eliminate workplace exposures to toxic substances. It is highly disturbing to observe this trend. For example, Kotin [29] wrote recently, "What should the regulatory agency do about situations like asbestos where smoking can transform a nonhazardous exposure into a

Table II.   Estimated Age-Adjusted Prevalence Rates for
Different Chronic Conditions for 100 Moderate and
Nonsmoking Males and Females [27]

Condition, by Sex	Never Smoked	Half Pack or Less[a]	Half to Full Pack[a]
Heart Condition (excluding chronic rheumatic heart disease)			
Males	4.1	3.8	4.4
Females	4.3	2.8	3.5
Hypertension			
Males	4.1	3.8	3.6
Females	8.6	7.4	6.5
Chronic Bronchitis			
Males	1.0		2.3
Females	1.2	1.6	4.0
Chronic Sinusitis			
Males	8.3	7.7	10.1
Females	11.1	10.8	14.0
Peptic Ulcer			
Males	2.4	3.0	4.6
Females	0.16	2.0	2.5
Arthritis			
Males	6.8	6.6	5.9
Females	12.9	11.1	12.5
Hearing Impairment			
Males	8.0	7.8	7.5
Females	5.8	4.2	5.3
All Other Chronic Conditions			
Males	72.5	65.4	73.0
Females	95.9	85.1	99.5

[a] By heaviest amount smoked.

hazardous one?" Not only is this point of view scientifically unsound, it represents a blatant attempt to shift responsibility for preventive health measures from industry to the worker.

Veatch criticized most persuasively this trend to blame the victim for disease:

> The correlation of disease, mortality, and even so-called voluntary health-risk behavior with socio-economic class are impressive. Recent data from Great Britain and from the Medicaid system in the United States reveal that these correlations persist even with elaborate schemes that attempt to make health care more equitably available to

all social classes. In Great Britain, for instance, it has recently been revealed that differences in death rates by social class continue, with inequalities essentially undiminished since the advent of the National Health Service. Continuing to press the voluntarist model of personal responsibility for health risk in the face of social structural model of the patterns of health and disease could be nothing more than blaming the victim, avoiding the reality of the true cause of disease, and escaping proper social responsibility for changing the underlying social inequalities of the society and its modes of production [30].

In addition, questions of fairness are raised by the almost exclusive focus on smoking as *the* voluntary action to be prohibited by management. Why shouldn't the failure to exercise regularly, the lack of a balanced diet, the consumption of alcoholic beverages, or residence in polluted areas be used as a condition to deny employment to a worker? Health risks have been associated statistically with each of these factors and, to a certain extent, the worker voluntarily makes these lifestyle choices. Further, the emphasis on smoking may represent unfair discrimination against blue collar workers, because they have a higher percentage of smokers than management [5,6].

Antismoking movements in the Western world have strong emotional aspects. These may, in part, be traced to strong convictions held by some that poverty and disease are caused by sloth and unhealthy lifestyles. Thus, a punitive aspect might underlie some hiring practices that exclude smokers. For example, some years ago, a scientist was asked about the advisability of employing a smoker as a parking fee collector in an enclosed garage. The scientist had previously established that the booth in which the collector was expected to work had excessively high levels of carbon monoxide (over 100 ppm). The scientist advised the employer not to hire the smoker because the workplace exposure to carbon monoxide would unduly increase the smoker's total exposure [31]. It is difficult to understand why the scientist did not recommend to the employer that the collection booth be properly ventilated so that it would become safer for all employees. A fair interpretation of this incident may be that the smoker has to be taught a lesson by the denial of a job.

The possibility that smokers may be treated in a punitive manner also arises in the context of compensation hearings. We discussed elsewhere [17] the case of a light smoker who had been employed many years in underground mining who was denied any compensation at all. The board said it was "speculative" that such exposure would aggravate his condition [32]. Michigan attorneys practicing in the compensation area are of

386   JOB DISCRIMINATION

the opinion that awards to smokers tend to be one-third less than those to nonsmokers [17]. A survey of compensation cases on appeal in Wisconsin indicated that smoking was brought up by the employer in the majority of cases, regardless of the disease.

The constitutional problem of policies that restrict hiring of smokers have been examined in Germany in much greater detail than they have been in the United States. Briefly, the constitutional questions revolve around the freedom of an individual to engage in acts that are of traditional social custom and that cannot be abridged without compelling reasons. The German Labor Council has dealt in some detail with the rights of smokers and nonsmokers and employers' actions to restrict smoking within the context of the new German constitution [33]. Perhaps such discussions should take place in the United States.

SUMMARY

Most common reasons for restrictions on smoking in the workplace are to avoid fire from flammable materials or damages to products or equipment, or for customer relations. Lately, there have been instances of policies that restrict employment to nonsmokers or exclude smokers from certain compensations and perquisites. These policies pose a new issue in public health that requires careful evaluation. Restricting employment opportunities and rewards because of smoking affects individuals who are not ill, have no criminal record and are in need of employment.

1. There is evidence contrary to the idea that smoking interacts with some substances to which workers are exposed on the job so as to heighten the risk of disease.
2. The question of whether other people's tobacco smoke is hazardous to workers on the job is open.
3. That smoking workers represent an extra cost factor to industry at best is conjectural.
4. Lifestyle characteristics are apparently acquired by individuals at an early age because of a variety of social and cultural factors. The attempt to blame the victim's lifestyle for disease often represents a diversion of investigation of environmental, occupational or social disease antecedents.
5. "No smoking" policies often appear to be punitive and may affect awards in workers' compensation cases.
6. Finally, there are constitutional problems that need to be examined, which revolve around the freedom of an individual to engage in socially traditional acts that should not be forbidden without compelling reasons.

REFERENCES

1. Bennett D, Levy S: Smoking policies and cessation programs of large employers in Massachusetts. Am J Public Health 70:629–31, 1980.
2. Sterling T: Smoking, occupation and respiratory disease. Prepared for the American Lung Association Occupational Health Task Force Meeting, April 9, 1980.
3. Bettelheim B, Janowitz M: Dynamics of prejudice. New York: Harpers, 1950.
4. Bettelheim B, Janowitz M: Social change and prejudice. New York: The Free Press, 1964.
5. Sterling T: Smoking characteristics by type of employment. J Occup Med 18(11):743–54, 1976.
6. Sterling T, Weinkam, J: Smoking patterns by occupation, industry, sex, and race. Arch Environ Health 33(6):313–17, 1978.
7. Axelson O, Sundell L: Mining, lung cancer and smoking. Scand J Work Environ Health 4:46–52, 1978.
8. Dahlgren E: Lung cancer, cardiovascular disease and smoking in a group of miners. Lakartidningen, 1980.
9. Axelson O: Effects of low level and background radiation from radon daughters. Presentation before Vancouver, B.C., Royal Commission Hearings, January, 1980.
10. Weiss W: The cigarette factor in lung cancer due to chloromethyl ethers. J Occup Med 22:527–29, 1980.
11. Pinto S, Henderson V, Enterline P: Mortality experience of arsenic-exposed workers. Arch Environ Health 3:325–31, 1978.
12. McDonald JG, Liddell FDK, Gibbs GW, Eyssen GE, McDonald AD: Dust exposure and mortality in chrysotile mining, 1910–75. Br J Ind Med 37:11–24, 1980.
13. Hammond C, Selikoff I: Relation of cigarette smoking to risk of death of asbestos-associated disease among insulation workers in the United States. In: Bogovski P, Ed. Biological effects of asbestos. Lyon, France: IARC, 1973, pp. 312–17.
14. Hammond C, Selikoff I: Multiple risk factors in environmental cancer. Fraumeni J, Ed. Persons at high risk of cancer. New York: Academic Press, 1975.
15. Hammond C, Selikoff IJ, Seidman H: Asbestos exposure, cigarette smoking and death rates. Ann NY Acad Sci 330:475–90, 1979.
16. Saracci R: Asbestos and lung cancer: an analysis of the epidemiological evidence on the asbestos smoking interaction. Int J Cancer 20:323–31, 1977.
17. Sterling T: Does smoking kill workers or working kill smokers? Int J Health Serv 8(3):437–52, 1978.
18. White J, Froeb H: Small-airways dysfunction in nonsmokers chronically exposed to tobacco smoke. N Eng J Med 302(13):720–23, 1980.
19. Adlkofer F, Scherrer G, Weimann H: Small-airways dysfunction in passive smokers. N Eng J Med 303(7):392, 1980 (Letter to the editor).
20. Aviado D: Small-airways dysfunction in passive smokers. N Eng J Med 303(7):393, 1980.

21.   Huber G: Small-airways dysfunction in passive smokers. N Eng J Med 303(7):392, 1980 (Letter to the editor).

22.   Testimony before the U.S. Congress, House Committee on Agriculture, Subcommittee on Tobacco, Effect of Smoking on Nonsmokers, Hearing, 95th Cong., 2nd Session, September 7, 1978 Washington , DC: Government Printing Office, 1978.

23.   In the matter of arbitration between Schien Body and Equipment Co., Inc. and United Steelworkers of America, Local 8557, FMCS No. 77K17279, Grievance No. 77-1, Decision of Arbitrator, Farmington, MO 63640.

24.   Fielding J: Preventive medicine and the bottom line. J Occup Med 21: 79–88, 1979.

25.   Kristein M: Economic issues in prevention. Prev Med 6:252–64, 1977.

26.   National Commission on Smoking and Public Policy: A national dilemma: cigarette smoking or the health of Americans. New York: National Commission on Smoking and Public Policy, 1978.

27.   National Center for Health Statistics: Cigarette smoking and health characteristics. (PHS Publication No. 1000, Series 10, No. 34) Washington, DC: Government Printing Office, 1965.

28.   U.S. Public Health Service: Morbidity, smoking and health. A report of the Surgeon General. DHEW Publication No. 79-50066, 1979.

29.   Kotin P, Gaul L: Smoking in the workplace: a hazard ignored. Am J Public Health 70:575–76, 1980.

30.   Veatch R: What is a just health care delivery? Branson R, Veatch R, Eds. Ethics and health policy. Cambridge, MA: Ballinger Publishing Co., 1976, pp. 127–53.

31.   Smith N: Community exposure to carbon monoxide in Calgary, Alberta. Oral presentation before the Annual Meeting of the Canadian Public Health Association, Vancouver, Canada, November 19–21, 1975.

32.   Workmen's Compensation Council, Decision No. 221: re bronchitis and emphysema, British Columbia, December 20, 1976.

33.   Schneider W: Rauchen am arbeitsplatz and mitbestimmung des betriebsrats. Die Quelle 4:226–28, 1980.

DISCUSSION: CHAPTER 30

Robert S. Bernstein, Theodor D. Sterling, Stephen P. Teret,
Marvin A. Schneiderman and Douglas G. Mortensen

DR. BERNSTEIN: I am personally strongly opposed to the philosophy of victim blaming, which I think often has been prevalent in the area of occupational health. But I would like to make the following comments with respect to smoking and some of the studies that you have presented here.

My comments have to do with what's been called the healthy worker factor — the strong survivors effect that one sees in working populations. The studies that you mentioned, most of which I am not familiar with in detail, may very well be biased in that cross-sectional studies of prevalent problems among the population will neglect the problem that comes about because working populations have to remain healthy in order to continue to work. The studies that you mentioned in which nonsmokers appear to have done worse than smokers could have been biased if the studies were ones in which cohort or population based data were not available.

And second, with respect to the National Center for Health Statistics [NCHS] data, I can't tell whether in fact any differences are there because there are no intervals for figures. It is not clear to me from use of prevalency data in this table whether we are talking about problems that have come about because of cross-sectional study as opposed to cohort study and whether the smokers, because of obstructive problems of hypersecretion of mucus, were no longer working in the settings when the studies were done.

DR. STERLING: Actually, you have asked two questions. I'll try to answer them. I said before that there are a large number of studies that have investigated conditions in industry and especially the prevalency of lung cancer and mesothelioma and other industrial-caused diseases. A number of these studies have compared smokers with nonsmokers. These are very good studies. They are accepted as such. They just are not quoted in the literature. I find it interesting that so few health scientists are familiar with these studies. I am not going to ask for a show of hands, but few people seem to know of the studies by Axelson or Dahlgren, or the recent followup lung cancer study of Weiss among chloromethyl ether workers. It is as if these studies disappear from the literature. Few, if any references to them are found. It is so much assumed that it is smoking that causes the lung cancer that no attention is paid to studies such as these. But there is data in some occupations the risk for lung cancer is higher in nonsmokers than smokers. These are viable data that need to be attended to.

As for the NCHS data, my attempt was to show that when one looks at the calculational problems that smoking workers lose more time than nonsmoking workers, many claims are based on selected information. This information is selected from various sources but these are often anecdotal sources. One could just as well try to make a case that smoking decreases cost to industry due to absenteeism. One could take data similar to that published by the National Center for Health Statistics and build a case on them. In short, I was trying to vividly demonstrate that there are data that can be and are being used according to the biases of the user.

Now as far as your other remark is concerned about the healthy worker. I find it rather curious that we explain the fact that some groups of people have less disease by the fact that they are healthy workers. To be sure, I think this should be explored. But I would also like to see studies that explore that some workers have more disease not because they smoke, but because they are sick workers.

MR. TERET: I'm confused by Table II in your presentation. You are using prevalence rates rather than incident rates. If the smoking had a killing effect on the people, would not that reduce the prevalence of the disease so that it would show the same relationship between figures as would be shown if you used people who had pancreatic cancer or people who didn't use seat belts in cars?

DR. STERLING: Those are figures taken from statistics published by the National Center for Health Statistics. They published these prevalence rates, not I.

MR. TERET: I understand that. Aren't they misleading, tending to show that one would conclude smoking is a protective effect, whereas smoking, if it had an adverse health effect, would in fact change the prevalence rates?

DR. STERLING: You are committing the same logic error that Weiss did. The data do not show that smoking has a preventive effect. Rather, the NCHS findings demonstrate that there are data existing showing that nonsmokers have a higher prevalence rate of certain diseases than do smokers. It does not say a bit more. Of course, one can further investigate this point that smoking has a preventive effect. But I think it would be premature for anybody to come to such a conclusion from these data without further probing.

DR. SCHNEIDERMAN: As you know, Weiss had an exchange with John Goldsmith in the *Journal of Occupational Medicine* in which they discussed the same particular issue, and Weiss and Goldsmith just don't seem to come to the same conclusion that you do with regard to Weiss' data. This is just a comment.

How do you account for the changes in the reported incidence or mortality from lung cancer in both men and women in this country? The rates for men at younger ages are declining, the rates for women are increasing quite rapidly, coinciding with the roughly 25-year lag from the time the different population groups took on relatively heavy smoking in the United States after World War II.

The second question is, as you know and as you scolded me for doing, I've estimated that in men, at least, roughly somewhere in the order of 80% of lung cancer could be attributed to cigarette smoking. I have a 0.83 and the people in the American Cancer Society have 0.838. I'm not going to quibble with them on that. What proportion of lung cancer would you attribute to cigarette smoking? And to what do you attribute that portion that you don't attribute to cigarettes?

DR. STERLING: Here's what Weiss says, "The data suggests that continued cigarette smoking entails a factor which partially inhibited carcinogenic effect of chloromethyl ether."

Many have always considered Weiss to be militant antismoking. He has a history of telling workers with lung cancer, some of whom may have been exposed to chloromethyl ether [CME], that their disease was due to smoking. He did so for many years until somebody else discovered these cancers were a result of chloromethyl methyl ether exposures. Even then he wrote a curious paper insisting that while there may be more lung cancer among nonsmoking CME workers, the smokers coughed a lot more than the nonsmokers. But I will give him credit to respond to data rather than his beliefs and quote him.

Now to your questions. First, how do I account for different changes in the rate of lung cancer among men and women. In fact, I published some relevant data on that in the *American Journal of Public Health* about 12 years ago. I pointed out that there had been a proportional decline in industrial work among men and an increase in industrial work among women. I think what we see today when observing a rising lung cancer rate among women is "Rosie the Riveter" who, in World War II, started to work in the shipyards and was exposed there to asbestos, or worked in other industrial shops. We can't however, prove it with good certainty because of the very few studies that have bothered to look at the occupation of the woman. There are many studies that ask, "does she smoke?" Almost none that ask, "where did she work?" The National Cancer Institute has been especially remiss in its obligation to pay attention to these facts. There has been an increase among women workers in certain industries and, more important, there has been a shift of women's work. For instance, we find proportionally six times as many women welders now than we did 30 or 40 years ago. I think such findings have to be attended to.

I'm not saying that this causes the disease. I'm saying that health scientists have been remiss in not looking at women's occupational effects.

Finally, questions as to how much lung cancer is caused by this or that is a game of numbers. There are very few data showing the mutual effects of smoking, air pollution and, especially, occupation. I have pressed for a review of this issue for many years now, but there is not enough interest. We have, however, developed some relevant and revealing data and are about to publish it. These are studies in which populations of lung and other cancer and heart disease patients are followed retrospectively to see how many were exposed to fumes and dust and how many of them smoked. We seem to find that as many as 60% of lung cancers may be of occupational origin.

To answer, then, applying numerical techniques that are standard and acceptable, we come up with possibly 60% of lung cancers due to occupation.

MR. MORTENSEN: I am intrigued by your suggestion that there might be constitutional issues in smoking. I would like to review briefly my understanding of the application of constitutional law and ask you whether there can be a constitutional issue in smoking.

First of all, your last statement or outline states in effect that constitutional problems revolve around the freedom of an individual to engage in acts that are traditionally of social usage and this freedom cannot be breached without compelling reasons.

My understanding is that the constitution was intended to prevent the government from taking away rights—assuming that smoking is a right of constitutional dimensions. Therefore, are there constitutional problems or issues involved if the private sector or industry restricts access of employees or potential employees who smoke? I think that constitutional rights only apply to governmental actions.

Second, you state that employers should not be able to breach this right without some sort of compelling reason. My understanding is that compelling interest relates to those factors that someone is born with or those that are immutable. My question is whether smoking is such a characteristic that the courts should prevent discrimination in the absence of showing a compelling interest?

DR. STERLING: I found this discussion on constitutional law in the German literature. In the American literature there has been no such discussion. I was intrigued by the fact that this question of individual rights was raised as a constitutional issue in German literature. I'm not a lawyer, I can't really answer your question. I think it would be an interesting concept for lawyers interested in constitutional law to take a look at.

CHAPTER 31

EXPOSURE CONSIDERATIONS AND HYPERSENSITIVITY

Otto P. Preuss, MD

Brush Wellman, Inc.
Cleveland, Ohio

Hypersensitivity can be defined as an altered reactivity of the body (allergy) to foreign proteins (allergens) or certain chemicals (haptens) that have a specific affinity to the body's own proteins. In the latter case, such a hapten-protein combination will be considered by the body as "foreign" and will result in the formation of antibodies and the development of allergies.

Depending on the antigen characteristics, four basic types of reactions can be seen.

TYPE I

Type I, also called anaphylactic or immediate reaction, will result in bronchoconstriction or skin eruptions. These reactions may occur only minutes after inhalation or ingestion or even mere contact with the responsible antigen, and are dependent on the development of immunoglobulin E (IgE) antibodies. The latter attach themselves to granulated mast cells and, when linked with specific antigens, stimulate these cells into releasing histamine, serotonin and other substances capable of triggering allergic reactions such as asthma and hives. Certain individuals have much higher than normal concentrations of IgE, and these individuals have a tendency to react much more readily to specific antigens. They are much more likely to develop asthma or eczema, and they are called atopics.

TYPE II

Type II, or cytotoxic reactions, will result ultimately in complete destruction of intruding foreign cells. The transformation reaction is the best known within this category.

TYPE III

Type III reactions take several hours to develop. They manifest themselves either in a generalized "serum sickness" or localized organ responses, called an Arthus phenomenon. They may occur after either ingestion, injection or inhalation of certain antigenic materials and are based on the formation of toxic antigen-antibody complexes. Farmer's lung, caused by spores from moldy hay and resulting in an allergic alveolitis, is one of the best known examples in this category. Immunoglobulins A and M appear to play a particular role in their development.

TYPE IV

Type IV reactions are based on the sensitization and activation of T-lymph cells and require at least several days for manifestations, which primarily involve the skin or lung. They are, therefore, referred to as delayed, cell-mediated hypersensitivity.

Types I, III, or IV reactions may be encountered at the workplace.

Of the immediate Type I hypersensitivity responses, occupational asthma from exposure to isocyanates or vinyl chloride vapors are the best known examples. Others of the category are listed in Table I.

Table I.   Type I Hypersensitivity (Occupational Asthma)

Provoking Agent	Examples
Organic Dusts	Feathers, wool, furs, wheat grain, avian proteins, *A. fumigatus*, red cedar wood
Chemicals	Toluene diisocyanate, complex platinum salts, aluminum soldering flux

Hypersensitivity pneumonitis from exposure to organic antigens or chemicals belongs in the Type III category. Table II lists various agents capable of causing such reactions and the specific diseases produced by them.

Table II.   Type III Hypersensitivity Reactions

Clinical Condition	Source of Agent	Agent
Farmer's Lung	Moldy hay	*Micropolyspora faeni; Thermoactinomyces vulgaris*
Bagassosis	Moldy bagasse	*Thermoactinomyces vulgaris*
Mushroom Worker's Lung	Mushroom compost	*Micropolyspora faeni; Thermoactinomyces vulgaris*
Suberosis	Cork dust	Cork dust
Maple Bark Disease	Maple bark	*Cryptostroma corticale*
Sequoiosis	Redwood sawdust	*Graphium; Pullularia*
Papuan Lung (New Guinea Lung)	Moldy thatch dust	Thatch of huts
Wood Pulp Worker's Disease	Wood pulp	*Alternaria*
Malt Worker's Lung	Moldy barley	*Aspergillus clavatus; A. fumigatus*
Dog House Disease	Moldy straw	*Aspergillus versicolor*
Bird Fancier's Lung (Pigeon Breeder's Lung)	Pigeon, parrot and other bird droppings	Sera, protein and droppings
Pituitary Snuff Taker's Lung	Bovine and porcine pituitary snuff	Pituitary antigens
Wheat Weevil Disease	Wheat flour	*Sitophilus granarius*
Paprika Splitter's Lung	Paprika	

Type IV reactions may be caused by infectious microorganisms (tuberculosis, histoplasmosis, coccidioidomycosis) or by metals, such as beryllium or cobalt. Other organic substances, such as poison ivy and penicillin, or chemicals, like hexavalent chromates, amines or epoxy resins, may produce dermal reactions.

All three hypersensitivity categories are quite selective and involve generally only a small percentage of all persons exposed. Unfortunately, there are few useful predictive tests that allow one to detect the individual most likely to become sensitized. Only for Type I reactions can it be said that the atopic group, namely persons with a history of true food allergies, asthma or eczema, will be quite prone to develop similar allergic responses to other sensitizing agents of this category. But even here it is quite possible that a person may become sensitized only to a rather specific group of agents. (A nonspecific test for bronchial hyperreactivity following the inhalation of methylcholine exists which, nevertheless, may be quite helpful.)

Skin tests and inhalation challenge tests can be conducted for the majority of sensitizing agents from the Type III category. However, because of the possibility of sensitizing a susceptible individual or of an unexpectedly violent reaction in a presensitized individual, such tests must be reserved for strictly diagnostic purposes, under well controlled conditions, and should not be used for prediction. The same applies to certain skin tests for Type IV reactions such as for beryllium.

For all these reasons, a thorough preplacement history is of great importance whenever a potential exposure to sensitizing agents is a possibility. If this history reveals previous reactions to related materials likely to be encountered at the workplace, reexposure should be avoided. Accordingly, the examining physician has no choice but to advise against placement of individuals on jobs in which the worker will use or be near the agent in question.

Because of difficulties in determining who is particularly prone to sensitization prior to job placement, exposure to materials capable of producing allergic reactions should be kept below sensitizing levels by safe operating procedures and exhaust installations. Nevertheless, some individuals are so hypersensitive that they may develop reactions in spite of the most effective precautions. However, if several workers show reactions, a most thorough check on all preventive procedures and exhaust installations should be undertaken, since in such a situation exposure control breakdowns could be the contributing factor.

CHAPTER 32

CONSIDERATIONS ABOUT
REPRODUCTIVE HAZARDS

M. Donald Whorton, MD
School of Public Health
University of California and
Environmental Health Associates, Inc.
Berkeley, California

Concern for solving the reproductive problems resulting from occupational exposures is relatively recent. Only within the last few years has anything constructive been done in this area. Historically, research into reproductive problems has been inadequate, reflecting this lack of concern. The policies set have been discriminatory, primarily aimed against one group—females, and have usually resulted in the exclusion of all females from certain jobs "for their own protection." For this reason, reproductive hazards are considered primarily a women's issue.

Reproductive issues in the workplace raise political, economic and equal employment opportunity questions. Are these questions justification for blanket discriminatory rules?

The incidents involving dibromochloropropane (DBCP) have clearly shown that males can also be affected by reproductive hazards. Prior to this finding, only Occupational Safety and Health Administration (OSHA) lead standard addressed reproductive issues for both sexes. There is considerable debate regarding the validity of the data in that standard. However, there is no question that DBCP is a very potent male testicular toxin that can affect the testes alone, without involving any other organs.

In 1980 the Department of Labor and the EEOC proposed guidelines for reproductive hazards in employment. These proposed guidelines stated that males and females must be treated alike, unless one sex is affected differently. If the effects are different, there must be an evaluation of the possible effects on the other sex as well. One cannot state that the agent only affects females without looking at males, both in animal and human studies.

The EEOC guidelines state that exclusionary policies will only be upheld if reliable scientific evidence confirms potential harm to female employees. Recently, the current Administration withdrew these proposed guidelines.

Let us review some of the scientific problems encountered in attempting to evaluate reproductive problems in the workplace.

There are two major categories of reproductive problems: (1) the inability to conceive, and (2) adverse pregnancy outcome, which includes spontaneous abortions or birth defects.

About 15% of all couples in the United States are childless due to infertility. Another ten percent have fewer children than they would like. The male plays a significant role in 30% of these cases and is a contributing factor in another 20%. The female factor is solely responsible in 50% of the cases and is a contributing factor in 30%.

Any evaluation of couples presenting for infertility must include the male. Historically, however, the husband refuses to go for evaluation. The simplest and quickest evaluation for infertility is a semen analysis of the male. Unfortunately, that examination is frequently the last thing done. The evaluation of the female is more time consuming and more difficult. The initial step in such an evaluation is to record morning basal body temperatures for three months. In comparison, semen analysis, from collection to analysis, can be done in three hours. In evaluating infertility problems in the female, one must investigate the function of the ovary and of the oviduct, in addition to determining the receptivity of the uterus. In the male, one must investigate testicular function, plus patency of the delivery system, e.g., epididymus, vas deferens, seminal vesicles and prostate. One must not forget to consider the possibility that the subject is impotent.

There are a number of factors that must be considered in a diagnosis of infertility in males (Table I). However, in some 65% of the cases, the diagnosis is idiopathic.

There are basically three types of screening procedures for testicular function that can be done at a workplace. The first is to obtain a semen analysis; the easiest parameter is a sperm count (density). The second is to evaluate hormonal factors, and the third is to obtain a reproductive history. Each of these types of evaluations poses problems.

Table I.   Factors Affecting Sperm Production

Anatomical	Vericocele, cryptorchidism, torsion, trauma (pressure necrosis), vasovasotomy
Heat	Role of scrotum, effects of prolonged fever (38.5° C), baths, shorts, work
Infection	Mumps, venereal disease
Drugs	Antimetabolites, hormones, nitro-amino compounds
Toxic Effects	DBCP, lead, alcohol, marijuana, radiation
Diseases	Endocrine

The semen analysis, by far the most definitive, is the most difficult to do. In order to collect the semen samples, the employees must be motivated and cooperative, as this is not a passive test. The employees must be convinced that such a test is justified.

There are numerous reasons for an employee to decline or to be excluded from participating in semen analysis. Men with vasectomies must be excluded. In some studies, the vasectomy rate has been as high as 40%. Men whose wives are beyond child-bearing age are generally excluded inasmuch as they are no longer interested in the outcome. The same may be true for men whose wives have been surgically sterilized.

Impotency is seen with increasing frequency in men past their mid-40s. Other medical disorders, such as prior prostatic surgery, can render a man incapable of producing a semen sample. Some men have refused to be tested out of fear of abnormal results. These men were afraid a low sperm count would somehow mark them for life and feel that, as a result, no woman would want to marry them. For others, religious beliefs and cultural influences dictate that they do not participate. Finally, a semen analysis is believed by some to be too personal a test and there is tremendous fear that the results will not be kept confidential. I canot overemphasize the fact that in many work places the employees have a tremendous amount of fear—sometimes justified, sometimes unjustified—that all test results are immediately known by everyone in the company. If one cannot guarantee confidentiality of results, few employees will participate.

If a study population can be identified and recruited, the next major problem is the logistics of obtaining the sample. The best method is masturbation. Unfortunately, some men will not masturbate. For some, a literal interpretation of the Bible prohibits masturbation, regardless of reason. Abstinence for 48 to 72 hours is important. The best environment in which to collect the sample is at home; however, one sacrifices semen motility. While one could collect the samples at the workplace, the problem of the stigma of the collection room could inhibit participation.

Before any interpretation of the results can be done, it is necessary to determine what is a normal sperm count, particularly with respect to fertility. Many medical texts or laboratory references give as a normal sperm count 60–150 million/ml semen. For fertility purposes, generally a level above 20 million/ml is adequate. For the results to be scientifically valid, one must identify a reference, or control population. However, obtaining semen samples from a nonexposed group is difficult to nearly impossible. Frequently, one must use other published results. In using published results, it is particularly important that one ensures continence time is equal in both groups.

Blood samples are easier to obtain than semen samples; thus, serum hormone tests would appear to be more advantageous. The first drawback is the cost of analyzing the samples, which can range from $50 to $100 per sample for FSH, LH and testosterone. In addition, since the tests are done by radioimmunoassay, not all laboratories can perform them. The accuracy of the test depends on the specificity of the antigen. The lack of sensitivity of the tests in relationship to the sperm count presents an additional problem. Only when the count falls below 1 million/ml will the FSH level rapidly rise. Thus, hormonal results can only be used as screening tests if a large percentage of the population shows zero sperm counts (azoospermia).

This brings us to reproductive histories as a screening test. If one had a situation in which a child was born to each couple every two years, it would be relatively easy to detect a drop in fertility or newborns. In this country, with the widespread use of birth control measures, family size has decreased. In addition, a percentage of the work force has raised its family and is not interested in additional children. The stability of the work force, the availability of spouses, etc. are also important factors; and, if one is utilizing personnel records to determine numbers of dependents, one must be careful to determine those that may be the wife's by a prior marriage.

Recently, Dr. Richard Levine and his colleagues from CIIT developed a method of monitoring fertility of workers. This method was verified for DBCP exposure group [1,2]. The sensitivity of this method for less dramatic effects is still unknown, but if it works in all groups, it will provide a very easy way to do surveillance, certainly easier than anything I have already described. Ideally, one wants a surveillance test to be easily administered, relatively painless, inexpensive and acceptable to the employees. Requiring repeated semen samples as a surveillance tool is not very productive for very long.

Table II [3] shows the differential diagnosis of amenorrhea and contains the major categories and each subcategory of amenorrhea. The

Table II.   Etiological Classification of Amenorrhea [3]

Lesions of Central Origin
  Neurogenic
  Pituitary disturbances
  Psychogenic amenorrhea
Lesions of Intermediate Origin
  Chronic illnesses
  Metabolic diseases
  Nutritional disturbances
  Excretory and metabolic disease
Lesions of Peripheral Origin
  Ovarian amenorrhea
  End organ cyptomenorrhea
Physiological Amenorrhea
  Delayed puberty
  Pregnancy
  Postpartum amenorrhea
  Menopause
Etiology undetermined

entire outline covers two pages in the textbook. In addition to amenorrhea, fertility evaluation in females includes patency of the ovaducts, uterine competence, etc. It is important to know if the uterus is capable of carrying a pregnancy. If pregnancy occurs, there are two possible major adverse outcomes of concern: spontaneous abortion and birth defects. Estimates of the "normal" background rate for spontaneous abortions range from 15 to 40% of all pregnancies. This wide range is the result of the varying definitions of "spontaneous abortion." If definitive medical proof of conception is required (positive pregnancy test or tissue products of conception), the rate will be lower than if one accepts a history of a late period by three days, five days, seven days, etc. Thus, the definition becomes critical in developing background rates when one is asked to evaluate a workplace in which there is a complaint of increased spontaneous abortions or miscarriages.

Traditionally, hospital medical records have been used to verify spontaneous abortions. However, since the early 1970s, there has been an increase in elective abortions with many gynecologists and other physicians performing these procedures in their offices. Thus, the hospital rates for spontaneous abortions may be lower in some communities now than before. Therefore, one must be careful in utilizing hospital records alone to determine a community's spontaneous abortion rate.

One must also consider how the frequency of elected abortions has affected the spontaneous abortion rate. In a community with a high frequency of elective abortion, the spontaneous abortion rate may be falsely depressed. One of the scientific problems in evaluating the complaint of increased spontaneous abortions in the workplace is that you are generally working with a small number of employees who are complaining of a relatively common event. It is difficult to determine how one decides whether or not there is an increase, especially a statistically significant increase, without adequate numbers. An apparent increase can easily be due merely to chance.

A definition problem also exists with defects, as the estimated background rate ranges from 1 to 14%. The problem lies in the definition of "birth defect." Traditionally, birth defects are considered structural changes, both major and minor. Common major birth defects include abnormalities of the cardiovascular system, the nervous system and the musculoskeletal system. Minor birth defects can be skin colorations, webbing between two toes, extra toe, etc. In most cases, major birth defects are readily identified and brought to the attention of the parent. This is not true of minor ones; skin changes are frequently called birthmarks.

There has been recent interest in the concept of functional birth defects, first described in the children born to female heroin addicts. While there may be some structural abnormalities associated with these functional birth defects, they are not currently identifiable and, if present, may be at the subcellular level. These birth defects may not be detected at birth but only later in life. If a birth defect is not identified at birth, its presence will generally escape the usual reporting system via the birth certificate. Since there are so many types of birth defects, any is considered an unusual event, although some are more common than others. Since the reporting system utilizes birth certificates, only birth defects that can be identified within the first two or three days of birth are generally reported by the physician who certifies the birth.

If, in attempting to determine cause for birth defects, one asks mothers of children with birth defects and mothers with normal children what occurred during their pregnancy, the responses would differ considerably. A woman with an abnormal child has mentally reviewed all that occurred during the pregnancy, looking for a reason; a woman with a normal child has not considered the conditions surrounding her pregnancy. Thus, in any study of birth defects among working women with subjective exposures, this recall bias must be considered. In the evaluation of exposure, the timing and amount of exposure is critical for teratogenic effects.

Finally, it is highly unlikely that a worker had only a single type of exposure to toxic agents; thus, interaction must be considered.

In 1979 Goldberg reported on the differences in birth defect rates based on source of the data [4]. From 1968 to 1975 the Metropolitan Atlantic Birth Defect Program registered 209,668 births, in which 31,105 major birth defects were identified, for a rate per thousand live births of 14.8%. The National Center on Health Statistics, over a two-year period (1973-1975) in selected cities or areas, registered 8,409,277 births, in which 29,839 major birth defects were found, for a rate per thousand live births of 3.6%. There is a fourfold difference between these two data sources. The National Center relied entirely on reported defects from birth certificates, whereas the Atlanta program obtained data every possible way, including birth certificates, interviews with parents, interviews with the physician, school records, etc. These data show that the rate depends on the diligence used in gathering data, plus the definition used of "birth defect."

When considering workplace exposures, it is important to understand the stages of embryogenesis and susceptibility to teratogens. Teratogens are agents that cause structural changes or abnormalities, clinically called birth defects. In the early predifferentiation period (weeks one and two of conception) the embryo is not susceptible to teratogens. Adverse events in this period will kill the embryo.

In the period of early differentiation [weeks three to eight (ten)] of conception, the embryo is most highly susceptible to the effects of teratogens. With advancing age and development of the embryo, there is an increased resistence to teratogens. Since ovulation occurs 14 days prior to onset of expected menses, the period of highest susceptibility to teratogens begins the day after the menstrual period is expected. If a woman waits until she misses a second period to confirm pregnancy, a large portion of the critical susceptibility period to the effects of teratogens has occurred.

Currently known causes of developmental defects in humans include: genetic transmission, 20%; chromosome aberrations, environmental causes, including radiation, infections, drugs and chemicals, 7-10% and maternal metabolic imbalances, 1-2%. The remaining 65-70% of all birth defects are due to unknown causes.

Certain chemicals are logically suspected to be teratogens. This group includes many drugs, especially antibiotics and alkylating agents, almost all categories of pesticides, heavy metals and fat-soluble solvents. This group includes thousands of chemicals that one could assume, in the appropriate dose and at the precise time, could be teratogenic in humans.

This chapter was designed to point out the absence or paucity of scientific data that are needed to make policy decisions. Any policy must be very carefully and narrowly tailored to the issue at hand. For example, the removal of all women of reproductive age is discriminatory and short-sighted and is a disservice to the individual woman, as this policy basically implies she is not responsible for her own actions. One must realize that most couples in the United States who do not desire children utilize some type of birth control. Is it responsible to remove a woman whose husband has had a vasectomy? I do not think so. One also must remember that 90% of all women in the United States have all their children by age 25.

One cannot just remove women or men from their jobs without ensuring that they understand and concur in the action. Sometimes, one must remove a worker from his or her job against his or her wishes to protect the health of others, but this should only be done for sound reason.

While I can describe the scientific problems associated with policy decisions in this area, I have few answers other than prudence. Policies that are developed must be devoid of rhetoric and must protect the rights of those employed, while protecting the future generations of this country.

REFERENCES

1.   Levine RJ, et al.: A method for monitoring the fertility of workers. I. Method and pilot studies. J Occup Med 22:781–91, 1980.
2.   Levine RJ, et al.: A method for monitoring the fertility of workers. II. Validation of the method among workers exposed to dibromochloropropane. J Occup Med 23:183–88, 1981.
3.   Novack's textbook of gynecology, 9th ed., 1975.
4.   Goldberg, et al. J Am Med Assoc 242, November 23, 1979.

DISCUSSION: CHAPTERS 31 AND 32

T. Lamont Guidotti, M. Donald Whorton, Robert L. Jennings, Jr.,
Otto P.Preuss, Michael Holthouser, MacDonald Caza, Larry C. Drapkin,
Kenneth S. Cohen and William N. Rom

DR. GUIDOTTI: Dr. Whorton, you pointed out the difficulty in comparing rates for congenital defects between populations and that is, in fact, quite a serious epidemiologic quandry. Pediatric oncologists frequently study their incident rates for cancer, such as leukemia, comparing them in ratio with cancers that have a fairly constant rate, such as retinoblastoma and Wilms. Could a ratio for reproductive outcomes be determined by comparing the rates for diseases, such as cystic fibrosis and sickle cell anemia, and by selecting the appropriate population with ratio characteristics?

DR. WHORTON: I think so, but one of the problems is how do you handle certain diseases like those you described. One of the biggest problems is that our databases for most outcomes are so poor. I know some of the regulatory agencies and NIOSH have been struggling to develop a better database. It is unclear to me, since we do not know our own rate of adverse birth outcomes and since we live in a very heterogeneous society, what other societies could be used for comparison and how we would deal with those other factors in societies that are different from this one? Should we compare with only Western European countries, or maybe should we compare with the whole world? The question then is what do we do with their data? It is an issue that definitely needs to be addressed.

MR. JENNINGS: Recently, there has been increasing attention paid to the issue of prescreening of workers due to the possibility they may have an immunological, genetic or other predisposition to a given disease or reaction to an agent used on the job. The dilemma seems to derive from the fact that it is efficient for a company and there is some rationale in attempting to minimize the occurrence of disease in the workplace. Nobody wants a person to get a disease.

And it may—I'm not accusing anyone of doing anything—but it may well turn out that it is a lot cheaper or more economical and efficient for an enterprise to ensure that it hires a very healthy population without much tendency to disease and thereby reduce the need to control possible hazards.

But, on the other hand, the consequence of management's program, which is economically efficient, is that people's job opportunities and their opportunities to partake in the American way of life and economic benefits are dependent on things over which they have no control.

Thus, society may arrive at the point where opportunities are very limited by factors over which individuals have no control. Is that the kind of society we want to have?

Do the doctors on the panel view this as a problem and, if so, how should companies and society approach it?

DR. PREUSS: I pointed out that there is a considerable dilemma, inasmuch as there are prescreening tests and some of them identify definite hazards. Others, however, are inconclusive. There has to be an awful lot of research done before we can use prescreening tests routinely.

Second, even in those instances in which you would use these tests, it would certainly not relieve the employer of the obligation to keep the exposure to potential antigens or allergens to a minimum. On the other hand, however, a medical history points out that an individual is extremely sensitive. I think, as a physician, you have no choice but to advise him not to work in a particular area.

DR. WHORTON: I agree with Dr. Preuss. The other issue that comes up is how good are the data we are using to make decisions—what we are screening for? As an example, let's consider the sickle cell trait. Historically, until ten years ago or so, it was thought that if one had a sickle cell trait, one would never have a sickle cell crisis. However, three recruits in basic training with sickle cell trait at Fort Bliss became severely dehydrated on a hot summer day and severely stressed, precipitating a sickle cell crisis. From this episode, some people in industry have concluded that anyone with sickle cell trait should not be hired for any stressful or hot job. That makes no sense at all. Maybe there are very peculiar situations in which if one has some genetic problem he or she should not be in a particular job, but there would be very few of those situations, and they should be very carefully selected.

At one time many companies maintained a policy basically like the Army—we only want healthy people. But time and the laws for equal employment are changing this. If this policy were not carefully thought out and very well documented scientifically, we would be in great danger of displacing a large number of individuals for very false reasons.

DR. HOLTHOUSER: Dr. Preuss, the more I read about isocyanates, the more confused I become. John Peters' work indicates that a safe exposure level should be much lower than the current standard. However, the recent work of Hans Weill and his group indicates that perhaps the current standard is acceptable, assuming that one is dealing with an immediate type of reaction where almost any subsequent exposure will trigger a response in an individual. Is there such a thing as a safe level when you are working with sensitizers and subsensitizing level and, if so, what is your opinion of the level for TDI as the case in point?

DR. PREUSS: I do not have any specific experience with TDI. I would have to say there have always been some individuals who will be sensitized at extremely low levels.

For instance, in a Type III reaction, you may show precipitating antibodies. You could also show the same antibodies in a lot of individuals who never developed a disease, so subsensitization may occur. However, subsequent exposures still have to be at a certain minimal level to produce disease.

DR. CAZA: Yesterday, we talked a little bit of lead and women, today of hypersensitivity. There seems to me to be a deeper consideration that we have not addressed at all.

From a medical point of view or an occupational health point of view, it may be very well for Dr. Preuss to advise an employee that he or she should not work in a particular department. However, this is a medical opinion only. It becomes an administrative decision to determine whether or not to follow the medical recommendation.

Now, in the context of what seems to be the direction of human rights where, seemingly, an employee can, if he is not a danger to others, make the decision to accept a risk or hazardous employment if the risks of his employment are explained to him, where does that leave management? They are somewhat in a "Catch 22" situation. If I discriminate against an individual and not offer a job, I am liable for discrimination litigation. If I do give him the job and something happens, I am then liable either for compensation or another type of litigation. I do not think this conference has addressed this subject and I believe it will be a major issue in the future.

DR. PREUSS: I strongly agree. This is a tremendous dilemma, but I certainly don't have an answer for it.

DR. GUIDOTTI: Dr. Preuss, you quite correctly pointed out that it is proving very difficult to distinguish between an allergic or an irritant mechanism producing bronchial constriction in individuals with elevated resting bronchomotor tendencies, such as asthmatics. These individuals may experience a greatly increased airway resistance with irritant anesthetics. This behavior is very important as a practical matter in setting standards because a large percentage of the population is, in fact, atopic and may have these airway characteristics. A few years ago, the excellent experimental work in nitrous oxide of Orehec and his colleagues was very heavily cited and, in my opinion, was construed to suggest an extraordinarily low threshold of bronchoconstriction. The experimental situation used was very artificial. My question is rather broad and rather sticky. Should standards for irritant gas, in this particular situation broadly defined, be based on the general population with safety factors that include significantly large minority populations such as asthmatics (which constitute a small percentage of the population) or on lowest demonstrable thresholds in sensitive populations?

DR. PREUSS: That is a million-dollar question. I think if we speak in general terms, we probably should take the average total population as a base, because there will always be one individual who will surprise us and who will show a reaction to concentrations from which, based on our past experience, we would never anticipate a response.

MR. DRAPKIN: It seems that in this conference we have been cautioned, when determining general exposure standards for workers, not to always be willing to use animal data to arrive at conclusions about human exposures. We often look at animal data in the case of reproductive hazards as well, but we always are very cautious. In other discussions, we talk about the issue of potential liabilities arising from a damaged child if we are not cautious. It seems to me that this contradiction in cautionary measures is not very ethical in that what we are really considering at the bottom line is potential liability. I would like to know if my impression is wrong and if perhaps more than economic concerns are involved. Are we only concerned with money here?

DR. WHORTON: I think in reproductive issues one has to look at animal data because there are so few human data. When you must transform or go from one species to humans, it would be nice if all studies used all nonhuman primates as the only animal data. However, using nonhuman primates in most situations is not practical. Therefore, one is really left with rodents or other species that multiply quickly and have relatively short life spans. In evaluating the current human data, one must be cautious for scientific, not monetary, reasons. The databases are too inexact to toss caution to the wind.

Another question in reproductive issues is the extrapolation from in vitro results. There are no simple answers; this is a very difficult area in which to make such assumptions.

MR. DRAPKIN: I did not mean to suggest we should not look at animal data. I was trying to suggest that oftentimes we gloss over these data as they relate to exposure problems; however, we do not when we are talking about reproductive hazards. I thought that contradiction should be pointed out.

DR. COHEN: Rather than eliminating certain portions of our working force, let's control the hazard and allow all people within our working environment the opportunity to earn a wage.

DR. HOLTHOUSER: It is very difficult for one person to get something done or for a small group of people to get something through government. But, if the general population or a large segment of it is sufficiently outraged about what is happening to them, they will get milestone legislation, like the Civil Rights Act, passed and enforced. Now, black and white people eat together, for instance; something almost unheard of in many areas 20 years ago. I keep hearing we can't do this or that. We are pushing for more and stricter regulations in all aspects of life. I think if a large, powerful group such as organized labor pushed a lot stronger, more would be done. After all, they represent those who are getting sick, at least much more so than the groups pushing currently for health and safety regulation.

Companies tend to do what they must do to remain solvent and profitable. For example, they worry more about EEO concerns than about health and safety standards because they stand to lose fat government contracts due to the teeth in EEO enforcement, which are not present in health and safety enforcement.

DR. ROM: There may be issues about informing workers of the results of medical surveillance data. Could you comment on collecting urine, for example, from women employees for testing, then including pregnancy tests on their urine and informing them when they are pregnant by removing them from a potentially hazardous exposure?

DR. WHORTON: If you are asking about collecting urine without telling the employees why you are doing it, I think that is unethical. Collecting urine for drug screens was very popular in some companies. I think that any time one asks an individual to participate in a test, especially to provide a biological sample, one is obligated to tell them what one is going to do, why one is doing it, and how one will report their results to them. It is poor practice to state, "If you don't hear anything, the test is okay." You have to respond back to them, preferably in writing.

DR. ROM: One of the real issues that we have in workplaces involves working with materials that we don't know too much about. An example is diglycidyl ether (DGE). OSHA and NIOSH put out a hazard warning about two years ago, saying that DGE may cause testicular atrophy in experimental animals. (It reminds one of the DBCP in 1961.) Yet, this chemical is used in a number of processes, particularly in epoxy, which is widespread in industry, and women may work in these areas. What ethical responsibilities does a company have if they are employing women in this situation? Should they remove them from exposure? Should they measure exposures? Do they have ethical responsibilities in investigating these chemicals? What would you recommend as a consultant to that company, for example?

DR. WHORTON: Well, first of all, it goes into the right of the employees to know what they are working with, as well as their right to know what the potential toxic effects may be. They should know what the employer is doing to decrease exposure, if it is excessive. One can work with most any dangerous chemical or agent as long as it is controlled; thus, there should be no ill effects. In a situation in which there is testicular toxicity, why would you be removing the women?

DR. ROM: Because it may affect both sexes.

DR. WHORTON: But you would remove the men before the women, wouldn't you?

DR. ROM: Somebody has to work there.

DR. WHORTON: In evaluating animal data for testicular toxicity, one needs to know what other effects, at that dose, are seen in the animal. If the animals are nearly moribund, then the testicular toxicity is the least of the animal's worries. On the other hand, if, like DBCP, testicular toxicity is shown in absence of other effects, the chemical is more ominous for male workers. In the latter situation, one has the obligation to inform employees and to develop a medical testing system for those who want to be tested. One must provide health monitoring for the workers, as this is the only method to assure that workplace controls are adequate.

DR. ROM: Dr. Preuss, we heard yesterday that KBI is out of business, so Brush Wellman is virtually our only beryllium industry, or at least must be the major one. There remain a number of scientific issues concerning etiology of beryllium disease; for example, hypersensitivity vs overexposure. Harriet Hardy's review published in *Environmental Research,* discussed this. As medical director of the last surviving beryllium company, Brush Wellman, could you respond to the following: What do you think are your ethical responsibilities both to yourself and your management in performing research to clarify the scientific issues relating to etiology, pathophysiology, and complications of beryllium disease?

DR. PREUSS: I think this is a scientific issue that really has no bearing on our responsibilities to protect workers and to adhere to the standards.

The hypersensitivity theory really has no effect on our efforts. Actually, there is only one very small group that is still insisting that the beryllium toxicity is based on a strict dose-response relationship. Every immunologist who has worked with beryllium feels differently, as do lung specialists experienced with allergic alveolitis. There can be no question that beryllium belongs in the category of those agents that are producing a hypersensitivity lung disease. It is clearly demonstrated by the fact that beryllium creates a granulomatous interstitial lung disease and that the granuloma is well recognized as a cell-mediated hypersensitive response. Epidemiologically, chronic beryllium disease is very selective, since only 1-2% of all exposed contracted the disease, even under the worst exposure circumstances. During the 1940s, people had fantastic exposures to thousands of micrograms and still we saw only 1-2% of those exposed — women had a slightly higher percentage — contracting the disease. I have found confirmation of these data in Japan, where exactly the same results were seen. However, as far as our preventive efforts are concerned, this really has no bearing whatsoever since, for lack of differentiating criteria, they must include all workers. The only area where the hypersensitivity factor has any importance is in the differential diagnosis. The lymphoblast transformation test is of great help in diagnosing the disease because, unfortunately, we have a host of other hypersensitivity lung diseases, type III or type IV. All of these can produce the same physiological and radiological effects, so the differential diagnosis is extremely difficult, and the presence of beryllium in the lungs doesn't mean anything, because most people who have high concentrations of beryllium in the lung do not have beryllium disease.

So the lymphoblast test, which indicates cell-mediated hypersensitivity, is a very valuable diagnostic tool, but that is about all. We still do not know enough about this test as far as its overall behavior in all workers is concerned, but it can be very meaningful if we have an individual who does have clinical evidence of interstitial disease and a clear history of exposure to beryllium. If I do have a positive lymphoblast transformation test in such an individual, I do not need an open-lung biopsy any more, which did not tell me anything specific anyway.

In such cases, the test is a tremendous diagnostic help. Whether it is really a very valuable surveillance tool remains to be seen and we will have to go on trying to find out.

I hope that, eventually, we will be able to do continuous testing prior to employment and periodically every year or every other year, so we really could learn more about the value of the test for the workers' medical surveillance.

SECTION 5

ETHICS

CHAPTER 33

ETHICS AND ENVIRONMENTAL HEALTH

Thomas H. Corbett, MD
Anesthesia Service Associates of Toledo
Sylvania, Ohio

Virtues are, in the popular estimate, rather the exception than the
rule — Emerson

- **ethics:** the rules of conduct recognized in respect to a particular class
  of human actions or a particular group or culture
- **morals:** principles or habits with respect to right or wrong

When discussing ethics and environmental health, we are really discussing a form of medical ethics, which traditionally has had proper moral connotations.

Formalized medical ethics are at least as old as the Code of Hammurabi, the ancient King of Babylon. In comparison, the Oath of Hippocrates is a latecomer. Yet, some aspects of medical ethics change and must be reconsidered in the light of advances in medicine and changes in law, custom and society's attitudes toward its ways of life and death. Ethics may also change in light of economic considerations.

Kant, in his Lectures on Ethics, had this to say:

Ethics studies the intrinsic qualities of actions; jurisprudence whether
they are lawful.
Ethics is not, however, excluded from the legal field...but expects
that even those actions to which we are bound by law should be
performed by us not because we have to...but because they are in
themselves right and we are so disposed.

It is important to note what he said about "the intrinsic quality of actions"—in other words, their essence. Essences of right and wrong can be hard to calibrate, hard to determine. However, there is a gauge that not only evokes the best in human spirit, but is as practical as any ethical gauge can be. It is called the Golden Rule. Essentially, it tells us to treat others as we would want them to treat us or our loved ones. It tells us not only to do our level best, but also to assert ourselves if we find someone else who has done ill.[1]

When, as in medical ethics, the Golden Rule applies to a person to person situation—the doctor and his patients—it is a simple relationship. However, the ability to adhere to its principles becomes more difficult and complicated when considering this code of ethics in relation to large groups of people interacting at different levels. Group interests come into play both in terms of power and economics, the two usually being closely related. Furthermore, it is very easy for certain groups, when economic considerations are involved, to look at the public in coldly abstract terms rather than in intimately human terms, and this can also lead to breaches of ethics.

A case in point is the environmental contamination disaster in Michigan involving flame retardant polybrominated biphenyls (PBB). In 1973 sloppy packaging procedures at the Michigan Chemical Company resulted in the shipment of bags filled with PBB instead of magnesium oxide to the Michigan Farm Bureau Services, Inc., which mixed the chemical into dairy cow feed that was distributed to hundreds of dairy farmers throughout the state. Although mixed only in cow feed, the PBB contaminated hoppers also used to mix feed for numerous other types of livestock. Millions of chickens, pigs, goats, sheep, horses and other animals were thus contaminated as well. The nine million residents of Michigan were subsequently contaminated through the ingestion of these contaminated animals and their products. The health effects on humans were unknown, but millions of farm animals were rendered ill from the disaster.

Sick farm animals were sold through routine channels for nine months before the illness was diagnosed as chemical contamination with PBB. Neither the companies involved nor the Michigan Department of Agriculture was able to diagnose the problem. The diagnosis was made by a persistent, independent dairy farmer with a degree in chemical engineering. Furthermore, the Michigan Department of Agriculture, more concerned with the potential economic loss to the state than with human health, allowed millions of these sick and dying animals to be sold for human consumption on the grounds that they did not know what the trouble was!

Now once the diagnosis was made, the Golden Rule would dictate that since the animals were ill and the chemical would be passed to humans eating the animals or their products, the ethical thing to do would be to take the contaminated animals off the market, exterminate them, dispose of the carcasses in an environmentally safe manner and then decontaminate the farms. This, however, was not done. Again, because of economic considerations, these sick and dying animals were still sold for human consumption. Some farmers, concerned about human consumption of diseased animals and their products, shot entire herds of animals rather than have them sent to market, thus incurring great personal financial losses. It took several years for public outrage to force state and federal bureaucrats to finally take definitive action to end the contamination cycle. During that time, Michigan residents continued to accumulate PBB in their body tissues. The long-term effects of these concentrations of PBB are unknown, and the jury is still out in this regard. But there is cause for concern. Like polychlorinated biphenyls (PCB), PBB is known to cause cancer and birth defects in laboratory animals.[2]

Wherein lies the fault? It lies with the Michigan Chemical Company, whose sloppy practices showed a total lack of concern for the prevention of potential contamination episodes and, coincidentally, a total lack of regard for protection of its workers; and it lies with the Michigan Farm Bureau Services, which tried to cover up the contamination episode. With regard to the Golden Rule, their ethics were despicable. However, in either case each company was obeying its primal instinct, that of survival. What may be ethical in terms of environmental health may not be good for business. Either company might have been forced out of business by the disaster. The potential economic losses were staggering to the imagination. The fault also lies with the Michigan Department of Agriculture (MDA), whose dual purpose for existence is both to promote Michigan agribusiness and to protect consumers, the former purpose obviously being considered the more important of the two. The MDA was very lax in its attempts to diagnose the problem of sick and dying animals, and it set a high tolerance limit once the PBB contamination had been discovered. The MDA tried to minimize the economic losses to the state. Fault also lies with the Michigan Department of Public Health which was lax in determining potential health hazards in humans, and it lies with the Michigan legislature and the governor, who did not take the necessary political steps to stop the pollution of the state and the contamination of its residents.

The economic considerations were paramount in the decisions taken by these groups. Initially, the costs of cleaning up the mess were staggering, but it cost far more later because of the delay.

The bottom line in effective environmental health programs is money. It costs money to prevent environmental health problems, both in terms of averting disasters and in controlling them once they have occurred.

The greatest current economic threat to environmental health is the budget squeeze caused by inflation. We are now faced with the very serious dilemma of determining how to control mounting environmental pollution within a strained economy.

We mentioned that the primal instinct of a corporation is survival, and this instinct determines the code of ethics necessary for this survival. The same primal instinct is true for governments and the people who run them. Again, the ethics of a government are those necessary to maintain it, and a nation's economy is its backbone. So, a government will do whatever is necessary economically to ensure its survival. The same holds true for its politicians.

The Reagan administration has recognized our current economic dilemma caused by massive deficit spending and is attempting to curb inflation before the dollar is totally destroyed. Hence, the massive budget cuts, many of them in the area of environmental health programs. We must make the administration see, however, that in terms of economics, an ounce of prevention is worth a pound of cure, and that investing in solid environmental health programs is a blue-chip investment, reaping billions of dollars in dividends in the form of health care cost savings and the continued productivity of hundreds of thousands, perhaps millions of people who would otherwise be disabled, hospitalized or die from environmental disease.

REFERENCES

1. Nesbitt, TE: Medical ethics. J Am Med Assoc 245:241–242, 1981.
2. Corbett, TH: Cancer and chemicals. Nelson-Hall, 1977.

CHAPTER 34

PROFESSIONAL LIABILITY IN OCCUPATIONAL HEALTH: CRIMINAL AND CIVIL

Kenneth S. Cohen, PhD, PE, CIH
Consulting Health Services
El Cajon, California

Legislative law changes very slowly, but changes in case law are rampant. Certain changes in case law are rapidly altering statutory law, particularly in the area of professional liability. Professionals in the field of occupational safety and health need to understand and be alert to these rapid changes. The references in this chapter are primarily to California law, but comparable statutes or case law are found in most states.

I should like to preface this chapter with several caveats. First, it is in the best interests of each professional to seek private legal counsel for review of the statutes or case law in your state, your educational background and your professional jeopardy (or vulnerability) with regard to the actions that you take in your professional activities. It is well worth the cost involved to exercise this reasonably prudent self-investigation.

Second, if you are employed by a corporation, company, or an organization—military, government or otherwise—check the specifics of the protection alluded to by your employer. I would advise you not to accept blanket determinations: the "general understanding" that you are covered by someone's professional liability insurance should be verified in writing, if possible. Professional liability should be documented in writing by your corporate legal staff as to your specific status as an employee in relationship to the employer. Finally, action on these recommendations should also be on the advice of your private counsel, as I am not attempting to practice law without a license.

I am not sure that many professionals in medicine or industrial hygiene appreciate something called "the adversarial role." This can be typified by describing my recent experience in court. I sat on the stand and four defense attorneys pinged away at me, while the plaintiff's attorney for whom I was testifying rigorously objected. On leaving the courtroom, the attorneys joked with each other as to how beautifully they each had defended their rights to argue. In a similar experience, I participated in a deposition session in which the attorneys went tooth and nail at each other, arguing the admissibility or inadmissibility of my testimony. At the close of the deposition, when we were off the record, the attorneys suggested we all go out to lunch and enjoy the rest of the day. The animosity disappeared almost instantaneously because this is the "adversarial system." I have begun to believe that this is part of the grooming or "care and feeding" of trial lawyers during their educational experience. In most cases, attorneys are friends or "brothers under the law," yet when in the adversarial position, they must take polarized positions. For most people, it is alarming to see the animosity flow between people in the courtroom who were potentially friends moments earlier. An appreciation of this "adversarial" relationship may assist persons in the health professions to better understand the conditions under which they are being required to testify or perform.

I would now like to focus my crystal ball, if I may, on a futuristic look at some of the potential or real situations of professional liability that I have seen coming up on the horizon in my modest experience.

DUTY

In its use in jurisprudence, this word is the correlative of right. Thus, wherever there exists a right in any person, there also rests a corresponding duty on some other person or all persons, generally.

We must examine closely the concept of duty. What is our duty to the injured person or the injured worker as it relates to our activities? It is very important to determine the nature of our agreement to enter into certain activities. If we are doing volunteer work, if we are doing a gratuitous level of activity, then it becomes a moral obligation to continue to perform the duties. However, if, in fact, we have a contractual relationship with remuneration involved, then the breach of duty is construed as negligence. This difference is not typically clear nor understood by most health professionals.

For example, most professional organizations have codes of ethics that include criteria for defining the responsibilities of individuals practicing in that field. Under the nursing code of ethics, the nurse has the responsibility to question a physician's orders if he or she feels they are not efficacious. If the nurse fails to do this, he or she is liable for personal malpractice separate from that of the employer. This "questioning" is a part of their oath, code of ethics and standards of care in practice. This duty to question is of particular importance in industrial practice situations, where independent nurse practitioners or nurses are located in dispensaries in which a physician is rarely seen and in which they are working under "directives" of a physician. These "directives" are "carte blanche" prescriptions that direct the nurse to use given medications, to issue medications or to perform certain procedures under specific circumstances laid down as "protocols."

Negligence

I would like to address the concept of breach of duty, or negligence, and workers' compensation as a sole remedy for injury. Baron [1] did an eloquent job of outlining some of the possible areas for recovery in lieu of or in addition to workers' compensation. These new areas relate directly to the competence of medical and health professionals involved in treatment or protection engineering. Hammer [2] also described some of the "chinks" in the engineer's armor. I became very uneasy as I noticed that Mr. Hammer was, in effect, feeding attorneys those questions that should, can and, probably, will be used in cross-examination to establish lack of training or lack of adequate preparation.

In the professions of industrial hygiene and occupational medicine, it is obvious on review of the certification and board examinations that questions in areas such as toxicology are not current and appropriate with respect to the new chemical environment in which physicians and industrial hygienists may have to work. And how sound and valid are the ABIH points for recertification? Do CME for physicians truly reflect on a credit-by-credit basis the continuing education and competence of the physician? Did the professional attend a conference, or did he go to a specific ongoing course in which he learned the techniques or "nuts and bolts" by which he can practice more adequately in the field?

Duty to Avoid Injuring Persons or Property

Every person is bound, without contract, to abstain from injuring the person or property of another or infringing on any of his rights.

The duty to avoid injuring persons or property is based on civil codes, which are laws of long standing that are binding on all individuals. Professionals in occupational safety and health have an overlapping responsibility not to deceive or withhold information that may result in injury.

INTENT TO DEFRAUD PUBLIC INCLUDES ANY DECEIVED PERSON

One who practices a deceit with intent to defraud the public or a particular class of persons is deemed to have intended to defraud every individual in that class who is actually misled by the deceit.

Class action suits are becoming common. The $100 million suit filed in Tyler, Texas, on behalf of the former employees of the Pittsburgh Corning Corporation in Tyler was such a case. The Tyler case was particularly significant because there were multiple victims of the same disease at several plants belonging to the same corporation. In these class action suits, we hear spokesmen from specialty groups such as unions alluding to the concept of multiple-occurring "coincidence" injuries; and we hear from the "shoeleather epidemiologists" such comments as "Joe died of leukemia at the refinery, and you know Sam died of leukemia, and old Charlie died of leukemia, too." This should be a warning to health professionals; if you don't make the associations between worker injury and disease similarities, workers will.

LIABILITY FOR NEGLIGENCE OR TORT

Everyone is responsible, not only for the result of his willful act, but also for an injury occasion to another by his want of ordinary care or skill in the management of his property or person, except so far as the latter has, willfully or by want of ordinary care, brought the injury.

In the practice of occupational health, it is encumbent on the professional to understand the term "negligence" in order not to be found negligent. I believe that a health professional must know the enemy and, in this case, he may be his own enemy by virtue of his errors of omission or commission. I would advise you to review with your own attorney your activities to determine those for which, in reference to current law, you might be found negligent.

Whether negligent or not, you may someday find yourself in the witness box testifying for someone with whom you are involved, for whom you have provided treatment, or as an expert witness against someone with whom you have had no contact. In any case, let us discuss protocol on the witness stand. Although many of you have served your apprenticeship in the witness box, you will probably be subpoenaed again at some future time. I believe there are important criteria that you should know with regard to the qualifications and testimony of an expert witness in court.

An expert, in a court of law, is someone who basically knows more than the lay person on a specific issue. If you are a specialist in any area, you can be called as an expert witness.

QUALIFICATIONS OF A WITNESS

Except as otherwise provided by statute, every person is qualified to be a witness, and no person is disqualified to testify in any matter.

The qualifications and weighting of an expert can be challenged by the opposing attorneys at any time during the course of testimony or deposition. In challenging a witness, the attorney asks the jury to weigh the level of "expertise" of one witness against a second expert witness. This "weighting" is based on the degrees, credentials, certificates, etc. of one side vs the other. For example, the testimony of a qualified physician in internal medicine or general practice can be completely overshadowed by the testimony of an individual in the same medical specialty who is, in addition, board-certified.

Another factor in the evaluation of qualifications is the manner in which the witness conducts himself in his professional activities. How does he handle the data that he collects, or the patients that he sees?

I offer the following brief checklist for those individuals who are or may be called as expert witnesses:

1. Keep a personal diary. There is nothing worse while sitting in a witness box than to be forced to sift through scraps of paper and charts attempting to identify something that you thought you would be able to reconstruct at some future date. Keep this diary to reconstruct meetings with management, meetings with your industrial hygienist and with safety persons, meetings with a manager or supervisor in the work area, or other significant or unusual occurrences.
2. Confirm conversations in writing and keep a copy. For those who are not aware of the California or San Diego jargon that is aptly expressed by the phrase CYA, I recommend that you cover your anatomy with paper. If you have a conversation with a manager in regard to a sensitive issue, such as carcinogen handling by workers, you may want to document that conversation and send a confirmation note to the manager. An example of this may be: "Dear Mr. Smith: In regard to

our conversations regarding Direct black-38, it is imperative that all
personnel contacting this material be informed in writing of its hazard
potential."

3. Review your corporate job description with both your private counsel
and with the corporate or company attorney and clearly define your
responsibilities and duties. A clear and concise job description becomes far more important as health professionals become increasingly
vulnerable.

4. Good conduct in the witness box. It is important to understand the
protocol for a witness, whether you are an expert or not. Know the
techniques that make for a good presentation in court. Review your
evidence in detail. Notes brought into the courtroom may be entered
into the court record and can be lost to you forever. Only bring into
the courtroom that which you know you can and will easily give up.
Do not bring into the courtroom documentation of information you
do not want revealed.

For those of you who have never participated in the drama of the
courtroom, let me introduce you to *voir dire.*

Voir Dire

To speak the truth: the preliminary examination that the court may
make of one presented as a witness or juror, when there is an objection to
that individual.

The process of *voir dire* occurs after an individual has taken the stand
and has been presented by an attorney. The attorney gives the individual's
qualifications as an expert witness. At this point, the opposing attorney
may object to the individual as an expert witness. For the sake of this
presentation, we will assume that you are the witness. The attorney calls
for *voir dire* and proceeds to interrogate you beyond your wildest dreams
of torture and pain. He may ask you the length of your toenails, the
thickness of the hair on your left eyebrow, or any other obscure pieces of
information from your background that he can, in some way, tie to your
expertise. He may question the score on your kindergarten examination
or ask if you passed wood shop in high school or any other deep questions
that may be used to impeach your testimony. The final determination of
your qualifications as an expert is made by the judge. If he so determines,
he will acquiesce to your competency as an expert witness and to the
admissibility of the evidence that you are about to give.

Oftentimes, we are anxious to offer our testimony and very rapidly
spew out things we may regret at some time in the future. Witnesses

should be warned against the dangers of speaking before the mind is in gear. Many years ago, when I first started testifying in criminal cases, I was coached by a very experienced criminologist who runs the crime lab in the San Diego Police Department. This gentleman was of Hispanic background and he said, "Ken, if you want to make it in court, the jury must think that it is in your head before you open your mouth and say it. Say to yourself, 'one taco, two tacos, three tacos; then open your mouth and start giving your opinion." This pause provides you with an opportunity to think about what you are to say and also impresses the court that perhaps you were contemplating the question as opposed to giving a rehearsed response. I would caution you in the courtroom, as a witness for whatever side—I hope you are always a witness for the right side—to be a teacher to the judge or jury. Explain your technical terms, since the judge or jury may not be familiar with "idiopathic pulmonary fibrosis." As a witness for whatever side, take the initiative and be humble. God help us, but when we attain the stature and position that accompanies the title of health professional, the hardest words to cross our lips are "I don't know." Say, "I don't know" if you do not know, because everything you say will usually be held against you. Opposing attorneys will drag out of the archives every deposition you have ever given, every court testimony you have ever occasioned, and they will go back and read and reread these transcripts to find any point that might be used at a later time in some negative way.

Above all, as a witness, be honest. Only in this way will you serve as a credit to yourself and the subject of the trial. One of the most difficult questions that a physician or a good scientist will address is: "Dr. So-and-So, beyond a shadow of a doubt, can you say that this was caused by that?" A good scientist, properly educated, generally believes that it is not possible to know "beyond a shadow of a doubt" because there is always a level of knowledge that is beyond his grasp. There will always be a more finely focused microscope that can see deeper and clearer; there is always a greater chemical explanation or other evidence just about to be unearthed. We do not always know beyond a shadow of a doubt and, as a result, most doctors and scientists make poor witnesses, because they are not willing to commit themselves. However, it is your best opinion that the court is seeking and all that can be contested is that at the time of your answer, you gave your best effort in forming your opinion.

Physicians must be particularly careful in the courtroom not to allow their early training to override the need for honesty. Some physicians will admit that one concept taught in medical schools as part of the bedside manner (or loving care) was "if you can't make it, fake it." The last thing the physician wants to do is to let the patient know that he may not know everything that exists in all of medicine and in all of science. Most astute attorneys now practicing in this area recognize this tendency.

Let's turn now to the subject of medical malpractice. There are a number of voluminous texts in print and others that are being written in all areas of malpractice. I find one of the best to be Dornette [3]. This text is particularly lucid in its approach to what goes into the malpractice action. There are similar texts in orthopedic surgery, and obstetrics and gynecology, and I hope someday one will be written on industrial hygiene, but I think that there are similarities in the concept of malpractice that overlap into all professions. Let us talk about the concept of malpractice.

What is malpractice? Baron [1] stated that, in occupational health and safety, malpractice is not limited to the practice of medicine in the workplace. It also includes "the doctrine of failure to warn" and a variety of other considerations related to the management of the workplace.

Insurance rates, which reflect the rising costs of malpractice claims (particularly in engineering fields), have increased dramatically. If we take the 1962 rates as 100%, the costs to physicians and surgeons had increased to 128% by 1972; and for architects and engineers, the rate skyrocketed to 493% by 1972. As you can see, those of us in the engineering fields are not lagging far behind the physicians, and will probably catch up at some point in the future. This becomes a significant problem, inasmuch as most industrial hygiene engineers, safety engineers or other allied health professionals do not currently carry any type of indemnification for malpractice or insurance related to occupational health. Since malpractice insurance is either nonexistent for those in allied health trades or extremely expensive, we end up going "naked." The law committee of the American Industrial Hygiene Association has been charged with finding some method by which industrial hygienists and/or safety engineers can be covered by some form of malpractice insurance. No such insurance yet exists.

In an evaluation of malpractice by any health professional, we must learn to judge "the standard of care" based on the practice of other professionals in that field. How do other industrial hygienists sample and analyze the levels of contaminants in the workplace? How do other physicians gauge the severity of an exposure? These are but a few examples in the health fields.

Acts of commission or omission are also considered in malpractice.

The Doctrine of *Res Ipso Loquitor*

The action speaks for itself.
Is what the health practitioner did or did not do directly attributable to

the injury of the patient? Could the injury or its treatment have been modified in some way if the activity was performed in some other manner?

How much training and education beyond the degree program has the practitioner had? How current is his information with regard to health contaminants within the workplace? These questions directly relate to the continuing education activities of the professional. Some experts in the legal profession are now suggesting that a professional is one who commits 10% of his time to those activities that will continually upgrade his level of knowledge.

I have discussed the general concept of duty and the obligations of voluntary versus contract services. With reference to malpractice, the professional must also be familiar with the statuses or local regulations or ordinances that directly affect the exercise of duty. The medical requirements laid out in Occupational Safety and Health Administration (OSHA) standards, such as the lead standard, are typical of trends in the law. Professionals are also expected to be familiar with the codes or white papers issued by professional societies that may directly affect performance of duty. Recently, a number of editorials in professional journals, such as the *Journal of the American Medical Association,* have dealt with professional relationships. These editorials are weathervanes of issues in the field.

Several recent rulings in malpractice law that seem to reflect a trend may relate to health professionals. A recent $2 million dollar judgment against the Church of Scientology alludes to problems in ministerial malpractice. The basis of this malpractice action was a charge of alienation of affections by families of parishioners of the church. Another case resulted in an $830,000 judgment against an Episcopal pastor and bishop in Dallas. The clergy were found guilty of negligence based on outrageous misconduct and clergy malpractice.

Engineers and architects are also finding themselves in the malpractice area in a variety of design defect cases. Design of industrial structures that provide for roof outlets of industrial effluent of a potentially toxic nature is an example of a malpractice risk area. Placement of intakes for mechanical ventilation systems in the immediate vicinity of roof outlets constitutes negligence, as it is reasonably foreseeable that the contaminant in question will most likely be recirculated. The use of exhaust ducting that specifically directs vapors to the roof line instead of into the wake of air over the building is another example. Cases have been filed in situations in which the architect, design engineer or contractor has modified the utilization of flameproofing materials specified for certain buildings as a cost-benefit replacement. Additional design judgment in the area of sprinklers will probably arise out of the MGM Grand fire in which the cost-benefit risk ratios developed precluded the installation of sprinklers in the original building. In fact, the cost would have been slightly less than the dollars expended on replacing the carpet in the building. The lives lost in this situation may justify a malpractice action.

An error in professional judgment can cost money and involve civil liability, and it can also verge on the criminal and be liable to prosecution under such sections as 17e of the Occupational Safety and Health Act (OSHAct) and other statutes. To appreciate the potential for criminal liability, we need only look at just one state—California. In one 15-month period, there were more than 100 investigations of criminal violations under Section 17e of the OSHA equivalent statute. Of those 100 investigations, 34 resulted in criminal proceedings, with 31 convictions. It is well established that California takes the lead in many areas such as this in exploring the potential for legal remedy.

A lead case in the area of criminal prosecution in California was *California v. Lockeed Shipbuilding,* etc. This case resulted from a tunneling disaster in California in which 17 men died as a result of a methane explosion. The project manager and safety engineer were sentenced to 20 and 18 years, respectively, and the corporation was fined $121,000. This case was reversed on trial error, but it could hold on the next review. To be prepared, we should review the criminal sanctions under the OSHA law. There is no "Miranda decision" in OSHA, and there is considerable controversy over the individual's rights in an OSHA investigation following a fatality. "You have the right to remain silent; you have the right to an attorney." I advise my corporate clients that if there is a fatality, not to speak to a compliance officer without an attorney being present.

There are instances in which Department of Labor and OSHA officials have been negligent. These include not only wrongful acts but the casual observations made as to how we ought to correct the health and safety of the environment. For example, a CAL-OSHA compliance officer issued a safety citation for a dual-bladed band saw. The officer casually recommended that the firm place the emergency disconnect switch between the two blades. As a result of this recommendation, the switch was installed. Subsequently, a worker cut his hand off reaching in for the disconnect switch. This is now being considered as negligence on the part of the OSHA officer.

I'd like at this time to take a small departure into the area of product liability litigation. I would recommend the following textbooks for your library if you are at all involved in these actions:

1.   *Toxic Torts* [4] is an excellent textbook on how to write up suits. It suggests references and contains a wide variety of theory on various product defects. The book is an introduction to the concept of inter-

rogatories. It explains how interrogatories are demanded and what the opposing attorney expects with each question.

2. *Handling the Occupational Disease Claim* [5] is another excellent textbook that directs the reader to the personal injury action and its many theories of remedy other than the simplistics of workers' compensation.

It is helpful to understand the sequence of a trial. When charges are filed, notices are sent to the parties of the suit that a case is pending. If you receive such a notice, it will be followed shortly by the "interrogatories," or lists and lists of questions directly related to the facts of the case, which you are required to answer under penalty of perjury.

If the opposing attorney fails to get sufficient information through interrogatories, he will continue his "discovery" by making specific requests for information or he will take depositions from you or other persons involved in the direct actions of the case. During this "discovery," the opposing attorney may demand specific information through a subpoena *ducas tecum* in which he may request copies of records or onsite examination of materials that are substantially related to the action in question or the injury. Expert witnesses often provide to attorneys a list of the appropriate questions to ask and the items to request during discovery.

Many persons involved in cases that affect some aspect of product liability have difficulty understanding the concept of defect. The legal definition of defect is "some aspect of a product design or action that directly relates to the injury in question." In the case of the Ford Pinto, it is interesting to note that the pursuit of information that eventually led to the decision that the product's defect had contributed to deaths and injuries resulting from Pinto crashes was effectively pursued by an Orange County attorney, Mr. Mark Robinson. He relates that he examined document after document before he found one buried memo that turned the entire case around. This memo was an obscure cost-risk benefit analysis developed by Ford's engineering staff that recommended changing the position of the gas tank in relationship to the bumper or the rear axle. These obscure memos are often the key to a case and a positive settlement or trial outcome. In some cases, attorneys instruct experts to go to a company with two secretaries, and document every copyable record in the research and development department of a chemical company. Occasionally these experts make visits to look at a piece of machinery. The attorneys must know the right questions to ask or the right answers will not be given.

The discovery process is much like a childhood game of "fish." The opposing side sits with a handful of cards, all of which you may need.

You ask, "Do you have any information on accidents?" And they respond, "No," and they are honest. The reason they can honestly refuse to provide information on accidents is in that particular company the term used to describe "accident" is labeled an "incident." Some companies may call them "happenings." The job of the opposing attorney is to ferret out these words, phrases and processes used to hide information which he needs to sue. An excellent coverage of this subject is also given by Peters et al. [6].

Product liability law states that: "A manufacturer who enters into commerce with his or her product holds himself out to be an expert in that product and, as that expert, must be responsible for all lay, technical, medical and engineering data on defect potential of that product." To keep abreast of this body of knowledge, the manufacturer must use computer searching on an ongoing basis so, as new information develops, he can be alert to the reasonably foreseeable consumer's need for warning. Another interesting and extremely important precedent of law developed by *Rawlings v. Bay City Marine* is that of design responsibility. If you manufacture a machine or a technical device for another division of your company, or if you make this machine designed for use of another but to their design specifications, the Rawlings case holds an important consideration for you. This case held that you as the engineer must institute all of the safety factors in that machine, whether or not your client will accept them, writes them into the specifications, or will even pay for them; or you will be held liable for that defective piece of equipment.

CONCLUSION

I would like to recommend to professionals that many potential liability actions are lurking in the workplace. They are more real than we would choose to admit, and the primary caution to all those involved in the industrial health workplace is "watch your step". There are many areas of professional exposure, but through academic preparation, continuing education and constant striving to understand the true education of practical experience, we can arrest many of these problems before they come to injury and court. If you do not know, get help. If you do not speak the language, make sure you get a translator who does. Did the worker get a warning in a language he truly understands? Occupational disease is being discovered today because we are keeping people alive longer, so that they can enjoy their tumors. Any material can be used safely—if you know the dangers.

REFERENCES

1. Baron FM: Alternative remedies for workers other than the Occupational Safety and Health Act (OSHAct) and workers' compensation. Chapter 27, this volume.
2. Hammer W: Design defects and workers' compensation. Chapter 26, this volume.
3. Dornette W: Legal aspects of anesthesia.
4. American Trial Lawyer's Association of America: Toxic torts.
5. Baron F: Handling the occupational disease claim.
6. Peters GA et al: Asbestos source book.

DISCUSSION: CHAPTERS 33 AND 34

Russell McIntyre, Thomas S. Corbett, Joseph T. Hughes, Jr., William N. Rom, Owen B. Douglass, Jr., Kenneth S. Cohen and Douglas G. Mortensen

DR. McINTYRE: Although Dr. Cohen has said that we are an audience of attorneys, physicians and industrial hygienists, I am an ethicist, and I am in the minority. I think that the rigor that is applied to philosophy and ethics could be very helpful in sorting out some of the issues we are dealing with here.

Dr. Corbett, I appreciate your attempt to begin by defining what you mean by ethics, but I would suggest that your definition of ethics is inconsistent with the intent of your paper. The definition you presented was "that rule of conduct which is particular to a certain culture." The "golden rule" that you presented is actually stated in the reverse by most of the religious communities in the world; that is, that "one must not do to someone else what one would not want to be done to him or herself."

In terms of the intent of your paper, I think that the negative statement of the golden rule is more appropriate. The definition you have chosen is a relativistic definition of ethics that says, in effect, that one must respect all value systems with equal weight; that does not allow you to really criticize in a productive way any value system at all.

The conflict exists in the illustration that you gave from Emmanuel Kant. Kant, of course, is known for developing what is called the "categorical imperative." That is really an attempt to identify some ethical principles that transcend any cultural, societal or, perhaps, local preference.

For example, in the Nuremburg War Trials, when the Nazi generals were claiming that they were under orders and, if they didn't perform the tasks, someone else would be sent to replace them, the standard invoked by the court was that there was a higher standard of responsibility, a higher standard of ethics to be identified.

Now, what I think we need to do is to identify some of these absolute principles for doing ethics in the arena of environmental health, ones that would be consistent with the intent of your paper. I suggest that it may come from one of the formalistic theories of ethics called deontology One of the primary responsibilities a deontologist identifies in ethics is — as in the case of medical ethics in general — not to harm. It may be, as Dr. Cohen has suggested, not to injure or not to deceive.

A second principle might be "to respect the sanctity of human life," i.e., referred to as the principle of nonexploitation, or the idea that persons must always be regarded as ends in and of themselves and never as means to other ends. It seems that, if you could identify specific principles of "duty," you would go far in defining the ethical — and professional — responsibility of the physician and others working in the field of environmental and occupational health. Would you do this for us?

DR. CORBETT: In writing this discussion, I had the dilemma of, if you will, having a code of ethics for different groups, and I thought that the definition I gave at the beginning of the talk was suitable for this particular portion of the talk, where we are taking two groups of people with opposing points of view or opposing codes of ethics. The two are not totally compatible with each other in terms of their viewpoints. The point of the definition I gave was essentially a code of conduct for a particular group. What might be considered appropriate for one group is not appropriate for another. You may have two conflicting sets of ethics.

Now, what I tried to do is differentiate that from what, at least in medical ethics, would be construed as the "golden rule," whereby one would not want to do anything to others that is harmful and, again, if you see someone else doing wrongly, you would bring that to their attention and correct it. There is a dichotomy here of a code of ethics. One part lies with the opposing point of view. This is one set of ethics vs medical ethics, which have the inherent thought of being proper or appropriate in terms of not harming anyone or, again, bringing to one's attention the fact that he is harming another.

MR. HUGHES: We will probably have a hard time applying categorical imperatives to occupational health. My two questions, Dr. Corbett, are: (1) Could you give us a little summary of any health effect studies that have been done since the PBB disaster? and (2) Has there been any liability compensation resulting from the PBB disaster?

DR. CORBETT: Health studies have been performed on the victims who had the highest PBB exposure — the farmers and their families. However, they weren't conducted until two years after the PBB incident.

The major lawsuit was by a farmer. Since that time, Michigan Chemical has paid damages to a number of other farmers on different grounds from those brought up in the first case.

The first farmer had lower PBB levels, and the case was decided in favor of the chemical company. However, higher levels of PBB in the herds were considered an appropriate cause for damages to be paid, and a number of farmers have been paid, although the amount was considerably less than the total losses.

DR. ROM: The PBB litigation is still going on with some of the low-level people (below the action levels). They were trying to get state legislative support for their cause; however, most of the settlements were out of court by the company directly to the farmers and were generally for less than what they asked for. Most of the farmers were reasonably satisfied. We ended up with an environmental public health problem for which we did not have the legal or governmental institutions to handle. This all happened prior to the Toxic Substances Control Act (TSCA). The PBB case gave impetus to get that act passed. I suspect that many of the problems that were generated by this episode could have been prevented by TSCA.

DR. CORBETT: In terms of carcinogenicity, I might also add that we don't really know anything in terms of long-term exposure of PBB. The disaster started in 1973, which means we are only eight years postexposure for this large population group, which really isn't a long enough latency period to see if we are going to have an environmental incidence of cancer in the state of Michigan resulting from this exposure.

MR. DOUGLASS: I have a question for Dr. Cohen. During your presentation, you did use the phrase "beyond a reasonable shadow of a doubt" for a question to be asked by an attorney during a court case. Will you please expound on how to answer this?

DR. COHEN: The concept of "beyond a shadow of a doubt" was explained yesterday in far better legal detail than I could attempt at this point. It is mainly used in criminal proceedings when instructions are given to the jury. The concept of making a decision excluding any variability of data is the hardest thing for a scientist-type person to make. Posing a question such as, "Were all cattle in Michigan contaminated during this period?" presents the dilemma I described. Now, we all know there may have been a cloistered area of Michigan that didn't buy grain from the particular supplier contaminated with PBB and, as a scientist, you know that there must be some cow that didn't eat this material. Because of that scientific integrity, you are reluctant to say with ultimate finality that all cows in Michigan during that period were, in fact, contaminated.

MR. DOUGLASS: How do you respond to such a question?

DR. COHEN: I state my opinion, such as "In my opinion, to the best of my belief and understanding, cows in Michigan were exposed on an overall basis to the PBB material."

MR. MORTENSEN: What good does it do for an attorney to attack a health care professional or an engineer's lack of educational training in safety matters if it is the individual professional who is being sued, not the school from which he got his degree or his profession? Couldn't this backfire against an attorney in having the jury say he's not liable for designing the dangerous defective product because he didn't know any better?

DR. COHEN: There is a fairly well established legal precedent that you never directly attack an eminently qualified expert witness. I am referring to an indirect form of attack, where you're trying to indirectly characterize the background and training of that expert and not necessarily his professional liability action. In my opinion, it is important that "professionals" invest in continuing education throughout their careers, to preserve their image and capability. I suggest that attorneys do not directly attack expert witnesses since the judge and jury often become defensive. Rather, attorneys should outline how well the witness comes to the witness stand prepared by continual training and continual upgrading in his or her particular expertise. The last thing the opposing attorney wants is for the judge or the jury to consider that this poor witness has been picked on and, thereby, characterize him or her as the underdog.

CHAPTER 35

ETHICAL ISSUES OF
OCCUPATIONAL MEDICINE RESEARCH

Molly Joel Coye
   National Institute for Occupational Safety and Health
   San Francisco, California

The AOMA "Code of Ethical Conduct" [1] addresses the individual provider of occupational health services, as did the landmark New York Academy of Medicine Conference on Ethical Issues in Occupational Medicine held in 1977 [2]. The individual practitioner, in fact, has been the object of most discussions of occupational medicine ethical issues, both in professional journals and in public discussions such as ours today.

Much of the debate has centered on the problem of ethical conduct for physicians in those cases in which we do not have certain knowledge about the health effects of an occupational exposure. Because of the large—indeed, overwhelming—number of occupational exposures that have not yet been evaluated, this constitutes a major dilemma for the individual practitioner. Ironically, it is in the effort to research and evaluate these potential health effects that a further set of ethical issues has now emerged.

This is what I will address in this chapter—the ethical issues of research in an occupational setting. My comments stem from two professional experiences: as occupational medical consultant for several years to the Oil, Chemical and Atomic Workers International Union and, currently, as Medical Officer for the National Institute for Occupational Safety and Health (NIOSH) in Region IX. I will outline some of the reasons why ethical issues arise in occupational medicine research, and how they differ from issues of that more familiar research form, the clinical-therapeutic trial. I will conclude with a number of principles of ethical research that might supplement the AOMA "Code of Ethical Conduct" for practitioners and discuss the potential usefulness of a peer review system for such research.

Some unmeasurable, but probably significant, amount of worker suspicion regarding occupational medicine research is a "spillover" from the generalized suspicion of individual medical practitioners, particularly "company doctors." Several authors [3–5] have acknowledged this suspicion by workers of physicians either directly employed by or contracted to industry. The AOMA "Code of Ethical Conduct" was developed precisely because of this widespread and deep-seated problem.

Although far fewer workers are subjects of occupational medicine research than are seen by individual practitioners, this suspicion of medical practitioners clearly carries over to prejudice initial contacts between workers and researchers. Where workers have been involved in studies, their experiences have frequently been described to me in words approximating this: "Some guys came in—I guess they were doctors or something—and they said they wanted to get us breathing into this machine. I guess the chemical we use here, it hurts your lungs, maybe it causes cancer even—I don't know. So they sent us this letter, with some numbers, but I never did hear if there was a problem here in the plant. We never saw them again." These complaints are, I believe, an eloquent indictment of research as it is commonly conducted in our field.

I would like now to list, in fairly abbreviated form, some of the most important criticisms that organized labor and individual workers make of occupational medicine research.

1. First to be questioned are the motives for undertaking research in this field and for the choice of specific research topics. Workers suspect researchers—and their funding sources, which are still primarily industrial firms or trade associations—of a greater interest in discovering means to reduce absenteeism or compensation premiums than in preventing long-term or delayed-onset diseases. More simply, they believe that much research has been undertaken to disprove occupational etiologies.

2. Workers find that the risks of research are frequently not explained or are inadequately explained to them. Because these are not clinical trials in which there is a possibility of direct benefit from an experimental therapeutic, such risks are usually undertaken with only indirect societal or work-group benefit to be hoped for.

3. The individual results of medical examinations or laboratory tests frequently are not provided to the worker or to his or her physicians.

4. Perhaps most important—I will discuss this later—is the question of bias in interpretation of the results of an investigation.

5. Where the question of bias fails to arise, it is often because workers have not been informed of the broader conclusions of the investigation. (For example: Were there any elevated mercury levels found in the work group? Did the investigators decide there was a relationship between silica exposure and lung disease in this setting?)

6. In other cases, workers are frequently ignorant of the results of an investigation because these results are disguised in our own medical language. I use the word "disguised" on purpose. No physician who has ever treated patients could fail to realize how impenetrable this language is, and the failure to translate to lay language for workers is a deliberate sin of omission.

7. Most workers have good reason to question the usefulness of occupational medicine research as a whole. What use, they argue, is an increase in scientific knowledge or understanding of health effects if the workers who are actually exposed to this physical or chemical agent never hear about the newly elucidated effects? The field of occupational medicine is replete with cases of occupational diseases — known or soundly suspected — that never were publicized or acted on by responsible medical groups.

8. Finally, a particularly difficult problem: the failure to disclose a health hazard or occupational disease found incidental to the course of a separate piece of research. As Morton pointed out in his discussion of the responsibility to report occupational health risks (Chapter 6), it is frequently difficult for industrially employed physicians to communicate their recognition of such hazards and risk even among themselves, much less directly to the workers, when management fails to support them.

This is a brief list, but it is disturbing. Some of these problems, of course, are much easier to deal with than others. If we want to make the results of individual tests available to workers or their physicians, that is not difficult, and we could even learn to make the results of our examination and studies intelligible to lay persons.

There is another more difficult issue, however, that I would like to spend more time on. This is the claim of objectivity that we make to the subjects of our research. This claim is valued as a way of describing as "nonpolitical" or "unbiased" the conclusions of a study and as a "nonpolitical" way of resolving conflicts between labor and industry. However, our practice is science *and* art. The goal of our science is to diminish the realm of uncertainty, the gray areas where only by art and our informed personal convictions can we make a decision. In these areas, where we must decide the line dividing normal from abnormal populations — defining our own tradeoff between false negatives and false positives — we do ourselves and our patients a disservice to claim an unpassive, impartial objectivity. Workers know quite well — and this has been repeatedly confirmed by health researchers from Yerushalmy's early studies [7] — that individual physicians have characteristic patterns of decision-making in these gray areas.

For example, several workers [8–12] have all reviewed inter- and intra-observer variability in readings of chest films for coal workers' pneumoconiosis, finding marked variability (or put more significantly, lack of agreement) among even expert radiologists trained in the reading of pneumoconiosis films. Substandard films were shown not to be a significant explanatory factor, and lack of experience led most frequently to underreading of disease. This brought into question the competence of medical diagnostic techniques in deciding compensation cases on which the livelihood of many workers and their survivors would depend.

In 1973 Morgan et al. applied the techniques of decision analysis to this problem in medical decision-making [13]. They found that the index of detectability curves for all readers were substantially parallel, indicating that the same criteria were used to distinguish between the two populations; only the placement of the actual cutoff — that is, the tradeoff between false positive and false negative — differed. For each reader there was a characteristic placement of this cutoff. The decision between false positive and false negative, then, is a pure value decision; we weighed the risk of disease missed against the risk of unnecessary job loss, the risk of groundlessly provoking anxiety against the lost chance to warn and educate workers about an unsafe or unhealthy working condition. No random variation, but our values and ethics determine where we draw this line. It is crucial here to understand that we are not talking of "bias" in the more common or direct sense, in which a researcher would knowingly alter his or her data or interpretations to conform to external pressure. The process is more subtle: first, in the selection of a physician or research group according to their particular pattern of scientific interpretations (the powerful effect of selective funding of research) and, secondly, in peer pressure to bring one's interpretations into conformity with "current wisdom."

Although workers are not aware that as elegant a study as Morgan's exists to document their suspicions, they are fully cognizant of the phenomenon. Yet, researchers and medical practitioners continue to claim impartial, scientific objectivity — and deny the phenomenon we have just described. Understandably, workers are left with a troubling and generalized suspicion of occupational medicine research as well as practice.

What ethical precepts can we suggest for application in the area of occupational medicine research? There are, I believe, three major principles that we can use well in this situation:

First — and I hope, least controversial — is the ethical duty of professional competence. Very simply, if we undertake to research the health effects of occupational exposures, we owe it to the workers potentially at risk to perform that research with state-of-the-art excellence in epidemiology, biostatistics, and general preventive medicine skills, including worker education.

Second, I would argue that we have an ethical duty to reconsider and redefine the concept of "proof" that has evolved with the clinical therapeutic trial. In clinical trials of new drugs, surgical procedures, and so on, the principle of *primum non nocere* dictates that we be most careful not to make the mistake of falsely concluding that our therapy has improved the patient's outcome. This, the Type I error, consists of falsely rejecting the null hypothesis, i.e., the hypothesis that the treatment made no difference. We all learn in medical school to hunt for the "p value," but the "p value" tells us only about the probability of making a Type I error. This use of the "p value" stems from our determination as physicians not to accept or use an inefficacious or dangerous form of therapy. In the field of occupational medicine research, however, our primary concern must be exactly the opposite—to avoid the Type II error, or to falsely conclude that there is no difference between two groups (i.e., falsely accepting the null hypothesis).

Practically, what this means is that in a clinical trial a 10% chance that the observed difference between study and control groups is due only to chance—there is no real difference between the therapy and its absence; and this is often very good grounds for rejecting that therapy (particularly since most therapies have appreciable side effects). In the occupational medicine setting, however, we are usually trying to find out whether a particular disease or symptom complex is associated with one or more given exposures. Our greatest concern must be that we not falsely conclude that there is no association when, in fact, an association does exist; and the "p value" does not tell us what our chances of making this error are. If we should commit the Type I error of finding an association where none exists, there will be far less immediate threat to our patient's health than if we fail to find a true association and to urge appropriate protection.

Third, I believe that we should borrow from our colleagues in other fields of medicine a major institutional means of defining and maintaining ethical standards in medical research: the process of peer review. Until the 1960s, there were no regulations in this country that protected the rights of the subjects of medical research. In 1966 the National Institutes of Health (NIH) issued guidelines requiring prior review of the judgment of principal investigators regarding the welfare of human beings used in chemical and behavioral studies.

As a result, for the past 14 years medical research funded by NIH has been routinely subjected to early review by peers, primarily for what we would consider ethical problems. Board members are publicly identified by name, occupation or position, and relevant background. A major study of the performance of the Institutional Review Boards (IRB) was reported by Grey et al. in 1978, in which they found that board members and investigators were "virtually unanimous in agreeing that the review procedure helps to protect the rights and welfare of human subjects" [14].

Because much of occupational medicine research is conducted outside of university settings, an equivalent board might be established by the profession to approve study protocols developed in industrial, private consulting and other settings. I should point out, however, that most university medical center review boards are, in fact, quite ill-prepared to deal with issues particular to occupational medicine research, and therefore, even research conducted in a university setting could benefit by review of a more specialized board, such as that which I have proposed here.

I would like to conclude with a very informal proposal of some further ethical principles, perhaps better called operational guidelines, which could form the basis for an ethical code for occupational medicine research and for review and approval of research projects. Where labor is unorganized, some of these suggestions would be more difficult to implement; in such cases, at least full posting of the relevant information should be required.

1. Whenever possible, there should be joint labor-management agreement on the nature of the research project, on the protocol, and on the planned use and distribution of the study results.
2. All research proposals should include informed consent protocols providing for the written, knowledgeable and voluntary consent of the workers involved, and including the right of the worker to obtain answers to any questions he or she may have about the study or about known effects of the hazards to be studied.
3. All research protocols should disclose the source and amount of proposed funding.
4. All research proposals should outline a means of providing test results to the individual worker or his or her designated physician only; and of providing the group results without identifiers to the employer and to the employee representative in lay language. All researchers must be explicitly free to and encouraged to publish their findings in professional, peer-reviewed journals.
5. All longitudinal studies should be reviewed periodically, and protocols should include criteria for the possible termination of studies when significant health risks are identified in the early stages.
6. All research protocols should explicitly include a plan for the education of potentially affected workers about potential health hazards identi-

fied in the design protocol. Others have discussed the nature of the physician's legal duty to inform workers of health hazards on the job [15]; this is also an ethical duty and should be treated as such by the review boards.

7. All research protocols should explicitly outline the intention of and procedures whereby the investigators will notify both the employer and workers of any potential health hazards discovered incidentally to the investigation. Wegman [16] and Scala [17] discussed the physicians' and toxicologists' duties to warn at the New York Academy of Sciences Conferences in 1977, and Morton has called for the establishment of a review and mediation program to which occupational physicians could turn for professional assistance when they wish to report occupational hazard information but encounter constraints by management [6]. The peer review system suggested here might, therefore, serve a second important function—as a source of professional guidance and advice when confronted with these difficult situations and of support of actions taken with their concurrence.

REFERENCES

1. Code of Ethical Conduct for Physicians Providing Occupational Medical Services: J Occup Med 18(8), 1976.
2. Conference on Ethics of Occupational Medicine: Bull New York Acad Med 54(8), 1978.
3. Colwell MO: Elements of credibility. J Occup Med 18(9):591–94, 1976.
4. Ackerman R: The organizational environment and ethical conduct in occupational medicine: A perspective. Bull New York Acad Med 54(8)707–14, 1978.
5. Tabershaw I: Whose "agent" is the occupational physician? Arch Environ Health 30(8):415–16, 1975.
6. Morton W: The responsibility to report occupational health risks. J Occup Med 19(4):258, 1977.
7. Yerushalmy J: Reliability of chest radiology in the diagnosis of pulmonary lesions. Am J. Surg 89(1):231–44, 1955.
8. Williams J, Moller G: Solitary mass in the lungs of coal miners. Am J Roent 117(4):765–70, 1973.
9. Williams R, Hugh-Jones P: The radiological diagnosis of asbestosis. Thorax 15:103–08, 1960.
10. Fletcher CM, Oldham PD: The problem of consistent radiological diagnosis in coal miners' pneumoconiosis. Br J Ind Med 1949:6, 1968.
11. Trout ED, Jacobson G, Moore RT, Shoub EP: Analysis of the rejection rate of chest radiographs obtained during the coal miners' black lung program. Radiology 109(1):25–27, 1973.
12. Reger RB, Morgan WKC: On the factors influencing consistency in the radiological diagnosis of pneumoconiosis. Am Rev Resp Dis 102:905–16, 1970.

13.   Morgan RH, Donner MD, Gayler BW, Margulies SL, Rao PS, Wheeler PS: Decision processes and observer error in the diagnosis of pneumoconiosis by chest roentgenography. Am J Roent 117(4):757-64, 1973.
14.   Gray B, Cooke RA, Tannenbaum AS: Research involving human subjects. Science 201:1094-1101, September 22, 1978.
15.   Weiner P: Workers' rights to health and safety information. Oral presentation for the Department of Industrial Relations, January 1979.
16.   Wegman D: The duty to report hazards. Bull New York Acad Med 54(8): 789-94, 1978.
17.   Scala R: The duty to report hazards. Bull New York Acad Med 54(8): 774-81, 1978.

DISCUSSION: CHAPTER 35

T. Lamont Guidotti, Molly Joel Coye, Leslie I. Boden,
Russell McIntyre and Victor E. Archer

DR. GUIDOTTI: I think you made a very important proposal on the peer review system, and I'm wondering if you could elaborate on how it would work in terms of review of proposals for occupational health research within the private sector.

DR. COYE: I would imagine that the ERC or, in the case of California, the state-supported occupational health centers that are already active in occupational medicine research could offer this service. They might offer to review protocols and, also, to review final reports before publication as well. This would be an opportunity for academically based groups to be more involved and learn more about the practical problems of field trials and, in many cases, it would offer industry groups an opportunity to initiate their projects with less chance of future charges of bias based simply on criticism of their industry finding.

DR. GUIDOTTI: Would this peer review board be exclusively composed of academicians, or would nonacademicians serve on the board as well?

DR. COYE: The institutional review boards under NIH include lay members. For the specialized boards I am recommending, I think it would be useful to also have at least a couple of labor and industry representatives. There are problems, among them obviously that of confidentiality, but I think one purpose of these boards is to defend protocols against future criticisms, because the broader the consensus initially, the better your chance of successfully defending it.

DR. BODEN: I've talked with a couple of people today about what I think is a particular dilemma in a certain kind of research. For example, say someone is investigating the relationship between exposure to a potential carcinogen and the presence of urinary mutagens by the Ames

447

test. Positive findings would indicate the possibility of some prior exposure to carcinogens with a further possibility of cancer developing sometime in the future. The dilemma researchers then face—and maybe these projects should not be undertaken in the first place for this reason—is when you find a group of people with abnormal results, how do you, or do you, inform those people? Do you suggest that they avoid this exposure, even though you don't know what it means, or do you just abandon this kind of test? It seems to me a very serious dilemma.

DR. COYE: Speaking more from my experience with OCAW than with NIOSH, I don't see any reason not to go ahead with those types of studies; but I do believe strongly that, before you initiate the study, you should discuss the interpretation of possible outcomes with the workers very thoroughly.

In my experience, if workers are told, "We think this is an important research problem, but we don't know for sure what the results will mean in terms of your own future health," they are usually willing to enter into the study. You will also have done a great deal to allay their anxieties. If you wait until after you have determined mutagenic results, for example, and then try to tell them this while telling them that you don't know what it means, you are going to have a hard time. So I don't see any reason not to do these studies, but I think you have to have full discussions of possible results with the workers ahead of time.

The second point is that one group of workers who learn, for example, that they have a higher incidence of urinary mutagenesis than another group should make their own decisions about whether they want to seek protection from exposures. It should not be a question of our decision to get rid of workers who have mutagenic results. The workers have to decide whether to seek more protection, given that we don't know what potential health effect may exist in the future.

DR. BODEN: I try to imagine myself as an informed lay person in this area being given results that show some abnormality was found. I would be faced with tremendous uncertainty about whether I ought to try to seek protection that might prevent my having cancer at some time in the future or be more concerned about the immediate effect threatening the security of my job.

DR. COYE: That's exactly why my first suggestion was that there should be joint union-management agreement on the protocol, because the union should have the chance to say, "We don't want these answers. We don't know what to do with them. We don't want to live with them." They should have the opportunity of rejecting a research project looking at the incidence of mutagenesis in their urine; but, as a researcher, you should not deny a group of workers the right to decide this for themselves. They should have the chance to decide whether they want to assume the burden of knowledge.

DR. McINTYRE: I've been on our Institutional Review Board for three years now. One of our concerns, although we do not want to deal with it, is monitoring the research as it goes on. The only thing we do is review the reapplication that comes in on an annual basis. We have really no way of knowing what is going on at the site.

In your suggestion of developing some basis for longitudinal review and interventions, would you see it as the responsibility of the review board to do some kind of monitoring or policing and then having the right to interrupt the protocol if, in fact, they are able to determine that it might be hazardous to the workers?

DR. COYE: One of the elements you might include in such a system would be a staging system for longitudinal studies; in other words, the identification of a series of nodal points. For example, when the first baseline studies are completed—even though in a longitudinal protocol these results would not be published until the entire study is completed—the researchers might be asked to make a summary report to the Institutional Review Board. This would not require continual monitoring on the part of the IRB, but would enable them to review the study periodically.

In the lifetime of most studies, we are only talking about two or three nodal points. I feel that this is a fairly reasonable request to make of a research group.

DR. ARCHER: If I were to plan an experiment on human volunteers to test the human carcinogenicity of a new chemical that has a positive Ames test, the experiment would almost certainly be unethical. We would not be permitted to do it. Yet, if you were a chemical manufacturer, you could expose hundreds of workers and perhaps thousands of consumers to the same chemical, and the question of ethics would not even be raised. Now, if this appears to be somewhat of a divergence here in attitude, how would you interpret this difference?

DR. COYE: Unfortunately, what I have suggested only covers cases of industry-supported research, not what industry routinely exposes their workers or the consumers to. I don't think that the purview of an Institutional Review Board could go further than to review studies. Of course, this does not touch on what you have pointed out. What you are describing is a major contradiction, and we do need to begin to address the problem of ongoing routine worker or consumer exposure to potential carcinogens or other toxic effects.

CHAPTER 36

ETHICAL ISSUES IN INDUSTRIAL HYGIENE IN THE 1980s

John D. Yoder, ScD
ITT
New York, New York

Ethical issues for industrial hygiene in the 1980s — a difficult topic.

Before I go any further, I would like to emphasize that I am not a specialist in ethics. I am not a philosopher in ethics. I am an industrial hygienist in industry who is attempting to do his job: to protect the health of employees; to prevent occupational illnesses; and to ensure that these things are done in an ethical way.

How do we define ethics? Webster [1] defines ethics in various ways, such as:

2a: a group of moral principles or set of values
2c: the principles of conduct governing an individual or profession

Ethical is defined as:

3b: conforming to professionally endorsed principles and practices

These are broad definitions, open to too many interpretations.

Do ethics change with time? When I brought the subject of this presentation up with a number of industrial hygienists, this was often the question that was asked — "Do ethics change with time?"

Any group that calls itself professional must be concerned with ethics and must attempt to define ethical conduct for the group. And it seems that we need a more definitive concept of ethics than the dictionary definitions just mentioned. This is particularly critical for industrial hygienists, inasmuch as the health of entire work forces depends on their decisions and recommendations.

People in the field of industrial hygiene generally understand ethics as a concept, but few can define ethical conduct. One thing is clear, however; industrial hygienists are very much concerned with ethics and with what is ethical conduct.

Just as in all fields, industrial hygienists have found that the more one studies ethical concerns, the more avenues one sees for consideration and exploration.

However, before discussing ethical issues for industrial hygiene in the 1980s, I believe it is essential to attempt a bit of crystal-balling as to what is forthcoming in the 1980s.

1.   What will be the trends in industry and industrial technology?
2.   What will be the advances in the industrial hygiene field?
3.   How will all these factors affect the industrial hygienist and the industrial hygienist's work?

INDUSTRIAL DEVELOPMENTS OF SIGNIFICANCE TO THE INDUSTRIAL HYGIENIST

There are a number of important developments that we are likely to see in industry over the next few years. Many of these will be of major significance to the industrial hygienist. I will review just a few.

A dramatic increase in the use of robots or machine-assisted work will have a major impact on industrial technology in the next few years. This is of special interest to the industrial hygienist in that it is expected these robots will first be used in the more onerous and undesirable jobs. The obvious consequence is that some of the worst jobs will no longer be of concern to the industrial hygienist.

A second major development will be the revolution in microelectronics, which will result in many more closed-loop, automatic types of processes run from environmentally controlled, computerized control rooms. This will ensure cleaner operations with limited direct labor involvement that will be of less concern to the industrial hygienist.

It is predicted that almost all industry will become more quality-conscious as a result of the magnificent achievements of the Japanese and their quality programs. The end products of the increased attention to detail, better machinery, and improved processes will be better-kept, cleaner units and fewer occupational exposures.

But there is a "sleeper" in robotics and computer-controlled high quality processes. While fewer individuals will come in contact with the actual processes, the equipment will still need to be maintained and, as is true in many situations, the exposures of most concern are often those associated with the maintenance operations. It is in maintenance that workers are most heavily exposed to the materials used in the process. Here is where a host of exposures, often to unknown mixtures, develop.

There is one additional factor that should be mentioned and that is the influence of the increased concern for energy conservation. This concern will result in, again, better quality equipment and better process control on the units consuming energy sources and, thus, fewer toxic exposures. However, there is a potential hazard inherent in these energy conservation measures. Energy-conserving buildings are designed to be much tighter. Where in the past contaminants were likely diluted by infiltration of air from other areas, this will be less and less likely to occur in energy-efficient buildings. As a consequence, although we can expect better process controls and better equipment, we will find that the materials that are released will tend to accumulate unless attention is given to this point.

As indicated, these are but a few of the pending changes in technology, many of which are likely to result in industrial operations of less gross concern to the industrial hygienist. Will industrial hygiene become an obsolete function? Not on your life! There are too many factors with health implications and there are too many coming down the road. What will be needed, however, will be increased professionalism on the part of industrial hygienists and better training in these areas.

INDUSTRIAL HYGIENE IN THE 1980S

I believe it goes without saying that there will be better, more sophisticated (and, of course, more expensive) instrumentation available to the industrial hygienist in the 1980s. This will allow a better, more complete, more accurate evaluation of employee exposures and should be accompanied by better techniques for developing and retaining necessary records.

Some industrial hygienists foresee increasing use of fixed-location samplers to provide a continuous real-time indication throughout the day of workplace contaminant concentrations. I say, why not provide these improved continuous monitoring devices in a personal monitor on the employee? This should give us better estimates of exposures than do the fixed samplers.

In another area, I am encouraged that (although admittedly there will probably always be an infinity of things that sorely need to be considered) we are making major advances toward perfecting our techniques for discovering and evaluating occupational health hazards. The techniques are becoming more sophisticated, more sensitive and more accurate.

In the 1980s we will see refined epidemiologic techniques that will enable us to pinpoint earlier and with greater reliability potential new areas for concern in occupational health. However, to maximize the use of these techniques, the industrial hygienist will have to record and store the data in such a way that the information gathered can be optimally useful in these studies.

As discussed previously, many of the obvious health hazards will be designed out of industrial processes or will be handled by robotic types of equipment. Our attention will focus on evaluation of more subtle types of exposures. This might include exposures to carcinogens, exposures to microwave and RF radiation, and exposures that cause behavioral toxicology problems—those exposures that may not cause outright illness but can reduce efficiency and attentiveness.

Many industrial hygienists anticipate a larger role in management in the future. With the increasing concern and attention to health hazards by company management, there will be an increasing realization of the need for and the value of industrial hygiene input into a wide variety of business decisions. The industrial hygienist will have a more significant role in the planning stages, which will ensure that facilities are designed to eliminate or minimize potential health problems. Control will be obtained by system design. It will be through industrial hygiene expertise that we will determine the most effective combinations of various control techniques and the levels to which we will need to control exposures.

Through this expanded role, we will be able to engineer out as many of the human factors as possible, thus eliminating the need for workers to take active steps, including the use of bulky personal protective equipment, to protect themselves.

Will our role change? It may. It probably will. For example, I would expect that the fully trained industrial hygienist will be less and less involved with field monitoring. This work will be done primarily by technicians, while the fully trained professional hygienist will be involved in the management decision-making process relating to the health of the employees and to compliance programs to meet government requirements.

ETHICAL ISSUES IN THE 1980s

Now then, what of the ethical issues facing the industrial hygienist in the 1980s?

It seems clear that we will be facing fewer gross overexposures than we have experienced at times in the past. Ethical action in these past situations was relatively easy to define. Many of the issues facing us in the 1980s, however, will be more subtle, and it will be more difficult to clearly define ethical boundaries, and there will probably be more disagreement on what is ethical. It is difficult to foresee all the challenges that are forthcoming in this area.

One issue that is clear, however, is that of the worker's need and the worker's right to know what he or she is working with and the health consequences of exposure to those materials. Health and safety professionals agree that workers can best protect themselves if they are informed of the possible hazards. It is in this area that I predict there will be some major changes in the *perception* as to what constitutes ethical conduct.

When I first entered the field of industrial hygiene, it was felt that the company could best look after the health of the worker, and that it was not necessary for the worker to know all about the materials with which he or she was working. It is still true that the company can do the best job of protecting the health of the worker because the worker cannot be expected to understand some of the technicalities involved in providing optimum protection, and it is clearly the responsibility of the company to provide this protection. However, it is also clear that the worker should be informed of the material with which he or she is working and the consequences of exposure to these materials. The question, which I feel to be more practical than ethical, is how much information does the worker need and how should it be provided? There is an additional dilemma when materials or processes are trade secrets.

There will be significant new findings with respect to health aspects of occupational exposures over the next decade. How do we handle this information? As scientists, I am sure we would prefer to wait for additional data to enable us to make better decisions; however, for a scientist, are there ever enough data?

Furthermore, OSHA and other regulatory agencies may be slow in reacting to new information on health concerns. In such instances, we cannot wait until all the data have been compiled and analyzed. We must use our professional judgment to decide when action must be taken and to decide what actions to take. This is an area in which some of the biggest differences of opinion as to what is necessary and, consequently, what is ethical will surface.

A related major, but more specific, area of concern involves reproductive hazards. Many of the decisions as to how to handle materials that may cause reproductive disorders will be made by others, i.e., the toxicologist and the physician; but it is essential that the industrial hygienist be part of the team that considers these hazards.

Certain unions have claimed that they want these hazards eliminated — not the job and not the worker. However, while this is certainly desirable, it is not realistic in all cases. In the case of the female worker and her unborn child, there must be determinations made as to what levels of various toxic substances cause reproductive hazards that will harm the fetus. In some cases, a few molecules of a particular substance are thought to cause an effect during the most sensitive stages of the growth of the fetus. In all probability, we will not, in the foreseeable future, be sure of safe levels for many of the presumed toxic materials for either females or males.

Is it ethical then for certain individuals to say, "Let these people work," if we don't know the levels at which problems occur? To adequately control exposures in such situations, we are forced into a zero exposure concept, and a zero exposure concept is unworkable in most cases. Thus, it seems to me inevitable that, for good medical reasons and to protect the unborn and future generations, certain workers will have to be excluded from work with certain materials.

In still another area, there are those in the legal profession who have discovered the attractive and profitable possibilities associated with litigation for health effects arising from employment, whether or not these effects are actual, alleged, or imagined. Clearly, there will be far more litigation during the 1980s. Consequently, the industrial hygienist will need to be more aware of and familiar with legal considerations. As a corollary, the hygienist must also be concerned with ethical aspects of conduct in litigation. This is not at all new for many industrial hygienists, but the type of litigation and the extent of litigation may cause new concerns.

I am sure I have not considered nearly all the issues that the industrial hygienist will face in the 1980s. I am afraid my crystal ball just does not have that range.

Of course, in many areas we do not need a crystal ball. In the 1980s, industrial hygienists will face many of the same ethical issues they have been facing for years. For example, how do we handle those people who refuse to protect themselves from health hazards or who will not take medical examinations? When does the industrial hygienist "blow the whistle" on the company for which he or she is working? Should you leave a company that is not doing what you feel it should, or is it better to work within the company in a position in which you feel you can be most effective in the long run in protecting the health of the employee?

HANDLING THE ETHICAL ISSUES

Now that we have considered several of the major issues that the industrial hygienist may or will be facing in the 1980s, let us look at the reference points we have to handle those ethical questions that arise.

A code of ethics is a starting point. Several associated groups have codes of ethics — physicians, engineers and, I am sure, many others. But the code of ethics that applies for the practice of industrial hygiene specifically is that developed several years ago by the American Academy of Industrial Hygiene and is included here as the Appendix.

The code includes a purpose and four areas of responsibility — professional responsibility, responsibility to employees, responsibility to employer and clients, and responsibility to the public. It clearly states that the primary responsibility of the industrial hygienist is the protection of the health of the employee.

Earlier, I noted that it appears the simple dictionary definitions of ethics and ethical are too broad, that they are open to too many different interpretations, and that we need a more specific delineation of ethical conduct. In addition, throughout this chapter, I have asked — but not answered — a number of questions as to what is appropriate ethical conduct.

I believe that the Academy's code of ethics gives us the more specific delineation of ethical conduct that we need.

And I believe that the code, if not supplying the answers to all our questions, will provide at least a good starting point in dealing with ethical considerations in the practice of industrial hygiene.

SUMMARY

I have not presumed to define what is ethical in specific situations. I have attempted to project what may be coming in the 1980s for industry and for industrial hygiene. I have reviewed some of the issues in which ethical considerations will be involved. Some of these will be the same issues we have faced for years; others will be new and will probably generate considerable debate.

Furthermore, it seems realistic to accept that the concept of what is ethical—the perception of what is ethical—may change. However, even though many things about us will change—industrial activities, industrial technology, the equipment we work with, the demands on our profession—I believe the question posed earlier, "Do ethics change with time?" can be answered that, in reality, they do not. In other words, our perception of what is ethical may change; we may have to address new problems; but the basics of ethical conduct will not change. And, as for the industrial hygienist, I feel the code of ethics of the American Academy of Industrial Hygiene gives us (with some exceptions that require special consideration) the guidelines we need in attempting to do this most important work in an ethical manner.

REFERENCE

1. Webster's third new international dictionary of the English language, unabridged. Springfield, MA, G. & C. Merriam Company, 1971, p. 780.

APPENDIX: THE AMERICAN ACADEMY OF INDUSTRIAL HYGIENE CODE OF ETHICS FOR THE PROFESSIONAL PRACTICE OF INDUSTRIAL HYGIENE

Purpose

This code provides standards of ethical conduct to be followed by Industrial Hygienists as they strive for the goals of protecting employees' health, improving the work environment, and advancing the quality of the profession. Industrial Hygienists have the responsibility to practice their profession in an objective manner following recognized principles of industrial hygiene, realizing that the lives, health, and welfare of individuals may be dependent upon their professional judgment.

Professional Responsibility

1. Maintain the highest level of integrity and professional competence.
2. Be objective in the application of recognized scientific methods and in the interpretation of findings.
3. Promote industrial hygiene as a professional discipline.

4. Disseminate scientific knowledge for the benefit of employees, society, and the profession.
5. Protect confidential information.
6. Avoid circumstances where compromise of professional judgment or conflict of interest may arise.

Responsibility to Employees

1. Recognize that the primary responsibility of the Industrial Hygienist is to protect the health of employees.
2. Maintain an objective attitude toward the recognition, evaluation, and control of health hazards regardless of external influences, realizing that the health and welfare of workers and others may depend upon the Industrial Hygienist's professional judgment.
3. Counsel employees regarding health hazards and the necessary precautions to avoid adverse health effects.

Responsibility to Employers and Clients

1. Act responsibly in the application of industrial hygiene principles toward the attainment of healthful working environments.
2. Respect confidences, advise honestly, and report findings and recommendations accurately.
3. Manage and administer professional services to ensure maintenance of accurate records to provide documentation and accountability in support of findings and conclusions.
4. Hold responsibilities to the employer or client subservient to the ultimate responsibility to protect the health of employees.

Responsibility to the Public

1. Report factually on industrial hygiene matters of public concern.
2. State professional opinions founded on adequate knowledge and clearly identified as such.

DISCUSSION: CHAPTER 36

Victor E. Archer, John D. Yoder, Michael Holthouser, Jon R. Swanson,
Joseph T. Hughes, Jr. and Marvin A. Schneiderman

DR. ARCHER: I have been acquainted with several corporate efforts to inform workers of hazards of their employment. After explaining the biological and health effects that might occur, these statements always end with the comment that exposure at this company is so low that none of the bad effects will occur. This statement was made regardless of the actual level of exposures.

I can understand that company lawyers would not permit companies to predict harmful effects, but what would you say about the ethics of such statements?

DR. YODER: I think such ethics are very poor. There's no question about that. If you have exposures that can cause problems, you have to state that and define what you are going to do about it.

DR. HOLTHOUSER: Buildings are currently being constructed with energy conservation in mind, which may be affecting occupational health. For example, in colder climates, heated air is being recirculated rather than exhausted. Do you have any comments on the implication of such practices?

DR. YODER: In many industrial situations, we find poorly designed ventilation systems; ventilation systems that use much more air than they should be using. One way of getting around this is to have a more efficient design; for example, local exhaust as opposed to some general ventilation techniques. Local exhaust uses much less air to control hazards and, thus, reduces the amount of air needing to be tempered. There are many situations in which we used to say we will not recycle air. Now, however, because of energy considerations, we are reconsidering this stand. A number of studies have been done on how we can recycle the air and still avoid adverse health effects. There have been many improvements in techniques to ensure that the contaminants we are concerned with will be removed prior to recirculation.

DR. SWANSON: We have maybe ten or so substances to which we have employees exposed to only trace quantities—well below current standards. In these situations, I think the industrial hygienist should ethically do his best to quantify exposures to such trace substances for use in the future by other researchers. Do you agree?

DR. YODER: Yes, I do. I think we should be reporting all data developed because we never know what future research will show. We might need these data for future epidemiological investigations.

DR. SWANSON: In 1966 Henry K. Beecher, in a very important article on clinical research, identified some of the coercive pressures brought to bear on those in research positions, reflected in the accuracy of reports. There certainly have been a number of suggestions by occupational physicians about the kind of pressures brought to bear on them due to their being employed by industry. I'm wondering, in terms of the code that you have presented, has there been any track record in terms of how industry has received this? How insulated are you now as an industrial hygienist because of this document, or is your own position and that of others still as vulnerable to those coercive pressures and to the possibility of being discharged by the corporation for the whistle blowing that has been alluded to?

DR. YODER: I have not heard of any situations in which we have had problems with the code of ethics. There have been a couple of inquiries from people who were questioning the conduct of some industrial hygienist, but it has never gotten to the point of review by the American Board of Industrial Hygiene. The code is relatively new. I know a lot of the people use it to define to others what their conduct will be, but I have not heard of any problems.

MR. HUGHES: I find it interesting that you said that ethics never change. I wondered what you think about that in light of the "right to know" principle. It is my feeling that corporate ethics in general have undergone a real change in the last ten years in the whole concept of worker exposure, monitoring, and notification as to what is going on. There is a long history of some guinea pig–like experiments in industrial settings, which may have been done from lack of knowledge, but it appears that there has been a real change in ethics with respect to letting people know what they are exposed to.

DR. YODER: I consider that a change in the perception of what is ethical conduct. I think it is well agreed now that the worker should know these things, but a while back it was just not considered necessary to inform the worker.

The company was considered to have the best capabilities for controlling exposures and went about its business of doing that.

MR. HUGHES: Do you think in hindsight we would consider that unethical?

DR. YODER: Yes, in terms of today, but no in terms of that time and situation.

MR. HUGHES: That seems unethical to me, but that is a difference of opinion.

DR. YODER: I have no problems with the way things were handled back then, but things have changed.

DR. SCHNEIDERMAN: In view of Dr. Swanson's comment and your reply, I would like to read something from this morning's newspaper if I may. It is a very short item.

It is from Baltimore from the Associated Press, and it says, "The state hearing examiner has fined Bethlehem Steel Corporation $400 for violating safety rules by failing to test worker conditions of 12 employees who handled asbestos, an agent that has been linked to cancer for several years. Mr. 'So-and-So', the hearing examiner, wrote in his decision that the company failed to monitor the employees' working conditions. Without the monitoring, he said, 'There appears to be a substantial probability that death or serious physical harm could result.'"

CHAPTER 37

ETHICAL ISSUES FOR OCCUPATIONAL HEALTH NURSING

Mary E. Rahjes, RN, MPH
Colorado Department of Health
Denver, Colorado

Professions develop codes of ethical conduct to indicate to their members and to society that they accept responsibility for the regulation of their professional practice and behavior. These codes are based on the individual and collective philosophies of past and present members of the profession, and they outline what the members are able to practice, what they are expected to practice and how they prefer to practice. The ability of the individual to practice depends on his or her knowledge and professional competence. How the individual is expected to practice is determined by social, moral and political climates. However, because all of these factors are constantly changing, professions must periodically reexamine their beliefs and revise ethical standards.

It is difficult to differentiate between ethical, moral and legal responsibilities. The ethical and moral issues affect one's conscience, and legal issues affect one's liability.

State nurse practice acts are examples of the translation of ethical codes of conduct into statutes governing the practice of nursing. These practice acts are more prescriptive than are ethical codes in the regulation of personal and professional behavior. Codes of ethics in health professions focus primarily on the professional's accountability to the patient, to the client or employee, to the profession, and to the public. Nurse practice acts attempt to define the practice of nursing and set parameters, including minimum expectations, by which practice can be measured and judgments made.

It is difficult to define the practice of occupational health nursing and to achieve a consensus for that definition. The role of the occupational health nurse is rapidly and constantly changing, thereby creating new and increasingly complex demands. The evolution and expansion of the role results in a continuous process of decision-making. There is also a demand for an increasing level of intellectual capacity in the decision-making process. In addition, there are inherent risks that accompany change. Risk-taking is almost as common as decision-making, although one may make a decision *not* to take a risk. The continuing demand for decision-making and performance with increased intellectual quality provides the basis for one of the most important ethical considerations in occupational health nursing—the need to acquire adequate knowledge and skills and to maintain professional competence. We as nurses must be prepared to make the best decisions and to conduct our practice in the most competent manner. We must be able to decide when we need help to perform our duties and when to share the responsibility. We must also be able to assess the limits of our knowledge and skills before accepting responsibility.

It is difficult to achieve and maintain professional competence in occupational health nursing. The first problem is in basic nursing education. There are several levels of basic education that cause fragmentation in the nursing profession and create confusion among other health professionals. Any person who completes an accredited associate degree nursing program of two years average length of time, a hospital-based diploma program for three years average length, or a baccalaureate degree nursing program of four to five years average length is eligible to take the state examination to become a "registered nurse" (RN). It is also extremely difficult for employers of occupational health nurses to understand the different educational criteria for these various RN programs and, therefore, to know what they can expect from their nurse employees.

The basic coursework in the associate degree and diploma programs is often weak in science, public health and mental health, areas that are so important in the practice of occupational health nursing. Very few basic programs include courses or information on occupational health. Most baccalaureate programs do offer courses in the sciences, in public health and in mental health.

Many practicing occupational health nurses completed basic nursing education in hospital-based diploma programs 20–30 years ago [1]. These nurses need opportunities to upgrade or enhance their basic education, and to continue education for their changing and expanding roles. To be adequately prepared to practice in the occupational health nursing field, the nurse should have preparation in occupational health, industrial hygiene, safety, toxicology, physiology, epidemiology (with a special emphasis on worker populations), public health statistics, physical assessment, research, health education and health administration.

It is difficult for a nurse to be adequately prepared because of the variation between industries and occupational health settings. Each industry has unique characteristics that require different or special knowledge and skills.

The recent and continuing explosion of technology and the increase in the knowledge and science of health care affects our ability to maintain professional competence. It also increases the importance of existing ethical issues and creates new ones.

An increase in health care technology usually results in a decrease in direct personal service to patients and, in occupational health, to workers. In the future, occupational health nurses will spend more time in scientific measurement and monitoring and less in personal service. Nurses will have to work diligently to offset the depersonalization that can accompany increasing technology. Ashley referred to this ethical dilemma:

> Nurses in industry need to join the movement to bring humanism back into health care. We need more humanistic approaches in both nursing and medicine. Nurses can lead the way to reforms by examining their own history and the idealistic and humanistic aspirations of early practitioners in nursing. Historically, the focus of nursing practice was in the health of the total human being—the physical, the emotional or mental, and the spiritual or value aspects influencing human responses to unhealthy levels of existence. If industrial nursing has moved away from this focus—"the health of the total human being"—it is now time to move back to it! [2]

With the increase of technology in health care, we can expect increased specialization and, in some cases, the creation of "superspecialists." Increased technology and specialization is very expensive, and large populations are needed to support these services. In the past, this cost dilemma has often been solved by the regionalization or sharing of specialized services and personnel. Only large industries can afford a full complement of special occupational health services and personnel. Most industries will have to look outside their organization for these services.

The occupational health nurse may continue to be the only or major in-plant health professional with direct and continuous contact with the worker and the environment. The worker will expect the nurse to be able to interpret what has happened or what may happen in the work environment that affects his or her health and safety. This points out another ethical issue. How much can the nurse tell the worker about known or potential hazards in the work environment? Even scientists cannot agree on the degree of potential harm and risk in many situations. Misinterpretation of information may confuse the situation for the worker and place management in unwarranted legal jeopardy. On the other hand, information withheld from the employee may place that worker in danger of losing his or her good health or life. Employees need to have all the available correct information about the potential for exposure to risks in the work environment. This will enable the worker to make decisions or choices about how much risk he or she is willing to accept.

The practice of occupational health nursing could be compared to the interesting dilemma of having two lovers. In this case, the lovers are labor and management. The situation is further complicated by the fact that each lover knows of the existence of the other. Both lovers must be convinced that, when you are with the other lover, you are always thinking of them and have their welfare uppermost in your mind at all times. Some think that it is impossible to love two lovers equally—some do not. Like all situations in life, some of us are better at it than others, and some only think they are.

Again, the nurse must have adequate knowledge and the professional competence to make personal decisions about what is important for the health and safety of the worker, to interpret that belief to management, and to gain the cooperation of the worker and management in developing policies that are comfortable for all parties concerned. If the dilemma is not resolved, the nurse must decide whether to remain in a situation that she or he considers to be unethical. The ultimate responsibility is the protection of the health of the worker.

Occupational health nurses must take extra care in protecting the confidential nature of their work. In the industrial setting, there is a greater potential for disclosure of information to those not directly involved in a patient's care or in the conduct of company business.

Not only are occupational health nurses responsible and accountable for their own judgments, decisions and actions, but they must also protect the worker from the illegal or unethical practices of other persons. This is further complicated by the fact that nurses practice in a setting that is not in the business of health. Management must be made aware of the level and quality of practice that can be expected of licensed health professionals. They must also be instructed about the conditions under which paraprofessionals, unlicensed professionals and nonprofessionals may legally or ethically practice. Nurses must report new and continued incompetent, illegal or unethical practices to appropriate authorities.

There are numerous ethical issues that have not been discussed. I have directed my presentation to the ethical responsibilities of occupational health nurses to workers, employers and other occupational health professionals, and have emphasized that nurses must constantly change and develop new ethical codes based on the changing environment in which we practice. I have reviewed the special problem of increasing scientific knowledge and technology, and our responsibility to acquire and maintain the highest level of professional competence. I have not even touched on the ethical responsibility of occupational health nurses to the nursing and occupational health professions. I would like to state that everything that affects the health of the worker affects the ethical practice of all occupational health professionals. We share a joint responsibility for protecting workers from incompetent, illegal or unethical practices by any person. We must also assist management in the development of policies for assuring a healthful working environment. These policies must protect the worker and be acceptable to the occupational health professional and to the employer.

REFERENCES

1. U.S. Department of Health, Education and Welfare: Occupational health nurses: an initial survey. Washington, DC: Department of Health, Education and Welfare, 1966 (PHS publication no. 1470).
2. Ashley JA: Reforms in nursing and health care: industrial nurses can lead the way. Occup Health Nurs 23(6):11–14, 1975.

DISCUSSION: CHAPTER 37

Joyce A. Spencer, Mary E. Rahjes,
T. Lamont Guidotti and Russell McIntyre

MS. SPENCER: I appreciate what you said about the two lovers. I think this is truly the dilemma that any of us in occupational health face. There have been some interesting alternatives to the current system of providing health care given during this week. I'm wondering if you have given any consideration to the possibility of the nurse (and perhaps this could apply to any health professional working there) being jointly hired by both labor and management, perhaps through a contractual arrangement whereby the nurse is responsible to each party equally.

MS. RAHJES: I think you could eliminate some of the dilemma by both parties sharing the salary and responsibility for employment. The dilemma, however, would still be very real with respect to whom do you owe your emotional stability and your job stability.

MS. SPENCER: I agree with you. However, I think there is collective bargaining precedence where the company picks up the tab by collective bargaining, but the union has equal say in how it is managed.

MS. RAHJES: It may change the situation because it would define the fact that you are responsible to both, but I don't think it is going to make it any easier for the nurse.

DR. GUIDOTTI: At our institution, we are establishing a new occupational health program, and I was wondering if I could perhaps ask for your views on the interpolations among occupational health programs that tend to be dominantly medicine and engineering programs and the nursing school program and, specifically, those institutions operating within an era when nursing is undergoing review of its professional roles and incentives. How does an occupational health group that does not now include nursing education awaken interest in a more traditionally

oriented nursing program that does not at present recognize the opportunities of occupational nursing?

MS. RAHJES: Having developed an occupational nursing program in
a setting that did have at least a clinical specialty in public health nursing,
I was able to incorporate occupational health into the public health curriculum. That makes it much easier. I see a relationship between the practice
of public health and occupational health that makes it easier to develop
these kinds of programs. The problem with nursing education based on
pure clinical practice for individual clients or employees is that it does not
have the focus on the total issues and problems of the population that it
should have. For instance, nurses and nurse practitioners who see only
individual employees with their individual problems, whether they be in a
clinical setting or wherever, do not have the global view that they should
have of the total problem for the worker, including his environment, his
family and many other things that affect the situation. To answer your
question about developing programs for more traditional nursing schools,
I have yet to see many nursing institutions that have similarities to another,
so it is very difficult to make a general statement about how you would do
this. It is dependent on the situation, i.e., the particular university and
where you find yourself in the scheme. I also think it's much easier if there
is a school of public health where interdisciplinary courses and experiences
are available. It is also difficult to convince some medical schools of the
need for occupational health in their curriculum.

MR. McINTYRE: Early in the 1970s there was a discussion in the
nursing literature as to whether or not nursing was a profession. If you
use the definition that Elliott Friedson used in his book *Profession of
Medicine,* he suggests that it is questionable if nursing is a profession that
controls the contents of its own work.

I'm wondering, in light of this and in terms of what happened in the
past ten-plus years, has occupational nursing evolved to the kind of
professional status that Friedson was looking for; and, second, has it
been or is it becoming so recognized in industry?

MS. RAHJES: Well, you have just asked for a course in nursing that
would take about two semesters to present. It is a dilemma and depends
on how closely you define the word "professional." Needless to say,
nurses consider themselves professionals. They accept the responsibility
for their practice, they accept the accountability for the quality of their
practice, and they also attempt to regulate the action of members of their
own profession.

I think that by establishing good nursing role models who are competent, well educated, and have appropriate experience, occupational health
nurses are able to demonstrate to management and labor their profes-

sional abilities. I don't say that in the case of every occupational health nurse this is true. I'm sure that this is true in all professions. For instance, I could also defend the practice of occupational medicine where people have different views about professional competencies and acceptance. I can remember at one time explaining to a public health physician who came from a country in the Middle East about the medical practice of public health and occupational health in this country. I explained that, on the "totem pole" of medical practice, very often the public health physician is one of the lowest, and occasionally there is one lower, the occupational health physician. This applies to occupational health nursing as well. But that is changing—and I think it is because nurses are making an attempt to change and control their work and be responsive to the needs of the workers, to the environment, and to the public. There is a good case for better public relations also.

CHAPTER 38

RESPONSIBILITIES OF THE COMPANY PHYSICIAN

Richard W. Prior, MD

General Motors Corporation
Detroit, Michigan

My task in this section on ethics is to discuss the role of the company physician and his/her responsibility to ensure that the practice of occupational medicine is done in a quality manner with strict adherence to traditional medical ethics. To the public, medical ethics is also closely associated with availability, performance and who is paying the check.

None of us would deny that the public's opinion of the typical company physician has changed greatly, particularly during the last generation, from one of almost total disdain to a position of respect. This change has come about by a process of evolution that is still going on. If we think about it, there are many subtle, and some not so subtle, examples of this evolution. When did industrial medicine and surgery become occupational medicine? The latter is most certainly a name with more dignity. However, because of inherent organizational delay, some of this change obviously took place before the Industrial Medical Association (approximately 1970) became the American Occupational Medical Association—again, a name with more dignity and more representative of the ever-widening body of knowledge involved in a practice of medicine that deals with occupations. Another sign of the long evolution is the fact that the American Board of Preventive Medicine in Occupational Medicine came into being in the 1950s, when it was finally agreed by the founding fathers that a body of medical knowledge specific to occupations did indeed exist and deserved recognition.

Did you ever stop to think about the changes during the past 30 years in the names used to describe physicians who did medical work in industry? Most of us who have been in the field for a number of years started out as what the public called "Shop Doctors"—a term that implied a "suture and cast man" and did not connote either knowledge or respect. The fact that we were paid by the company was often interpreted as indicating we were not trustworthy. I distinctly remember the pained facial expressions on my favorite aunt and uncle when they learned, following my Army release, I was going to work full-time for General Motors rather than become again, to them, a respected general practitioner in our home town of Saginaw, Michigan.

Anyway, we've all seen a lot of these names come and go—from shop doctors, or factory doctors, to industrial physicians and surgeons to the present-day occupational physician. Medical Director is another commonly used term to denote the health team leader, although the latest development is to use a title such as "Director (or Vice President) of Occupational Safety and Health." This latter title recognizes the extent of growth of the entire field of occupational medicine.

Fortunately, the evolutionary process was aided by the development of several fine residency programs, which have trained a new breed of occupational physician—well-trained, scientific, objective and career-oriented. The impact of this group in upgrading the image of occupational medicine has been great as individuals, but numerically there is still a long way to go.

It was only a few years back that one of our leading labor leaders was quoted as calling industrial physicians, except for a few notable exceptions, the "crud" of medicine. There must still be some residual doubts, because it is very interesting to note that in one of our largest unions, the head physician, who in my opinion is an exceptionally talented and ethical physician, is still called a "full-time medical consultant for the union" rather than a medical director of that union. I must assume the "consultants" are more acceptable than "directors" to their public.

Getting back to ethics, the fact is that the majority of occupational physicians become such as a result of a second or midcareer change. For a number of reasons, the formal residency programs cannot numerically supply the need. This leads to the call for postgraduate type programs or continuing medical education (CME) programs which can somehow fill the gap to ensure that all physicians in industry—whether formally or informally trained, do have available more or less standardized training that ensures a high level of competence—and ethics.

Good training in medicine does not just happen. With the advent of the annual licensing requirements, CME became big business—you *must* sell tickets to pay for *good* teachers. As seminar chairman for the 1980 American Occupational Health Conference in Detroit last year, I know of the necessity to offer not only good technical papers, but also papers that will market well. However, we must not let our need for ticket sales completely determine the choice of subjects taught—even though it may indicate the relative scale of importance of certain subjects in the public eye. For example, as important a subject as ethics is, there are not a lot of seminars on this subject. As a matter of fact, it is interesting to note that "ethics," titled as such, is the last section of this four-day conference on Legal and Ethical Dilemmas in Occupational Health. I am sure none of us here would put ethics last—it was probably just the luck of the draw.

To ensure that the ethical background of physicians in industry was not left to the luck of the draw, the AOMA, to their everlasting credit, developed in 1976 a "Code of Ethical Conduct for Physicians Providing Occupational Medical Service" [1]. This is *the* premier document, and truly sets the standard in this field. While some in the audience may be fully familiar with this code, no presentation on ethics for occupational physicians would be complete without a brief review of the twelve principles of this truly outstanding document.

The opening paragraph sets the tone:

> These principles are intended to aid physicians in maintaining ethical conduct in providing occupational medical service. They are standards to guide physicians in their relationships with the individuals they serve, with employers and workers' representatives, with colleagues in the health professions, and with the public. . . . Physicians should:

1. accord highest priority to the health and safety of the individual in the workplace;
2. practice on a scientific basis with objectivity and integrity;
3. make or endorse only statements which reflect their observations or honest opinion;
4. actively oppose and strive to correct unethical conduct in relation to occupational health service;
5. avoid allowing their medical judgment to be influenced by any conflict of interest;
6. strive conscientiously to become familiar with the medical fitness requirements, the environment and the hazards of the work done by those they serve, and with the health and safety aspects of the products and operations involved;
7. treat as confidential whatever is learned about individuals served, releasing information only when required by law or by over-riding

public health considerations, or to other physicians at the request of
the individual according to traditional medical ethical practice; and
should recognize that employers are entitled to counsel about the
medical fitness of individuals in relation to work, but are not entitled
to diagnoses or details of a specific nature. . . .

With the advent of the OSHA Standard on Access to Medical Records,
Principle No. 7 is currently probably the most actively discussed. It precisely defines what information is released to whom. Notice the last line
or two, which refer to the information to which an employer is entitled.
He can know about the medical fitness of individuals in relation to work,
but he is not entitled to specific diagnostic details.

Principle No. 8 tells us to gather and report scientific knowledge in a
timely fashion.

Principle No. 9 deals specifically with the issue of ensuring that employees receive full information and counsel about their health.

Principle No. 10 says physicians should — "seek consultation concerning
the individual or the workplace whenever indicated."

Principle No. 11 directs cooperation and ethical relationships with
government and other health professionals.

Principle No. 12, of course, deals with the sticky issue of advertising
techniques.

The American Occupational Medical Association Code of Ethics certainly sets a fine target for all occupational physicians to reach. The
question becomes one of how to apply this philosophy to the everyday
world of medicine in the factory. There are many approaches, but allow
me to describe some of the concepts we try to instill in General Motors.

First, it is imperative that you hire a quality physician with good references and an inherent set of excellent ethics. We must be sure, particularly in the midcareer changes, that the physician does not allow the
typical philosophy of a middle-aged, conservative physician to adversely
influence his/her method of practice in the plant. A patient seen in a
factory setting is entitled to receive at least the same courteous, prompt
and quality care that a patient expects from a physician in a private
office — in spite of who is paying the bill. As we said earlier, the public
often associates ethics with availability and performance as well as who is
paying.

Even the most ethical physician will have trouble establishing that type
of reputation in a plant if he/she does not have a desire to give good
service.

The "Golden Rule" is nowhere more appropriate than in the plant
medical department. Promptly greet the employee as entrance to the

department is made. If you are busy, a smile and a simple "Good morning, Mr. Jones, I'll be with you in a moment," will set the tone for a friendly, productive visit. Do not be afraid to use the titles of Mr., Ms. Miss or Mrs. It connotes a degree of respect for the employee as a person.

One of the most common difficulties a physician has when changing from private to a factory setting is adapting to work on a regular schedule. Private practitioners work long hours—like 16 today, 12 tomorrow, but then only 3 on Wednesday. Nine o'clock tomorrow morning is often just sometime between 6:00 a.m. and noon to them. But the factory is different—five, 8-hour days are a tradition, with firm starting and ending times. Believe me, it's virtually impossible to develop a positive relationship with an employee-patient (who gets fired for being late) when the doctor is an hour late for that 9:00 a.m. appointment in the plant medical department.

It is important for the plant physician, early on, to establish that you "call 'em as you seem 'em"—regardless of the issue involved. Both the unions *and* management will respect a physician who does this. It is likewise important that an occupational physician who maintains an outside private office does not "recruit" patients from his factory contacts. In General Motors, we have a policy letter that notifies all that we highly discourage this practice—to prevent any appearance of conflict of interest.

Upper management must, likewise, be made aware of the constraints that ethics place on the plant physician. I personally have made many presentations to plant managers in this regard and use the same AOMA code of ethics we discussed earlier. I also show them this paragraph in our Instructions to Plant Medical Directors booklet.

> The Medical Director is the health officer of the plant location and, as such, has the prime responsibility of the prevention of illnesses and injuries of the employees. The Medical Director must be cognizant of the hazards that exist at the location and be prepared to make recommendations that will lead to control of these hazards. Notification and education of management in regard to these items is paramount, but the responsibility of the Medical Director is not fulfilled if rebuffed by management. Consultation with the Divisional Medical Director should be obtained.

Obviously, finances are of concern. We at GM are very concerned about the health of our employees, but in a company that lost $750 million dollars last year, our task is to make certain each of our health dollars is wisely spent.

Yes, ethics are the backbone of our American system of medical care. Today, we have explored the many things that really are perceived by the public as "ethics." It has been fun to reminisce where we came from, where we are, and, now, where and how we are going. If we are to reach the heights that the current concepts of occupational medicine promise, then all of us must join together. Whether we represent government, labor, or industry, we must look to the future. Adversarial positions between these groups must cease to exist. To accomplish our goals, constructive cooperation is and must be the watchword of the future.

REFERENCE

1. American Occupational Medical Association: Code of Conduct for Physicians Providing Medical Services AOMA July 23, 1976.

INDEX

A copy of the banquet address, entitled "Occupational Health and Ethics," by Paul Brodeur, can be obtained from the co-editor, William N. Rom.